MW00560842

American Dollhouses and Furniture

From the 20th Century

Dian Zillner

With Price Guide

Schiffer Publishing Ltd
77 Lower Valley Road, Atglen, PA 19310

Dedication

Dedicated to my son, Jeff Zillner, who has had a lifelong love affair with houses — both large and small. May he find as much joy in the pursuit of his hobby as I have in mine.

Photographs by Suzanne Silverthorn

(Except Where Noted)

Library of Congress Cataloging-in-Publication Data

Zillner, Dian.
American dollhouses and furniture from the twentieth century/Dian Zillner.
p. cm.
"With price guide."
Includes bibliographical references and index.
ISBN 0-88740-768-4 (hard)
1. Dollhouses--United States--History--20th century. 2. Doll furniture--United States--History--20th century. I. Title.
NK4894.U6Z55 1995
688.7'23'09730904--dc20 94-42165
CIP

Printed in Hong Kong.
ISBN: 0-88740-768-4

Published by Schiffer Publishing, Ltd.
77 Lower Valley Road
Atglen, PA 19310
Please write for a free catalog.
This book may be purchased from the publisher.
Please include $2.95 postage.
Try your bookstore first.

We are interested in hearing from authors with book ideas on related subjects.

Notice

All of the items pictured in this book are from private collections. Grateful acknowledgment is made to the original producers of the materials photographed. The copyright has been identified for each item whenever possible. If any omissions or incorrect pieces of information are found, please notify the author or publisher and they will be amended in any further edition of the book.

Contents

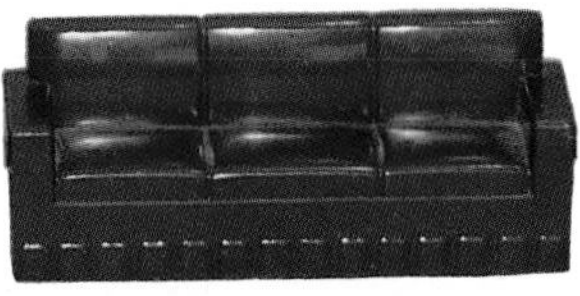

The author, Dian Zillner, has been interested in dolls and dollhouses for many years. She received her first dollhouse as a Christmas gift at the age of five in 1939. The inexpensive cardboard house was furnished with Strombecker wood furniture. Pictured here is the author, at the age of ten, with a favorite cloth doll.

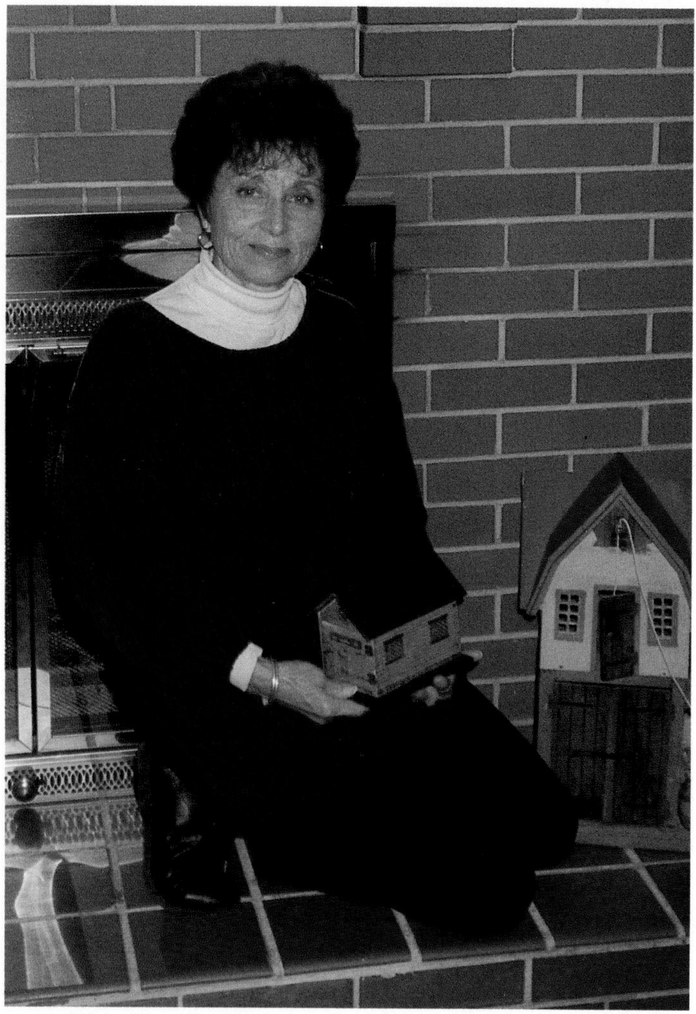

Marilyn Pittman has a varied collection of dollhouses and furnishings. Although she loves the early Bliss, Schoenhut, and Converse houses, she also makes room for the later Rich, Marx, and Keystone houses in her collection. Many of Marilyn's houses are pictured in this dollhouse book.

Suzanne Zillner Silverthorn, who took many of the photographs for this book, received this dollhouse as a Christmas gift when she was five. This home crafted house was taller than Suzanne as it was made to provide living quarters for the Mattel, Inc. Barbie family of dolls which were becoming so popular in 1964.

Introduction

More and more people enter the collecting field of miniatures each year. Some begin by making kit dollhouses and become addicted to old dollhouses and furnishings as well. Other individuals begin collections by accumulating older items and then also become interested in the new miniatures.

Although many books have been published on dollhouses and their furnishings, most of these earlier books focused on very old items. Although it is interesting to see these early museum type dollhouses, modern collectors also need information about the types of dollhouses and furniture they can add to current collections.

This reference book will concentrate on information about American dollhouses and doll furniture dating from the twentieth century. Although some of the items pictured were made in other countries, the pieces were manufactured for American based companies.

As with all collectibles, the value of a dollhouse or its furnishings depends on the condition, popularity, and rarity of the piece. A collector will pay more for the mint-in-box sets of furniture than for the same furniture assembled one piece at a time. As a rule, the cost of boxes of older wood Schoenhut and Strombecker furniture will be higher than for boxes of the more recent plastic furniture. An exception might be made in the case of Renwal plastic furniture, which is currently the most popular of the plastic furniture for today's collector. Other plastic furniture is also increasing in value, particularly the Ideal Young Decorator, Petite Princess, Princess Patti, and the Marx Little Hostess pieces.

The dollhouses in the most original condition will be the most expensive examples to purchase. Missing doors, windows, and chimneys depreciate the value of a house considerably. Damage to the house or redecoration also detracts from its value.

Although the older Bliss, Converse, and Schoenhut houses and furniture continue to be in great demand, newer Masonite houses have also become very collectible. The Rich and Keystone houses are especially desirable.

Along with these houses, the newer metal houses are also gaining strength as collectibles. The metal houses were made in such abundance that they are still very available and most of them are reasonably priced.

Probably the most amazing happening in the current doll furniture collectible field is the acceleration of interest in the plastic dollhouse furniture from the 1940s, 1950s, and 1960s. The quality furniture from Renwal, Ideal, Marx, and Plasco continues to be in much demand by today's collector. Perhaps this is because the furniture once belonged to the baby boomers who have now become collectors trying to recapture a piece of their childhood. This trend, apparently, will continue for some time.

Patty Cooper specializes in collecting the early wood dollhouses covered with lithographed paper. She is also very fond of the Schoenhut houses and furniture. As in most collections, other more recent houses of metal, cardboard, and hardboard have also found their way onto Patty's shelves. Although Patty photographed several of her older houses for this book, it is her sharing of her knowledge of the Schoenhut houses and their furnishings that will add needed information for today's collector.

Because of the assistance given by so many collectors, this book will provide much needed information on the dollhouses and dollhouse furniture from the more recent years. Also included in this publication is a list of addresses to help collectors in their search for dollhouses and furniture. The many pictures in this book are intended to demonstrate to collectors the diversity of items which are available. Also added to this resource book is a bibliography of the materials used to research this volume on dollhouses. It will provide helpful sources for those collectors who want to explore the field further.

I want to express my appreciation to the many individuals who answered questions, shared materials, and took photographs in order to make this book possible. A special "thank you" is extended to C. S. Olson, a long-time employee of Strombeck-Becker Mfg. Co., who was kind enough to loan several personal copies of Strombecker catalogs to be used in research for this book. Other companies also cooperated by supplying advertisements of dollhouses featured in their catalogs so that accurate research could be provided for this publication.

Thanks must also be extended to *Nutshell News* for their interest and help in locating copies of many of the late Dee Snyder's columns, "The Collectables," for use in research for this dollhouse book.

Special acknowledgment also goes to Cobb's Doll Auctions, Marian Schmuhl, George Mundorf, Pat Collins, Kathy and Bill Garner, Susan and Norman Lacerte, Judy Mosholder, Carl Whipkey, Mrs. David K. Large, Kathleen Neff-Drexler, The Toy and Miniature Museum of Kansas City, Barbara Staiger, Marcie and Bob Tubbs, Kay Houghtaling, Vilora Kergo, Gail and Ray Carey, Gordon and Judy Svoboda, Jackie Robertson, Betty Nichols, Donna Stultz, Rebecca Kepner, Leslie Robinson, Eleanor O'Neill, Leslie and Joanne Payne, Mary Louana Singleton, Paige Thornton, Patti Peterson, Clinton County Historical Society, Vernon M. Strombeck, H. B. Christianson, Arliss Morris, Marge Meisinger, Elaine Price, and especially Marilyn Pittman for their help in the preparation of this reference book.

In addition to these very helpful individuals, there are two "extra special" people who deserve added recognition. Roy Specht's wonderful photographs began appearing in my mailbox over a year before the publication due date and he was still fulfilling requests for additional photos until a few weeks before the deadline. Patty Cooper also supplied numerous photographs and shared many of her research materials in order to help make this publication possible. Patty has also been "on call" by phone to help solve a problem, share in a new discovery, or just to offer encouragement. These two collectors have added greatly to both the content and the "look" of this publication and to them I offer my deepest thanks.

Many of the people who are mentioned in this introduction have one thing in common: they are subscribers to the publication, *Dollhouse and Miniature Collector's Quarterly*, edited by Eleanor O'Neill (see sources). When I sent out a call for help while collecting material for this volume, these subscribers willingly volunteered. Although I haven't yet met any of these great people, I would like to personally thank them once again for their contributions to this book.

A special "thank you" also goes to members of my family who were so helpful with this "Dollhouse" project. To my daughter, Suzanne Silverthorn, who took many of the photographs, and to my son, Jeff Zillner, who was my "house repairman," and to my mother Flossie Scofield, who always offers support and encouragement, an extra vote of appreciation.

Acknowledgment and extra recognition is also extended to Schiffer Publishing Ltd. and its excellent staff, particularly to Sue Taylor, layout editor, and editor Leslie Bockol, who helped with this publication. Without their support and extra effort, this book would not have been possible.

Roy Specht is an avid dollhouse and dollhouse furniture collector. He has been interested in miniatures for over twenty-five years. He has a particularly exciting plastic dollhouse furniture collection. Pictured with Roy are some of his many furnished houses ranging from an original Strombecker house to a house furnished with the Marx Little Hostess pieces. Roy photographed many of the items from his collection for inclusion in this publication.

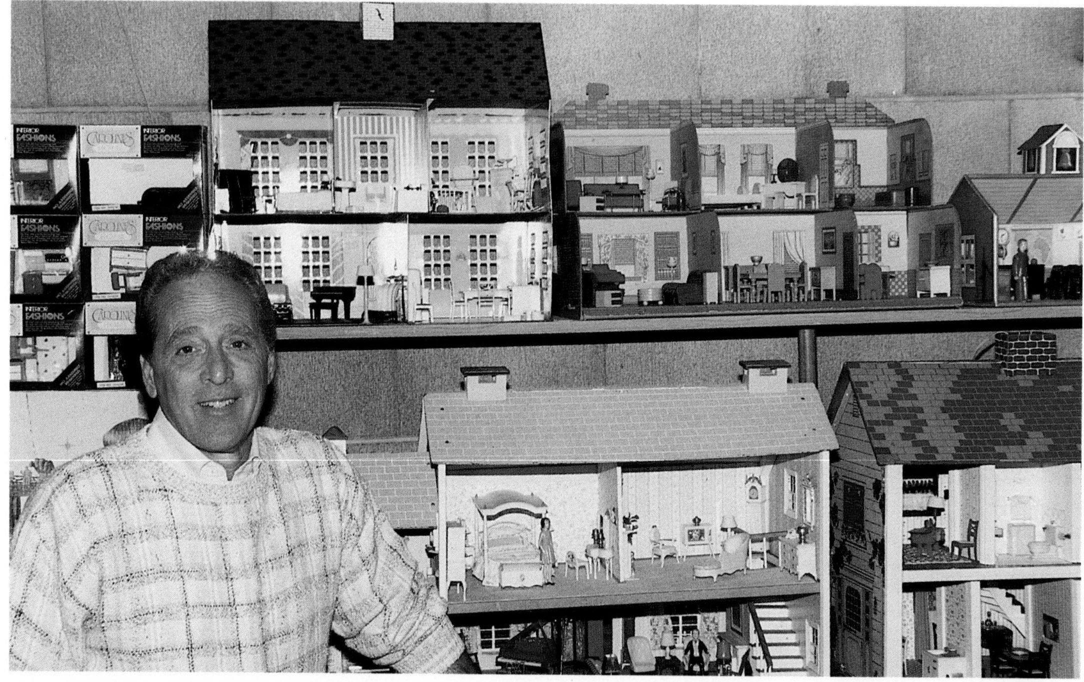

Wood Houses and Furniture

Bliss

The R. Bliss Manufacturing Co. was based in Pawtucket, Rhode Island from 1832-1914. The company was founded by Rufus Bliss in 1832 to make wood screws and clamps. By 1871 the firm was making toys and in 1873 games were added to their production. In 1914 the toy division was purchased by Mason and Parker from Winchendon, Massachusetts.

The Bliss products that are of great interest to today's collectors are the dollhouses and doll furniture. The first known dollhouse made by Bliss was "The Fairy Doll's House" advertised in 1889. The dollhouse was a much simpler model than the later houses. Basically it was composed of two boxes with a pediment top. The house had no front and could be taken apart with the smaller section (the second floor) fitting neatly into the larger section which made up the first floor. The house measured 20" by 12" and cost fifty cents. The house featured the Bliss method of decoration (lithographed paper over wood). The insides of the two rooms were complete with lithographed pictures, wallpaper, and some furniture.

The Bliss firm produced dollhouses through the early 1900s. Most of the dollhouses were made of wood with an overlay of lithographed paper used to show the details of the houses, both on the inside and outside of the rooms. Usually the dollhouses were marked "R. Bliss" above the front door or on the floor. Some of the houses were hinged to open on the sides, while others opened at the front. The smallest of the houses measured only 4 1/2" by 11" while others were nearly 30" tall. The two-story houses contained from two to four rooms with mica or printed windows. The outsides of the houses were quite elaborate and included architectural details like turrets, balconies, chimneys, and wrap-around porches with lots of gingerbread trim. Besides houses, the company also produced stables, shops, cabins, and churches. Sometimes dormer sections on the roof were made of cardboard instead of wood. The houses usually had a lithographed base and sometimes wooden steps were included to reach the porch. The roofs were usually painted red or blue.

The insides were not as fancy as the outsides of the Bliss dollhouses. The early houses were covered with lithographed paper that included fireplaces, doors, windows, curtains, wallpaper, and floor covering. Later, the interiors were covered in wallpaper instead of the lithography. Only a few of the two-story dollhouses had staircases.

By 1911 the dollhouses pictured in the R. Bliss Manufacturing Co.'s catalog had become much plainer than the earlier models. With the ending of the Gingerbread Age of architecture, the Bliss houses changed with the times and the towers and turrets were eliminated. Architectural details such as balconies and porches were retained.

One of the unusual designs done by Bliss was the two-story log cabin made with lithographed paper over wood. This Adirondack cabin, attributed to Bliss, contains three rooms in a two-story design. Part of the roof opens to the attic. Photograph and cabin from the collection of Patty Cooper.

The Bliss Company also produced folding dollhouses and these houses were featured in the company's catalog in 1911. Many of these house designs were the same as for those carried in the regular line. The folding houses were made of heavy cardboard, hinged with cloth.

Bliss manufactured dollhouse furniture from around 1888 to 1907. This furniture was not made to be used in their dollhouses as it was produced in a scale that was too large for the houses. Most of the furniture was made of wood that was covered with lithography paper. Many of these furniture pieces featured letters of the alphabet as decoration. Other furniture was decorated with pictures of children. The parts of the furniture that were not covered with lithographed paper were left unfinished. Items made by Bliss included: beds, dressers, rockers, cradles, tables, pianos, chairs, lamps, and settees.

Both the Bliss dollhouses and the furniture provide some of the most sought after prizes for today's collectors. The escalating prices on these items reflect this desirability and today's collector continues to hunt, often in vain, for a reasonably priced Bliss collectible.

Lithographed paper on wood Bliss dollhouse, circa 1900. The hinged sides open and the house contains four rooms. The two-story house is quite fancy with a large porch, balconies, and many added railings. The windows that are not printed are mica. The house is 29 1/4" tall, 20 1/2" wide, and 14" deep. It is marked "R. Bliss." From the collection of The Toy and Miniature Museum of Kansas City.

Bliss dollhouse made of lithographed paper on wood. Both sides open for access to four rooms. Mica windows where not printed. The house measures 20" wide by 24" tall. Also included in the picture are two bisque Armand Marseille dolls and three Steiff animals. From the collection of Cobb's Doll Auctions. Photograph by Darlene Cobb.

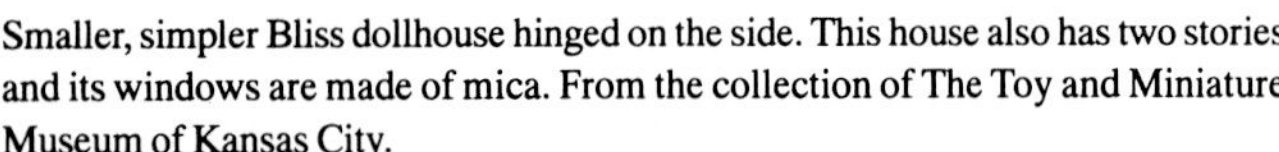

Smaller, simpler Bliss dollhouse hinged on the side. This house also has two stories and its windows are made of mica. From the collection of The Toy and Miniature Museum of Kansas City.

A smaller Bliss house is this two-story model that is 11" tall by 4 1/2" deep by 7 1/2" wide. The front of the house opens to reveal two rooms. Photograph and house from the collection of Marilyn Pittman.

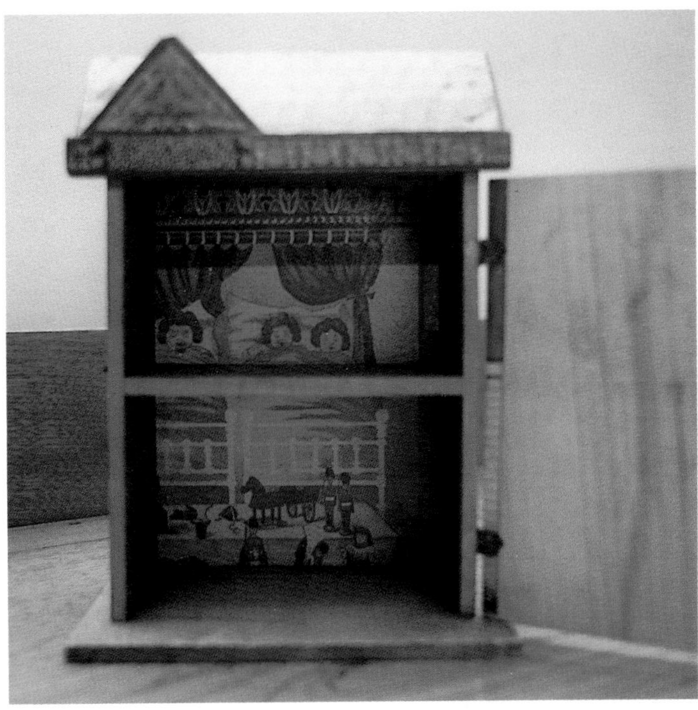

The inside of the small house shows the unusual lithographed walls of the house. Photograph and house from the collection of Marilyn Pittman.

This Bliss Garden House is marked above the porch "R. Bliss." The house measures 11 1/2" tall and the platform base is 15 1/2" wide by 12 1/2" deep. The house contains two rooms. Circa 1904. Photograph and house from the collection of Patty Cooper.

This Bliss lithographed paper on wood "Keyhole" house measures 9 3/4" wide, 20" high, and 7 1/4" deep. There are two rooms in the two-story house. Photograph and house from the collection of Patty Cooper.

This similar two-story, two-room Bliss house measures 18" high, 10 1/2" wide, and 8 1/2" deep. House and photograph from the collection of Patty Cooper.

This two-story, two-room Bliss house features overhangs above the upstairs windows as well as a fancy roof ornament labeled "R. Bliss." The house measures 20 3/8" high by 11 3/4" wide by 7 1/2" deep. Photograph and house from the collection of Patty Cooper.

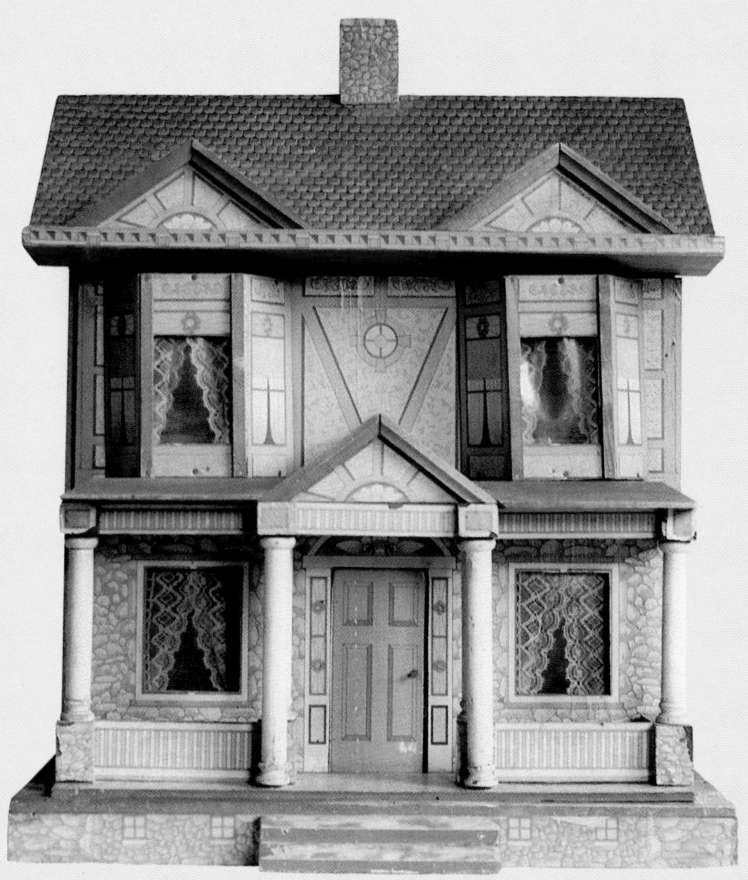

Pictured is a larger Bliss house which contains four rooms. It measures 24" tall by 20" wide by 11" deep. Photograph and house from the collection of Patty Cooper.

Bliss wood furniture with paper lithographed designs of the alphabet and children. This furniture, usually called the ABC furniture, was first sold around 1900. These chairs and sofa are part of the parlor set which also included a small table and a piano. Photograph and furniture from the collection of Patty Cooper.

Wood Bliss piano with lithographed paper designs made as part of the Parlor set circa 1900. Photograph and furniture from the collection of Patty Cooper.

Another parlor lithographed wood set of furniture dating from the early 1900s. It, too, has Bliss characteristics but it is not marked. Photograph and furniture from the collection of Marilyn Pittman.

Lithographed parlor wood furniture dating from the Bliss era of furniture production. The scale is 3/4" to 1". The manufacturer of the furniture is unknown although the pieces have Bliss characteristics. Photograph and furniture from the collection of Patty Cooper.

This set of wood bedroom furniture decorated with paper lithographed designs of children dates from 1895-1900. The furniture is not marked but is believed to have been made by Bliss. The company is known to have produced a set of bedroom furniture in 1895. The Bliss furniture came in sets with mixed scales. Photograph and furniture from the collection of Patty Cooper.

Converse

Morton E. Converse was born in New Hampshire in 1837. In 1878 he joined with Orlando Mason to make wooden toys, utensils, and wooden boxes. The company was located in Winchendon, Massachusetts and among their toy products were wagons and rocking horses. The company the two men founded became known as Mason and Converse. In 1884 Converse formed his own company under the name Morton E. Converse Co. In 1898 the name of the firm was changed to Morton E. Converse and Son. During the 1890s, the firm was the largest wood toy manufacturer in the world. By 1915 the company was making 3,000 toys in various styles and sizes.

Noting the success Bliss was having in manufacturing dollhouses using lithographed paper over wood, several other manufacturers began the process of making dollhouses using the colors lithographed directly on the wood. This technique was used as early as 1909. One of the companies trying this new method of making dollhouses was Converse. Another was N. D. Cass of Athol, Massachusetts. The Converse houses were not as elaborate as those made by Bliss. Most of the houses were of the bungalow type with a stone foundation printed on wood. In the 1913 Converse catalog, five sizes of these houses were listed. The houses ranged in size from 9" to 17". Some of the houses had the name of the firm printed on the border of the rug but others had no markings. Another identifying characteristic is the inclusion of a person or a cat printed on the windows as if they are looking outside. By 1916 the company was featuring dollhouses as tall as 26 3/4". These were two-story houses on a platform. The houses contained two rooms and a balcony. Most of the houses were only one story and measured around 7" tall. These smaller houses contained only one room. The basic Converse house was a bungalow with a porch decorated with two columns. The windows were printed on the wood and either the front or the side was hinged to open. Some of the houses even had printed furniture on the walls inside. This basic house was manufactured by Converse for many years. At first the two pillars were round but by 1926 they had been changed to square.

Besides dollhouses, Converse also produced wood buildings for boys. These included barns complete with farm animals and garages to house the new toy cars that were becoming so popular.

Around 1930 Converse changed their method of dollhouse manufacture and brought out a new house made of fiberboard and wood. The house was called the Realy Truly Dollhouse. It was 21" by 12" by 15" and contained four rooms. The house had a Georgian style door and shuttered bow windows as well as dormer windows. The company also provided Realy Truly furniture. The Sears, Roebuck and Co. catalog from 1930 featured the four rooms of wooden furniture at a cost of 89 cents for each room.

The living room set included a 5 1/8" wide settee, two lounge chairs, floor lamp, footstool, radio, and a long table. The dining room consisted of seven pieces of walnut finished furniture. These included a round table, four chairs, a buffet with a drawer that opened, and a server. The bedroom had twin beds (4 3/8" long with paper spreads), dresser with mirror and drawer that opened, a rocking chair, nightstand, chair, and a lamp. The kitchen furniture was painted green and included a stove, ice box, cabinet, sink, table, and two chairs. This furniture, particularly the kitchen, looks very similar to the Schoenhut pieces. The furniture is quite attractive and very collectible for today's collector. Since the furniture was made for such a short period, it is very scarce.

The new enterprise must not have met with much success as the Converse company was soon purchased by Mason Manufacturing Co. of South Paris, Maine. This marked the end for one of the most famous names in the history of dollhouses, that of Morton E. Converse.

Converse wood dollhouse with the colors lithographed directly on the wood. The house features a dormer on the roof. Photograph and house from the collection of Marilyn Pittman.

The inside of the small one room wood Converse house shows the fireplace and windows lithographed directly on the wood. Photograph and house from the collection of Marilyn Pittman.

Another small one room wood dollhouse with the colors lithographed directly on the wood. The house is attributed to Converse although it is not marked. It measures 11 1/2" deep, 12" wide, and 10 1/4" tall. Photograph and house from the collection of Marilyn Pittman.

Very small wood lithographed house with hinged front opening. Similar to the houses made by Converse. The house measures 7" tall, 5" wide, and 3 3/4" deep. Photograph and house from the collection of Marilyn Pittman.

Wood lithographed garage circa 1920s. Front opens to accomodate car. Probably made by Converse. Photograph and garage from the collection of Marilyn Pittman.

Small Converse house which has an open back. The one room house measures 8" tall, 9" wide, and 8" deep. The inside has no decorations. Photograph and house from the collection of Patty Cooper.

This wood house, attributed to Converse, measures 12 1/4" tall, 14 1/2" wide, and 11" deep. The house contains one room with printed windows and rug. It opens from the front. Photograph and house from the collection of Patty Cooper.

This interesting wood dollhouse is not marked and may have been made by either Cass or Converse. It measures 12" tall, 14" wide, and 12 1/2" deep. Photograph and house from the collection of Patty Cooper.

Advertisement in the Sears, Roebuck and Co. catalog for 1930 which features the Realy Truly dollhouse furniture attributed to the Converse company. There were four rooms of furniture pictured including the living room, dining room, bedroom, and kitchen. The kitchen furniture is very similar to the Schoenhut furniture being made the same year.

The front of the house opens to reveal one room. The walls and floor are decorated with windows, wallpaper, and a rug. Photograph and house from the collection of Patty Cooper.

The Realy Truly kitchen furniture as pictured in the Sears catalog for 1930. The cabinet is 3 7/8" by 3 1/8".

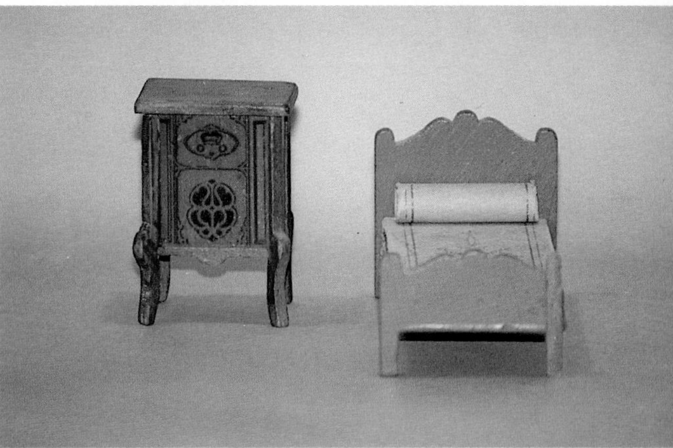

The Realy Truly bed and radio as pictured in the 1930 Sears catalog. The bed seems to be in smaller scale than the kitchen and radio pieces but most companies did not design all their furniture in true scale.

The Realy Truly living room sofa, chair, and floor lamp as pictured in the 1930 Sears catalog. Furniture and photograph from the collection of Barbara Staiger.

The Realy Truly dining room furniture included a table, four chairs, a buffet, and a server. Missing from the set pictured is one chair and the server. Photograph by Gail Carey. Furniture from the collection of Ray and Gail Carey.

Schoenhut

Albert Schoenhut emigrated to America from Germany at the age of seventeen. In 1872 he founded his own company in Philadelphia when he was only twenty-two. At first, the firm produced only toy musical instruments but by 1903 they were also making the famous Humpty Dumpty Circus.

The A. Schoenhut Co. manufactured their first dollhouses in 1917. The houses were made of fiberboard and wood. The houses had embossed siding and roofs which gave the appearance of stone walls and tile roofs. At first, all the Schoenhut houses opened from the sides.

The insides of the new Schoenhut houses were also quite attractive. The walls were covered with lithographs to represent wallpaper. The lithographing included doors which were partly open so it looked as though another room was beyond the door. The windows came complete with lace curtains.

The Schoenhut houses ranged in size from the simple one room bungalow to the largest two-story houses which contained eight rooms. Some of the larger houses had one dormer window, while others had two. These houses also included staircases.

The bungalows are especially interesting to today's dollhouse collector. This type of dollhouse was made by Schoenhut from 1917 until the late 1920s. The houses featured from one to four rooms. The bungalows all included front porches. Some of the models had dormer windows, others had porch railings and/or front steps.

In 1927 a new design was developed for the Schoenhut dollhouses. These models featured the Colonial look which would dominate Schoenhut's line of houses until the end of dollhouse production in 1934. These houses were described as being two-story models with attics in the Schoenhut advertising for 1928. The houses sold for $9 to $24 each. The old bungalow models were priced at only $3 to $10 each during the same year. The older interiors were discontinued in the Colonial designs and ordinary wallpaper or painted walls were used. All the two-story Colonial houses had staircases and most houses were attached to a platform foundation. Some of the new style houses even came equipped with a garden and a garage. A later Colonial model was made with a side porch.

Methods of opening the dollhouses changed with the years. Some opened on the front or the side, while other houses were produced with no backs. Many of the later houses came equipped with electric lights. The houses were made with a variety of outside decor including the look of brick, stucco, or clapboard siding. The houses were painted in bright colors and varied in the front door area designs. The houses all had shutters and most came complete with window boxes.

There were several different sizes in the Colonial houses. The rooms contained in the houses ranged from four to eight. The smallest of the houses measured 17" by 12" by 15" high while the largest was 26 1/4" by 25 1/2" by 23 1/2" tall. There were also different sizes in the same design of a house. By 1933 there were nine different models of the two-story Schoenhut Colonial houses featured in the company catalog.

A catalog picturing one of the last Schoenhut Colonial houses describes the house as having "brick finished walls, ivory window frames, green shutters, green roof, stairway, removable back, four rooms wired with five lights." The house was 26" by 17" by 24". It sold for $7.50. The roof lifted up in the front of these houses to reveal an attic that could also be used for storage.

In 1931 Schoenhut also offered dollhouses with a Tudor look. Three different models of these designs were featured in the catalog for that year. The two-story houses contained from four to six rooms. Each of these houses had a removable back wall. The houses ranged in size from 19 1/4" by 12 1/4" by 20 1/2" to 31 1/2" by 19 3/4" by 29".

The same year two Spanish-style houses were also offered for sale. These two-story houses featured staircases, four rooms, and were equipped with electric lights. The roofs remained similar to the ones on the Colonial models. The houses were made to look like stucco and the doors and windows had a different look than the those on the Colonial designs. The larger model measured 26" by 18" by 24".

A Dutch Colonial house was included in the Schoenhut catalog for 1931. This house also contained four rooms and featured a removable back. The house measured 20 1/4" by 15 3/4" by 21 5/8".

The company name usually appeared on a decal on the base of the house and read "Schoenhut Doll Houses." Sometimes a label was tacked onto the base which read "Manufactured by/ The A. Schoenhut Co./ Philadelphia, PA." At least eighty different models of Schoenhut dollhouses were made during their years of dollhouse production. Many of the variations from one house to another were slight (an added roof overhang, or a change in a front door arrangement) but the variations did make the houses all different. The houses were very well made and many remain in very good condition today. The Schoenhut houses continue to fascinate dollhouse collectors and they are still in great demand.

Besides the dollhouses, Schoenhut also made apartment house rooms in the early 1930s. The rooms had hinged walls so they could be folded for storage. The rooms were made of wood and were decorated on both the inside and outside. The rooms were sold with Schoenhut furniture. The various rooms included a living room, kitchen, dining room, bedroom, and bathroom. The five rooms could be joined together for play.

Surprisingly, it was not until 1928, over ten years after beginning production of the dollhouses, that the Schoenhut Co. began producing wood dollhouse furniture. The enterprise only lasted until 1934. It is hard to believe that in only seven years, a company could have made so many different designs of furniture in several different scales. According to their catalogs, at least some changes were made in the furniture designs nearly every year. The basic rooms of furniture always remained consistent. Included were a kitchen, bathroom, living room, dining room, and bedroom. Most of the furniture produced by the Schoenhut company was approximately 3/4" to 1' in scale. The furniture was designed to fit into the Schoenhut dollhouses.

Many pieces of the furniture shown in the 1928 company catalog, however, were close to 1" to 1' in scale. The kitchen pieces appear to be somewhat smaller. The kitchen set was produced in natural wood finish and included a table, two chairs, a cabinet (4 1/4" tall), and a two-door ice box. For the dining room there was a round table, four chairs, a buffet, server, and china cabinet. The dining room furniture was walnut in color. The living room set included a fireplace, two floor lamps (one in bridge style), table lamp, clock, mission oak table, rocker, chair, and sofa plus another unusual chair with sides in the shape of an "S." The bathroom came in white enamel and included both a tub and a separate shower (with shower curtain), a toilet, lavatory, and a stool. The bedroom was equipped with twin beds, a highboy dresser with a mirror, a mirrored vanity, and two chairs. According to 1928 Schoenhut advertising, the furniture was priced from $2.00 to $3.50 per room.

By 1930 a more modern living room had been added. With the exception of the fireplace and clock, the other furniture had been redesigned. The sofa and two chairs were painted and flocked to look like overstuffed furniture. A piano and bench were added which included the Schoenhut name on the front of the piano. The lamps were also updated.

Sears, Roebuck and Co. featured the new 3/4" to 1' line of Schoenhut dollhouse furniture in their catalog in 1931. Each room of furniture was priced at eighty-nine cents. The bedroom was light green and was made up of eight pieces of furniture. These included: 4 1/4" twin beds, 3 1/4" dresser with removable drawer, vanity, chair, rocker, table, and lamp. The living room included a sofa, two easy chairs, library table, piano, bench, and floor lamp. There was a seven-piece light green kitchen. Included were a 4" cabinet and refrigerator, 3 1/2" sink, table, two chairs, and a stove. The dining room also came in a seven-piece set. Included were a 3 1/4" buffet with removable drawer, 3 1/4" table, four chairs, and a 2 1/4" server. These pieces were all finished in brown enamel. The six-piece orchid bathroom set included a 4" tub, 5 1/4" shower stall, cabinet, stool, toilet, and pedestal sink. Most of this furniture also appeared in the 1932 Schoenhut catalog with slight modifications to the bedroom furniture and a smaller number of pieces of furniture included in each room. Perhaps this was because the company was also making a line of larger scale furniture that year. It was featured in the Sears catalog in 1932.

There were also five rooms of furniture in the larger 1" to 1' scale of Schoenhut furniture. Sears sold each room of furniture for only sixty-nine cents. The bedroom was orchid and included twin beds (5 1/4" by 2 7/8"), dresser (4" wide) with two opening drawers, two chairs, night table, and floor lamp. The living room included a green piano (4 1/4" by 4 1/2") and a bench. The sofa was red (6 3/8" by 3 1/4") as was the chair, ottoman, and end table. A floor lamp and a library table were also sometimes included in this set. There were seven pieces of kitchen furniture enameled white. These included a stove, table (5" by 2 3/4" by 2 5/8"), ice box (4 1/4") with top coil and opening door, sink, stool, and two chairs. The seven-piece dining room set was painted light green. Included were a table 5 7/8" by 3 1/4", a buffet with hinged doors, a server with a removable drawer, and four chairs. The seven-piece bathroom set was painted orchid. It included a bathtub, (5 1/4" by 2 1/2"), a separate shower stall, toilet, wash basin, make up table, bench, and medicine cabinet.

The Sears, Roebuck and Co. catalog for 1933 carried only the smaller 3/4" to 1' scale of Schoenhut furniture. The various sets came with six pieces of furniture and were priced at only forty-seven cents a set. The bedroom was the same as had been sold in 1932. It included a bed, colored paper spread, dresser (one drawer opened) with a metal mirror, two chairs, night table, and floor lamp. The kitchen furniture was painted yellow and included a table, sink, stove, ice box, (coil on top), and two chairs. These were all different designs than those from 1932. The dining room furniture was painted light green. There was a newly designed table, buffet (hinged doors), and four chairs. The living room furniture was also different than earlier years. The red sofa and matching chair were curved and featured embossed lines decorating the fronts of both pieces. A walnut radio on high legs, a green piano and bench and a floor lamp completed the living room pieces. The bathroom was furnished with a vanity, bench, tub, lavatory, medicine cabinet, and toilet.

The year 1934 appeared to be the last year of dollhouse furniture production for Schoenhut. Sears catalog again carried the furniture but did not identify it with the Schoenhut name as it had in earlier years. Several changes were made in the designs of the furniture. The wood frames were eliminated from the metal mirrors and the furniture took on a cheaper look. It was being sold by Sears for only forty-three cents for each six-piece set. The bedroom was blue and included a bed with a paper spread, a newly designed nightstand (pedestal), a mirrored dresser with one opening drawer, two chairs, and a lamp. The Schoenhut catalog also pictured a larger box of this furniture that included an additional bed, and a chaise lounge. The living room set contained a red sofa and chair, a green piano and bench, a floor lamp, and a newly designed radio which was flush with the floor. In the larger set featured in the Schoenhut catalog another chair, an ottoman, and a book case were added. The kitchen was again green with the same table and chairs but with a different design for the sink, stove, and refrigerator (no coil). The larger set of kitchen furniture also included a stool, a pantry, and a wastebasket. The dining room was buff enamal and included a different table and buffet (opening top drawer) and the same style chairs. The extra pieces added to the dining room set included a fireplace, clock, and a small table. The bathroom vanity no longer had a bench and there was no frame around the mirror. A stool was included with the tub, toilet, lavatory, and medicine cabinet. The bathroom pieces were painted orchid. The additional Schoenut pieces included in the larger set of furniture featured a shower stall, vanity bench, and a hamper.

Since Schoenhut furniture was only made for approximately seven years, this dollhouse furniture is much more scarce than the Strombecker line of furniture. With even the small approximately 3/4" to 1' line of furniture featuring opening drawers, many of the small Schoenhut pieces are better crafted than the Strombecker pieces, and therefore will bring higher prices. The 1" to 1' lines of furniture by both companies are very well made and the higher prices for the Schoenhut pieces would only come from their scarcity. Many dealers still confuse the furniture from these two companies and the collector should study pictures of both lines of furniture so they will be knowledgeable when making purchases of wooden dollhouse furniture.

The original A. Schoenhut Co. went into bankruptcy in 1934, ending their production of both the dollhouses and the dollhouse furniture. In 1936 O. Schoenhut, Inc. was founded in Philadelphia by Otto Schoenhut and his nephew George Schoenhut. The firm manufactured several lines of toys including "Pick Up Sticks" and toy pianos.

Another Schoenhut Company was also established about this same time. Albert F. Schoenhut and his son Fredrick were the founders of this enterprise called Schoenhut Manufacturing Co. This company carried on the dollhouse tradition of the earlier Schoenhut firm.

Their dollhouses were produced from 1936 to 1940. Several different models were made including a two story Colonial (with a covered porch or carport) which was named "Stuyvesant." The house measured 24" by 15 1/2" by 19". A small cottege called "Woodbine" measured only 16 3/4" by 10" by 12". Another house was made in a "Tudor" design. This "Cotswold Cottage" measured 22 1/4" by 11 1/2" by 14 1/2". A large Tudor was named "Cotswold Manor" and was 39" by 16 1/2" by 21" in size.

The most fascinating models of the new Schoenhut houses were the "Art Deco" designs. There were three flat roofed houses with this "modern" look. The houses were called "Long Beach" (22 1/4" by 13 1/2" by 15"), "Malibu" (29 1/4" by 14 1/2" by 17"), and "Beverly Hills" (41" by 21" by 19 1/2"). These houses all featured rounded windows. This company also produced toy house trailers during this period of time.

Two-room, one-story Schoenhut Bungalow featured in the Schoenhut catalog in 1923. The house measures 20" by 20 1/2" by 16 1/2". The wood and fiberboard house was embossed to represent stone walls and a tile roof. The house contains glass windows. A similar Schoenhut house featured three rooms and a porch railing.

The inside of the house is accessible from each side. Both rooms have lithographed walls to represent wallpaper and open doors.

Schoenhut also produced a large eight-room house with the look of stone. The house was pictured in their catalog for 1923. The two-story house measures 25 1/4" by 28 1/2" by 29 1/2". The house features a large porch, glass windows, and lace curtains. The company also made a similar house with six rooms. House and photograph from the collection of Patty Cooper.

This two-story house also contains two rooms. The house measures 11" by 14" by 18 1/2". It is embossed to give the walls and roof the look of stone and tile. The house also features a front porch. This house was shown in the Schoenhut catalog in 1923. Photograph and house from the collection of Patty Cooper.

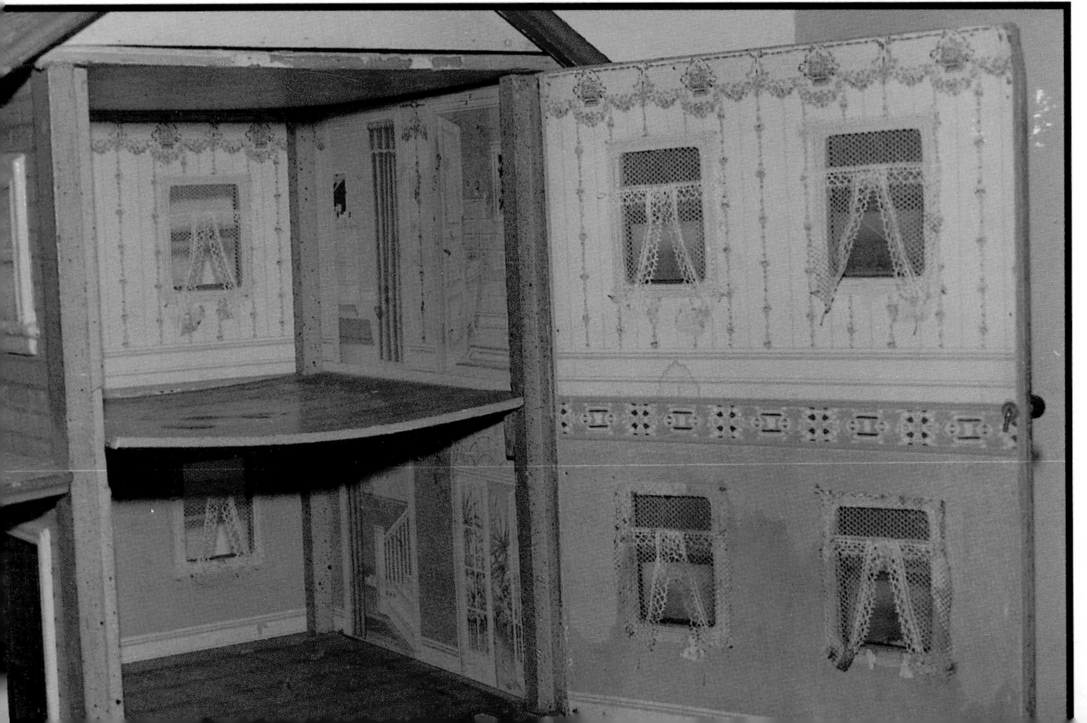

The house opens from the side to expose the fancy wallpaper on its walls and the lace curtains on the windows. The Schoenhut houses had floors covered with brown to represent hardwood floors. House and photograph from the collection of Patty Cooper.

The inside of the eight-room house featured the lithographed walls and electric lights which were a feature of many of the Schoenhut houses. Photograph and house from the collection of Patty Cooper.

Pictured in the eight-room house are various Schoenhut pieces of furniture. Photograph and house from the collection of Patty Cooper.

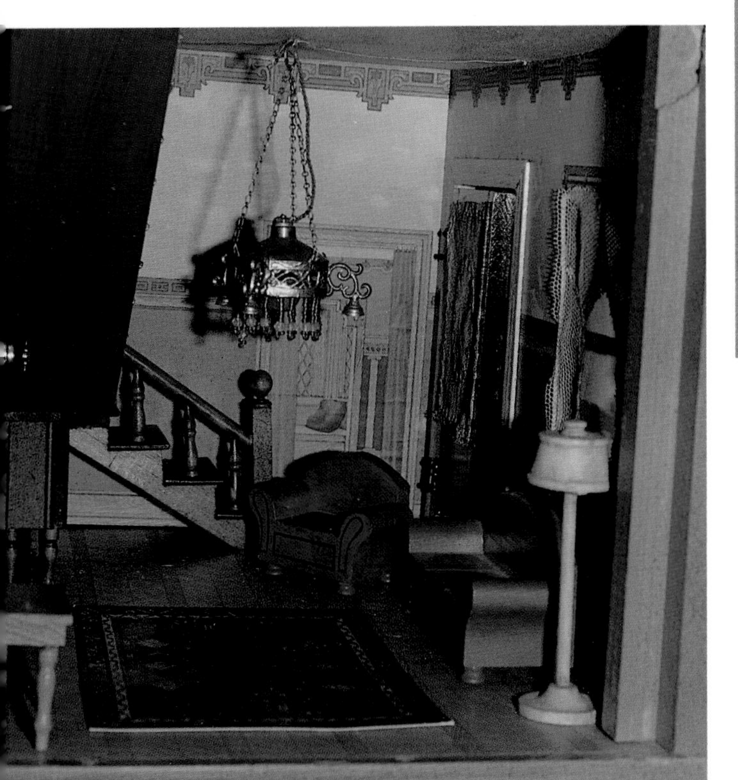

A close-up of the living room shows the stairway leading to the second floor in the eight-room house. Photograph and house from the collection of Patty Cooper.

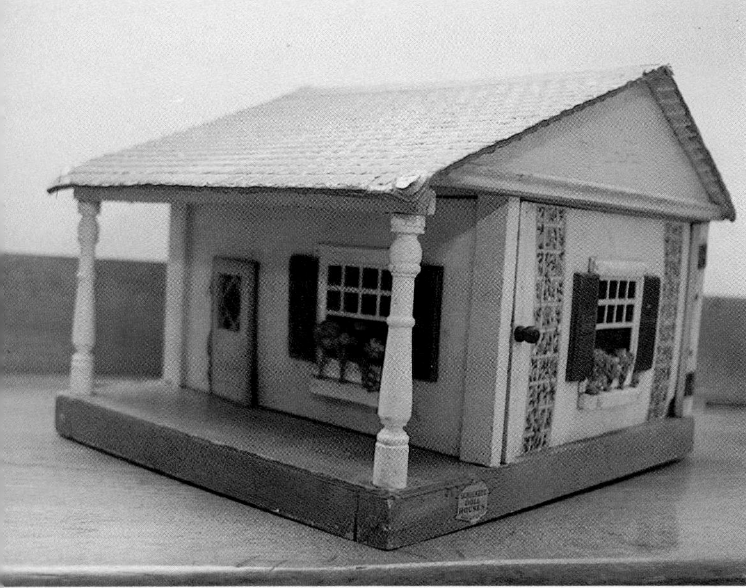

The Schoenhut catalog for 1928 pictured several models of the one-story bungalow. The houses ranged in size from one room and attic to four rooms and an attic. Pictured is a model with one room and an attic. The house opens from the side and measures 12 1/2" by 10 3/4" by 15 1/2". The old stone decor was replaced with a "stucco" look. Flower boxes added to the "new" look. Photograph and house from the collection of Marilyn Pittman.

In the 1927 Schoenhut catalog, the new line of Colonial dollhouses was featured. Through the years, these houses came in several different designs and sizes. All of the houses were two-story and contained from four to eight rooms. The houses ranged in size from a small four-room house 17" by 12" by 15" high to an eight-room house 26 1/4" by 25 1/2" by 23 1/2". Pictured is a four-room model from 1930 measuring 24" by 15 1/4" by 21". The shutters and window frames of the new models of Schoenhut houses were made of fiberboard. The shutters and window frames have been repainted and the doorway on this house has been restored. Window boxes that were on the house originally are missing.

Small Schoenhut houses were also produced to be used with train sets or in toy villages. Although these houses are not true dollhouses, they make an interesting addition to a dollhouse collection. The company also produced railroad stations. Photograph and house from the collection of Marilyn Pittman.

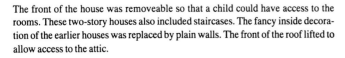

The front of the house was removeable so that a child could have access to the rooms. These two-story houses also included staircases. The fancy inside decoration of the earlier houses was replaced by plain walls. The front of the roof lifted to allow access to the attic.

The eight-room model of the Colonial house also was made in several different designs. Some houses featured the lower floor overhang roof while many did not. The door treatment also varied on the houses. The houses were painted cream, perhaps to represent stucco. Photograph and house from the collection of Marilyn Pittman.

This large eight-room Schoenhut house was two rooms deep, so the rooms were accessed through removable front and side panels. The two-room attic could be reached by raising the roof. The house was made on a very large platform, perhaps to provide space for a Schoenhut garden. The house measures 26 1/4" by 25 1/2" by 23 1/2". Photograph and house from the collection of Marilyn Pittman.

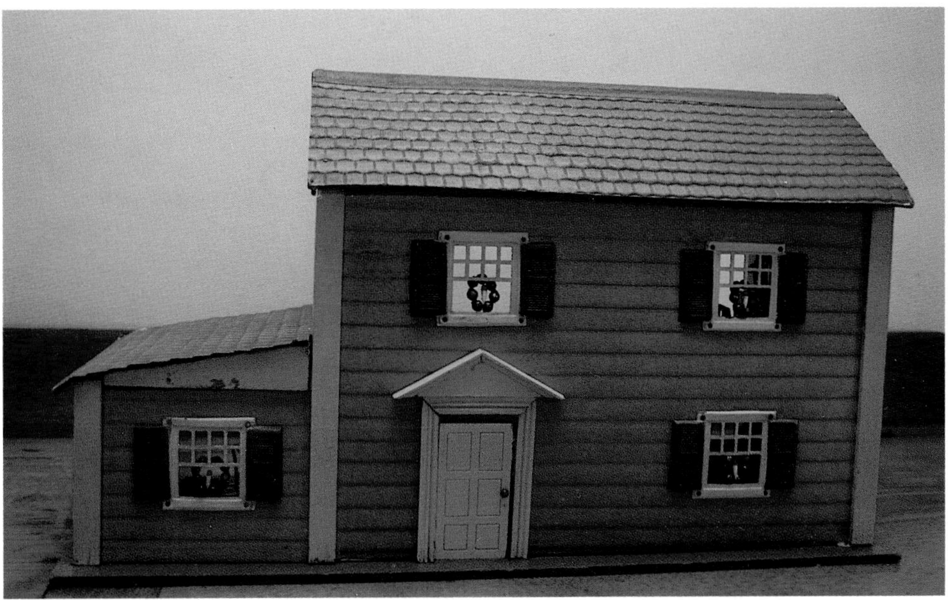

This five-room Schoenhut house was featured during 1934, the last year of dollhouse production. The house has an embossed clapboard finish and features electric lights in each room. The two-story house measures 24 1/4" by 9 1/2" by 14 1/4". There were two rooms upstairs and three rooms downstairs. The house contained no back for easy access to the rooms. Photograph and house from the collection of Marilyn Pittman.

Schoenhut's catalog for 1931 featured some new designs of dollhouses. This English style Tudor house measures 19 1/4" by 12 1/4" by 20 1/2". A similar house was also made which was 24" by 17" by 24 3/4". House from the collection of Kathy Garner, photograph by Bill Garner.

Another new model Schoenhut dollhouse from 1931 was this Dutch Colonial house. The two-story house measures 20 1/4" by 15 3/4" by 21 5/8". From the collection of Kathy Garner, photograph by Bill Garner.

The Tudor house contained four rooms and no stairway. The back was open for play. House from the collection of Kathy Garner, photograph by Bill Garner.

The Dutch Colonial house contained no staircase but the model did feature electric lights. The back was open for easy access to the rooms. From the collection of Kathy Garner, photograph by Bill Garner.

"Malibu" dollhouse produced by the "new" Schoenhut Manufacturing Company in 1937. The house measures 29 1/4" by 14 1/2" by 17". It contains five rooms and includes its original awning. Photograph and house from the collection of George Mundorf.

Schoenhut began production of wood dollhouse furniture in 1928. Most of this first line of furniture was nearly as large as the 1" to 1' scale. The kitchen pieces appear to be smaller. Pictured is the boxed bedroom furniture which included twin beds (5 1/4" long), two chairs, a vanity, and chest (both with mirrors). Several of the drawers opened. Furniture and photograph from the collection of Patty Cooper.

The same design of dining room furniture was used for several years. In 1928 it was produced in natural wood finish. Later it was sold enameled in green. Pictured is the entire set which includes a round table, four chairs, a buffet, server, and china cabinet. Many of the drawers and doors function. Furniture and photograph from the collection of Marilyn Pittman.

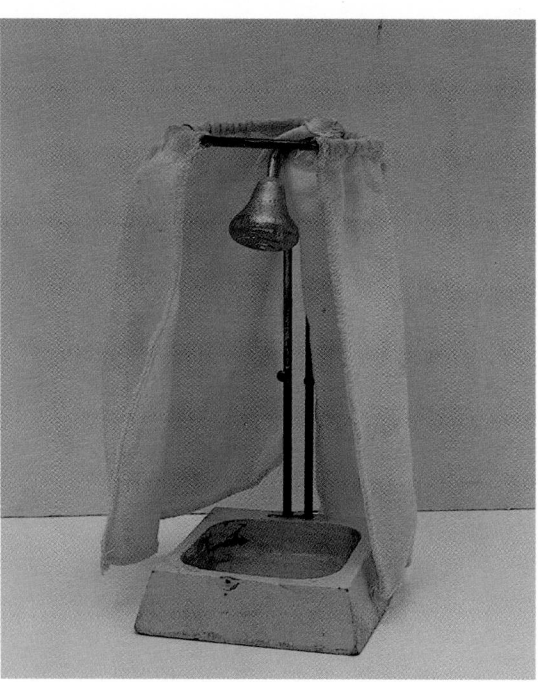

This Schoenhut shower stall is circa 1930. A shower curtain was needed to complete the item. Shower and photograph from the collection of Patty Cooper.

The Schoenhut bathroom pieces were also used from 1928-1930. Missing from the set pictured is the shower (with shower curtain) and a stool. Photograph and furniture from the collection of Marilyn Pittman.

The complete living room set from 1930 included a fireplace, grand piano, bench, floor lamp, clock, sofa, two easy chairs, and a bridge lamp. Photograph and furniture from the collection of Marilyn Pittman.

79F7109—Shpg. wt., 6 lbs. $1.89 SEE PAGE 2

Schoenhut's New Wood Doll House Furniture

The name Schoenhut means the maximum in quality. Far Superior to Metal Furniture Because It Is Practically Unbreakable and Then, Too, the Pieces Are Much Larger. Each piece made of nicely beveled and enameled or stained wood and glued and nailed together.

Any 1 Set Listed below.....89c | Any 2 Sets Listed below. $1.73 | Any 4 Sets Listed below. Postpaid.......$3.35

8-Pc. Light Green Bedroom Suite 89c

4⅛-inch Twin Beds, 3¼-inch Dresser with removable drawer; vanity, chair, rocker, table and lamp in proportion.

49F7142—Shpg. wt., 1 lb..89c

7-Pc. Modern Parlor Suite 89c

4¾-inch maroon color settee, walnut finish piano, 3x2⅞ in.; 4¾-inch lamp has gilt stand and **green** shade. Table, bench and two arm chairs in proportion and in colors to match.

49F7140—Shpg. wt., 1 lb....89c

7-Pc. Light Green Kitchen Outfit 89c

4-inch cabinet and refrigerator, 3½-inch sink; table, 2 chairs and stove in proportion.

49F7143 Shpg. wt., 1 lb...89c

7-Pc. Dining Room Suite 89c

3¼-inch buffet with removable drawer, 3¼-inch table, four chairs almost 3 inches high and 2¾-inch server. Finished in pretty glossy brown enamel.

49F7141 Shpg. wt., 1 lb....89c

6-Pc. Orchid Bathroom Set 89c

4-inch tub, 5¼-inch shower; cabinet, stool, toilet and pedestal sink in proportion.

49F7144 Shpg. wt., 1 lb....89c

The 1931 Sears, Roebuck and Company Fall and Winter catalog featured the Schoenhut wood furniture line. The five sets of furniture sold for eighty-nine cents each.

This blue wood sofa and chair set was part of the living room furniture issued by Schoenhut in 1931. In that year, the company produced its new furniture in the 3/4" to 1' scale. The other living room pieces included a library table, a piano, stool, and floor lamp. Photograph and furniture from the collection of Patty Cooper.

The Schoenhut kitchen furniture pictured in the 1931 catalog was painted green and included seven pieces instead of the five items from 1928 - 1930. The "ice box" was replaced with a new "refrigerator." A sink and stove were added, and a new design was used for the cabinet, table, and chairs. Furniture and photograph from the collection of Patty Cooper.

The dining room furniture was also redesigned in 1931 and included a rectangle table (3 1/4" long), four chairs, a buffet, and a server. The buffet is missing in the boxed set pictured. Furniture and photograph from the collection of Patty Cooper.

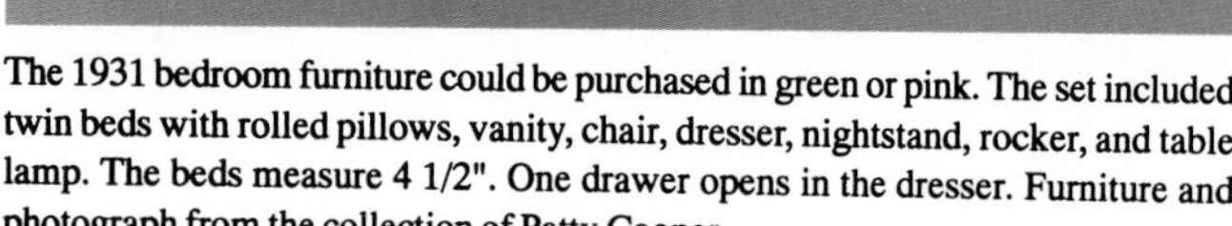

The 1931 bedroom furniture could be purchased in green or pink. The set included twin beds with rolled pillows, vanity, chair, dresser, nightstand, rocker, and table lamp. The beds measure 4 1/2". One drawer opens in the dresser. Furniture and photograph from the collection of Patty Cooper.

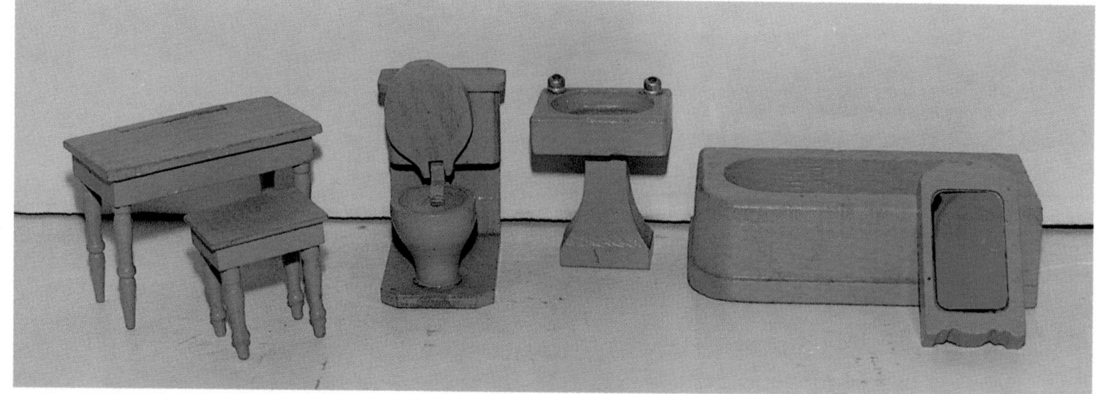

Pictured are various Schoenhut bathroom pieces. The toilet and sink appear to date from 1931. Also included in the 1931 bathroom set were a shower stall, a bathtub, medicine cabinet and a stool. The vanity (missing mirror), stool, medicine cabinet, and tub are from a later set produced in 1933. Photograph and furniture from the collection of Patty Cooper.

The 1932 Sears, Roebuck and Co. catalog featured the new Schoenhut large size furniture in the 1" to 1' scale. The sets included a bedroom, parlor, kitchen, dining room, and bathroom. Catalog from the collection of Patty Cooper.

Schoenhut's Latest New Large Size Doll House Wood Furniture—Many New Features—Larger Pieces—Finer Designs—Prices Greatly Reduced.

7 Pc. Orchid Enameled Bedroom Set 69c
Twin beds, 5¼x2⅞ in. 4 in. width dresser with two removable drawers. Two chairs, night table and floor lamp in proportion.
49K7152—Shpg. wt., 1 lb. 4 oz. 69c

7 Pc. Bright Red and Green Enameled Parlor Set 69c
Piano, 4¼x4½ in. and bench, enameled green. 6⅜x3¼ in. settee enameled red. Arm chair, ottoman, end table and floor lamp in proportion.
49K7150—Shpg. wt., 1 lb. 8 oz. 69c

7 Pc. White Enameled Kitchen Set 69c
Table, 5x2¾x2⅝ in.; 4½ in. Ice Box with hinged door. Gas range, sink, stool and two chairs in proportion. Shpg. wt., 1 lb. 8 oz.
49K7153...... 69c

7 Pc. Light Green Enameled Dining Set 69c
Table, 5⅞x3¼ in.; buffet has hinged doors. Server with removable drawer and four chairs in proportion. Shpg. wt., 1 lb. 4 oz.
49K7151...... 69c

7 Pc. Orchid Bathroom Set 69c
Bath-tub, 5¼x2½ in. Shower, toilet, wash-basin, make-up table, bench, chest in proportion. Shpg. wt., 1 lb. 4 oz.
49K7154...... 69c

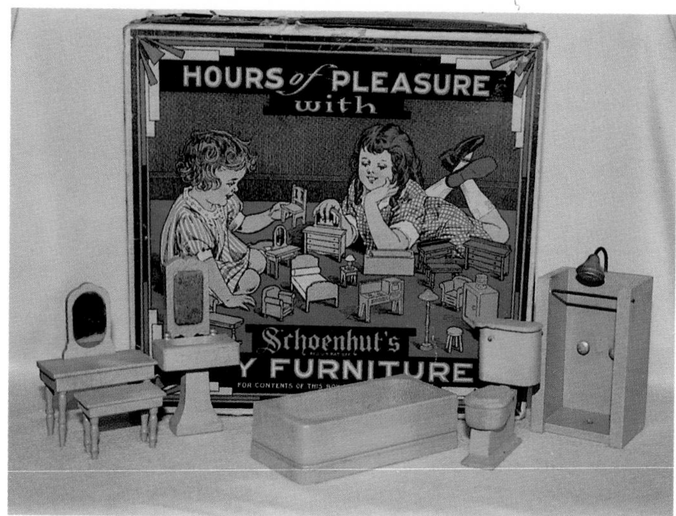

Pictured is the bathroom furniture from the 1932 line of 1" to 1' scale of furniture. The furniture pictured on the box, however, is in the smaller 3/4" to 1' scale. Photograph and furniture from the collection of Gail Carey.

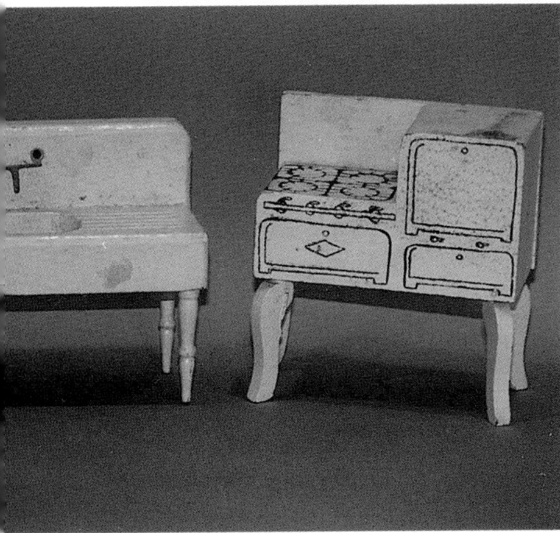

These Schoenhut kitchen pieces are also part of the larger 1" to 1' scale of Schoenhut furniture pictured by Sears in 1932. Missing is another chair, a rectangle table with turned legs and a stool.

The 1" to 1' scale of Schoenhut living room furniture marketed in 1932 looks very much like the furniture produced by Strombecker during this period. Although both sets of furniture are red, there are several basic differences. The Schoenhut library table has no stretcher and the footstool has an extra line around the bottom. Besides the items pictured, the Schoenhut set also included a grand piano, bench, end table, and lamp. See Strombecker chapter for more information. Photograph and furniture from the collection of Patty Cooper.

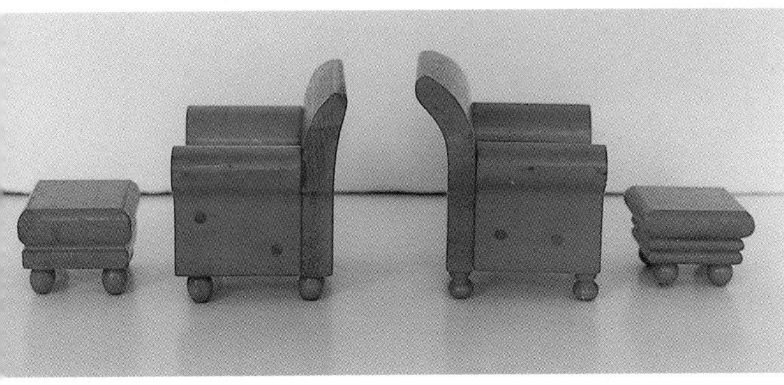

Other differences shown in this picture of the Schoenhut and Strombecker chairs include the placement of the nails that held the furniture together. The Schoenhut chair's nails are placed horizontally while in the Strombecker chair, one nail is at the top and the other is at the bottom on opposite sides. The Schoenhut chair leg also has an extra line and the back curves much more than that of the Strombecker piece. Photograph and furniture from the collection of Patty Cooper.

The Schoenhut furniture for 1933 was again in the 3/4" to 1' scale and was made more cheaply with less detail than the furniture from earlier years. The living room sofa and chair have rolled arms with lines embossed on the front. The sofa measures 4 3/4" long. Furniture and photograph from the collection of Patty Cooper.

The Sears, Roebuck and Company catalog for 1933 featured the new line of Schoenhut furniture at a low price of forty-seven cents for each room of furniture. Sets included a living room, bedroom, kitchen, dining room, and bathroom. Catalog from the collection of Marge Meisinger.

6 Pc. Light Green Enamel Wood Bedroom Set 47c

Wood Bed: 4⅝x2¼ in. Colored paper bed spread. 4 in. **Dresser:** Top drawer opens. Metal mirror. Two chairs, night table and floor lamp in proportion. Not Prepaid.
49 D 7162—Shpg. wt., 12 oz. 47c

6 Pc. Red, Green, and Walnut Enamel Wood Parlor Set 47c

Piano: 3¼x3¼ in. and bench, enameled green. Settee, 5x2¼ in. enameled red. Arm chair, radio cabinet and floor lamp in proportion. Not Prepaid.
49 D 7160—Shpg. wt., 12 oz. 47c

6-Pc. Canary Color Enamel Wood Kitchen Set

Table, 3¼x2⅛ in. Sink with drain board and dummy faucets. Gas range, ice box and chairs in proportion. Not Prepaid. Shpg. wt., 12 oz.
49 D 7163 .. 47c

6-Pc. Light Green Enamel Wood Dining Room Set

Table top, 4x2½ in. 4-inch buffet with hinged doors and 4 chairs in proportion. Not Prepaid. Shpg. wt., 12 oz.
49 D 7161 ... 47c

6-Pc. Orchid Enamel Wood Bathroom Set

Bath tub, 4⁷⁄₁₆x1⅞ in. Make-up table, bench, wash stand, medicine cabinet and toilet in proportion. Not Prepaid. Shpg. wt., 12 oz.
49 D 7164 .. 47c

The living room set also included a piano and bench, a radio on legs, and a floor lamp. The rocker pictured was part of the bedroom furniture set sold in 1931. Photograph and furniture from the collection of Patty Cooper.

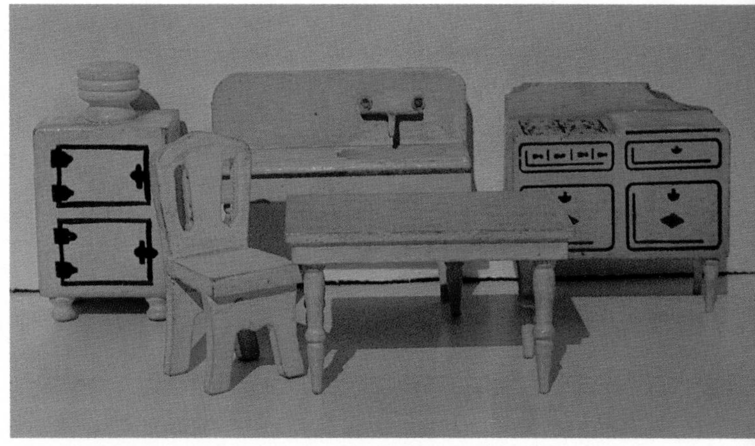

The refrigerator, sink, table, and chairs date from 1933, while the stove is pictured in the 1934 Schoenhut catalog. The same design of table and chairs was also used in the 1934 line of furniture along with a more modern refrigerator and sink. Photograph and furniture from the collection of Patty Cooper.

The 1933 Schoenhut kitchen included six pieces of furniture. There were no longer any working parts in the furniture. Photograph and furniture from the collection of Marilyn Pittman.

The 1933 bedroom pieces included a bed, floor lamp, two chairs, dresser with mirror, and nightstand. The dresser drawer opened. Photograph and furniture from the collection of Patty Cooper.

These kitchen pieces of Schoenhut furniture date from 1934. The table and chairs were the same as those pictured in 1933. Photograph and furniture from the collection of Patty Cooper.

The bedroom dresser for 1934 no longer had a wood frame around the mirror and designs for the nightstand, dresser, and bed were also changed. In addition to the items pictured, the set also included another chair and a floor lamp. Photograph and furniture from the collection of Patty Cooper.

Pictured is a boxed set of Schoenhut bathroom furniture from the last year of furniture production in 1934. The company appeared to be trying to cut cost by eliminating frills. Photograph and furniture from the collection of Patty Cooper.

This dining room furniture also dates from 1934. One more chair was originally included in the six-piece set. Photograph and furniture from the collection of Patty Cooper.

Strombecker

J. F. Strombeck, the founder of the Strombeck-Becker Manufacturing Co. was born on December 16, 1881 in Moline, Illinois. Strombeck attended the Moline schools until he dropped out after his second year of high school. He then began work for the D. M. Sechler Carriage Co. in Moline. After a few years he started his own air freight auditing business. J. F. Strombeck still felt he needed more education and he entered Northwestern University at Evanston, Illinois. He graduated in 1911.

Strombeck began a new manufacturing business that same year and was soon joined by R. D. Becker. The company produced many different kinds and sizes of wood handles. The business was incorporated in 1913 as the Strombeck-Becker Manufacturing Co. During World War I the firm helped the war effort by making tent poles for the army.

The company manufactured their first toys in 1919 (ten pins) but the venture was not successful and was discontinued. Several other attempts to produce wood toys were also made, through the years, but success was not achieved until 1928 when a ten cent airplane was marketed.

In 1931 dollhouse furniture was added to the toy line. The furniture was in the 1" to 1' scale. Furniture was produced for the living room, bedroom, dining room, kitchen, and bathroom. This painted furniture was made for several years. The fancier furniture had gold or silver swirls added to the finish. The bedroom pieces included twin beds (many contained three decorative holes in the footboard), a vanity dresser, bench, nightstand, and chairs. The vanity contained an opening drawer plus two small drawers that did not open. The vanity was made in two different sizes although the bench that accompanied both vanities was the same large size. The bathroom furniture included a bathtub (some models include a shower), toilet, lavatory, vanity and bench (different than the bedroom model), and eventually a medicine cabinet and clothes hamper were added. The kitchen furniture included a cabinet (with opening doors), table, two chairs, and refrigerator with coil top and opening door. A 1' scale kitchen stove on legs, a breakfast nook, and a sink on legs were also included in the set of kitchen furniture by 1933. The living room furniture was red and included a sofa, chair, footstool, library table, end table, radio on tall legs, and grandfather clock. By 1933 a grand piano, bench, floor lamp, and jardiniere were also being produced as living room items. The dining room pieces included a table, four chairs, a server (opening doors), and a buffet (opening drawer).

In 1934, Strombecker began the production of the 3/4" to 1' scale of furniture. The company catalog for the year stated that they were producing this new furniture because it was the size that best fit all the popular priced dollhouses. The new furniture was made for the bedroom, kitchen, bathroom, living room, and dining room. The bedroom pieces included twin beds, a nightstand, vanity, floor lamp, and chairs. This set of furniture came in walnut and was also later made in pink or green enamel. The kitchen pieces were green enamel and consisted of a refrigerator (coil top), table and four chairs, sink on legs, stove (on short legs), and a low cabinet. Later the kitchen pieces were made in a combination of white (sink, refrigerator) and green. The bathroom furniture was pink but would later also be made in white and other colors. The items sold for the bathroom included a vanity, stool, bathtub, lavatory (with two small nails for faucets), toilet, and hamper (the catalog calls it a clothes hamper but a wastebasket is pictured). The living room furniture included a walnut grand piano, bench, grandfather clock, radio, and table as well as a green sofa, chair, and a floor lamp. The furniture was also made in red. The sofa and chair were designed with molded lines on the arms and bottom and the legs looked like small beads. The dining room furniture was walnut and included a table, four chairs, a buffet, and a server. The set was also later made in blue enamel and the drawer in the buffet sometimes was made to open. Accessories were later added to the furniture. The furniture was sold in different size sets and even by the piece.

The 1936 Strombecker catalog also featured the 3/4" scale of furniture in unpainted sets. The consumer could buy the furniture for a cheaper price provided they did the painting themselves. Each set came with instructions along with sandpaper. These sets could be purchased for as little as twenty cents each.

The same catalog showed little changes in the 3/4" furniture. A table lamp, magazine rack table, and jardiniere were added to the living room set. The dining room included candlesticks and a wooden bowl. A clock, candlesticks, and a table lamp were added to the bedroom pieces and the beds and vanity included some decoration. The kitchen extras included a towel rack, bowl, stool, and wall clock. The table and chairs had added designs but the low cabinet was no longer pictured. The biggest changes were made in the bathroom pieces. Two towel racks, a bathtub seat, a scale, medicine cabinet, vanity bench, and portable heater were all additions to the bathroom set.

The 1" to 1' scale of Strombecker furniture pictured in the 1936 catalog was entirely different than the first line produced in the early 1930s. There were still five rooms of furniture but only the bathroom pieces retained the look of the earlier furniture. Part of the advertising stated that "The Museum of the City of New York has purchased for display purposes several complete sets of our Deluxe Doll House Furniture ... because of the accuracy of our reproduction." The living room furniture included a walnut radio (with bead type legs), six legged lamp table, magazine rack table, tan flocked sofa, chair, and footstool with bead type legs (molded lines on the arms and bottom), floor lamp, and table lamp. The dining room furniture was also made of walnut. The pieces had a Spanish look to their design. Included were a table, four chairs, a server, and a buffet. Two candlesticks were also included in the set. The new bedroom design was also made of walnut. Included were twin beds, a blanket chest with opening top, a vanity with rectangle mirror, bench, nightstand, two chairs, and a lamp. The kitchen was much more modern and included a table and chairs with a light oak look that featured added designs. The refrigerator was a larger model with a coil top, the sink was mounted on tall legs, and the stove was painted cream and green (flush with the floor). It contained a top that closed to cover the burners. A high stool and towel rack were also part of the kitchen set. The bathroom furniture was painted green and included the earlier hamper and medicine cabinet. In addition there were different designs for the bathtub (with faucets), lavatory (faucets), toilet, heater, stool, and towel racks.

The Strombecker catalog for 1938 pictured many new designs for both the 3/4" to 1' and 1" to 1' scales of furniture. The company was producing "New Modern Doll House Furniture Sets" in the 3/4" scale. This new modern-looking furniture is much sought after by today's collectors because it was not in production for as long as other designs and is harder to locate. The living room included a fireplace with round mirror, floor radio with rounded front, end table with open shelves, round ottoman, square backed sofa and chair with square footstool, table lamp, and unusual floor lamps with stacked open shades. The furniture was painted blue except for the lamps and the couch, chair, and footstool which were flocked in yellow. The dining room was painted coral and included a trestle table, six chairs (the backs and legs were one piece), hutch, server, and buffet. The bedroom furniture was painted yellow and consisted of twin beds with rounded head and footboards, dresser with round mirror, bench, two chairs, nightstand, table lamp, clock, and clothes rack. The bathroom was furnished with a bathtub, toilet, lavatory, medicine cabinet, wastebasket, dressing table, bench, scale, electric heater, and stool. The vanity also was equipped with a round mirror and the catalog pictures the lavatory placed in the middle of the vanity. The kitchen pieces were also completely different than the earlier models. The sink was contained in a "built in" cabinet that was flush with the floor. The upper part of the piece had a slight overhang. The refrigerator was a modern flush with the floor unit (no coil on top), and the stove was also flush with the floor. The table was a rectangle trestle model. The chairs were the same design as those in the dining room. The kitchen also included a high stool and a clock. The appliances were painted white while the other furniture was red. One of the favorite rooms of furniture for collectors is the "Modern Sun Room." This wood furniture included wire arms and legs in a modern design. These pieces were red and accompanied a yellow radio and end tables. Included were a davenport, two arm chairs, a table, end table, radio, floor lamp, and table lamp.

Even though Strombecker had come out with the new "Modern" design, the old models of furniture continued to be produced. A few changes had been made to this earlier furniture. The original sofa and chair were replaced by a newer design that featured a more square look with small legs which made the furniture nearly flush with the floor. A smaller kitchen stove was also produced which appeared to be an apartment size model.

Big changes had also taken place in the 3/4" to 1' scale of De Luxe furniture. All new designs were featured in 1938. The living room included a "Philco" radio (flush to the floor), occasional table (pedestal with legs), coffee table, sofa, chair (both with rounded arms), ottoman (all finished in "Dubonnet Izarine simulated upholstery"), floor lamp, and table lamp. The wood pieces were made of walnut. The dining room set was also made of walnut and included a more modern table, four chairs, buffet (opening drawer), and tea wagon. Accessories included candlesticks, and a fruit bowl. The bedroom furniture included twin beds, nightstand, vanity (round mirror), bench, chest of drawers (one opened), clock, and two lamps. The catalog pictured this set in white but it was later also made of walnut. The bathroom furniture included a more modern lavatory, toilet, hamper (opened), bathtub, towel rack, heater, and very modern vanity and bench. The catalog advertising states that the "major pieces follow the design of the Crane Company's Neuvogue fixtures as conceived for them by Henry Dreyfuss." The kitchen furniture includes a refrigerator without the coil (with opening door), a sink built into a cabinet (the top part overhangs the bottom slightly), a modern range, and a table and two chairs (these were the same as the earlier design with an orange finish). A wastebasket, bowl, and clock were also included in the set.

The new design of furniture also featured a nursery set. The original pieces included a youth bed, step up for bed, chifforobe, toy chest, shoofly, writing table, chair, lamp, and a clock. The door and one drawer of the Chifforobe opened as did the lid of the toy chest. The original color of this set was light grey with orange and yellow trim. Collectors are more familar with the furniture painted pink.

Besides all these new designs of furniture, Strombecker was also offering a line called "Custom-Built Doll House Furniture." This furniture is the best that was produced by the company and the current prices continue to escalate accordingly. This furniture sold by the piece instead of in a set. The 1938 catalog includes the following items:

Fireplace (No. 701E). Mantel and trim in hand-rubbed walnut. Embossed simulated tile floor and face.

Davenport (No. 701A). Three piece (divides into three separate chairs). Upholstered in two-tone blue simulated fabric or in golden rod and brown combination.

Lounge Chair (No. 701B). Matches Davenport.

Philco Radio (No. 701RP). Includes Swiss music box. Made of walnut. It could also be purchased without the music box.

Coffee Table (No. 701L). Walnut, three circular shelves with round central support.

Governor Winthrop Secretary (No. 701F). Walnut. Has doors which open with simulated leaded glass. Drop leaf desk with pigeonholes, and three pull out drawers. Also could be purchased complete with books (made of wood).

Governor Winthrop Desk (No. 701G).

Walnut. Just like the bottom of the secretary.

Grand Piano and Bench (No. 701CD). Walnut. Top lifts and closes. Could also be purchased complete with Swiss music box.

Tilt Top Table (No. 701K). Walnut. Top is hinged and can be fastened in a horizontal position.

Pier Cabinet (No. 701J). Walnut. Four shelves with a hinged door at bottom. Blocks of books included.

Grandfather's Hall Clock (No. 701H). Walnut. Stamped gold pendulum and weights and printed clock face.

Extension Dining Table (No. 702B). Walnut. Two pedestal legs with the Duncan Phyfe look.

Chairs (No. 702E and 702D). Walnut with leather seats.

Dining room chairs with or without arms.

Buffet (No. 702A). Walnut. Burled top and doors. Doors and drawers open.

Four Poster Bed (703A). Cherry.

Night Stand (No. 703D). Cherry. Drawer and door open.

Chest of Drawers (No. 703C). Cherry. Four large drawers and one small drawer all open.

Dresser (No. 703B). Cherry. Mirror set in cherry frame.

Three opening drawers.

One of the company ads from this period priced the Custom-Built furniture line at from 40 cents (three shelf coffee table) to $5.00 (grand piano and bench with music box) for each piece. The ad also stated that "Free Junior Play Rooms" were available if the furniture order was $1.00 or more. All of the walnut pieces of Strombecker furniture are stamped on the bottom in gold. The trademark reads "Strombecker/Playthings/Genuine Walnut" so the furniture can be easily identified.

Besides the furnishings for dollhouses, Strombecker also produced a set of school room furniture beginning in the late 1930s. The scale was supposed to be 3/4" to 1' but it appears to be a little larger. The boxed set included a teacher's desk (two real drawers), teacher's swivel chair, four students' seat desks with tilting seats, a cut-out blackboard, waste basket, and set of books.

In the 1942 Montgomery Ward Christmas Catalog, the 3/4" to 1' Strombecker furniture advertised for sale included many design changes from the "Modern" look of the late 1930s. The living room sofa and chair were flocked and made in an overstuffed design that was rounded instead of square as the "Modern" furniture had been. The new console radio resembled the popular "Philco" model. The bedroom furniture had been completely redesigned and featured a vanity dresser with a round mirror and an opening blanket chest. A chest of drawers was also sometimes used with this set of furniture. The head and footboards of the beds had also been changed. The new design of 3/4" walnut dining room furniture was also available at that time. The furniture included a table, four chairs, a buffet, and server. The kitchen furniture was also redesigned during the mid 1940s. The stove was doubled in size and was no longer an apartment size model. A modern refrigerator was also added to the kitchen line and the sink became a "built in" model flush with the floor. The kitchen retained the earlier "Modern" kitchen table but the chairs were redesigned. At about this same time, the grand piano produced by Strombecker for so many years was replaced by a modern "upright" model.

The late 1940s brought many more changes to the 3/4" Strombecker line of dollhouse furniture. The kitchen was modernized and the pieces produced and pictured in the 1950 Strombecker catalog are some of the nicest ever made by the company. The kitchen table and two chairs were made with wire legs and were painted red. The kitchen appliances were painted white. The built-in sink had open shelves on one end as did the stove. The refrigerator was also a new design. Accessories for the kitchen included a clock and a bowl. The bathroom furniture was also very unusual. The bathtub was attached to a back and side wall of imitation tile. A shower curtain rod was included in this piece of furniture. The toilet also was attached to a tile piece which contained a towel rack. The lavatory too was attached to its own piece of tile which featured a mirror. The catalog pictures two different lavatories. One is very modern with two small legs in front. The other model is a more ordinary design with a middle pedestal. Also included in the bathroom set were a hamper and a bench. The dining room pieces included a walnut Duncan Phyfe style dining room table with four matching chairs. Another dining room piece was a buffet with one opening drawer. This set is also extremely hard to find. The living room furniture included a davenport and lounge chair finished in green simulated upholstery with scroll lines on the arms. Also repeated as living room items were the walnut radio, coffee table, end table, upright piano, bench, table lamp, and floor lamp. The bedroom furniture was also made of walnut. It included twin beds, a vanity (round mirror), bench, nightstand, blanket chest, chair, two lamps, and a clock. This was the same design of furniture that had been made earlier in a painted finish.

Because the 1" to 1' scale Strombecker furniture was not as big a seller as the smaller 3/4" scale, the designs for this furniture were not changed as often as those for the smaller pieces. The larger furniture was more expensive and most dollhouses were not big enough to accomodate the 1" scale of furniture. The Rich and Keystone houses provide the perfect setting for this furniture.

The Strombecker dollhouse furniture box circa 1953 pictures six sets of "Deluxe" furniture in the 1" scale still available. Most of the items were repeats from the earlier line. These included: bedroom, living room, kitchen, child's room, dining room, and bathroom. The pieces found in each room were as follows: Bedroom: twin beds, vanity with round mirror and bench, chest of drawers (two drawers open), nightstand, and lamps. Living room: sofa, chair, footstool (all flocked), coffee table, occasional table, floor lamp, and television. Kitchen: sink, range, refrigerator, table, two chairs, and accessories. Child's room: youth bed, chifforobe (drawer and door open), blanket chest (opens), shoofly rocker, nightstand, and lamps. Dining room: table, four chairs, buffet, serving cart (walnut), candles, and bowl. Bathroom: tub, toilet, lavatory, clothes hamper, two towel racks, portable heater, stool, and wastebasket. This furniture was still being sold as late as 1954. Some of the 1" scale furniture was also available in "Do-It Yourself" assembly kits for much cheaper prices.

The Strombecker box for the 3/4" scale furniture circa 1953 pictured five rooms of furniture. The living room included an enameled sofa, chair, end tables, coffee table, occasional table, white television, two table lamps, and a floor lamp. The dining room furniture was the same walnut set which dated from the early 1940s. The bedroom furniture was also from the early 1940s and featured the vanity with the round mirror. The kitchen pieces shown on this box no longer included the open shelves of the earlier design although the stove and sink were exactly like the earlier models minus the open shelves. The refrigerator was the same design as those sold in 1949-1950 but was a little smaller in size. The table and chairs no longer featured the wire legs but were totally made of wood. The bathroom set no longer featured the tile design and looked more like the sets produced in the mid-1940s.

Besides the two scales of dollhouse furniture, Strombecker also sold several different designs of cardboard dollhouses. These houses, however, were not manufactured by the Strombecker Company. This firm made only wood products. They contracted with others to produce the dollhouses. Strombecker then marketed the houses furnished with their wood furniture. The first cardboard dollhouse is pictured in the Strombecker catalog for 1934. The house is called "Dream Home Doll House" and is No. 320F. It was a Dutch Colonial design measuring 19 3/4" in width, 14 3/8" in depth, and 16 1/4" in height. It contained five rooms and was made from cylinder kraft corrugated board. The house opened on two sides. The front and the back both dropped down and the roof could also be raised on the back of the house. The two-story house contained three rooms downstairs and two rooms upstairs. The house was licensed under Purdy Patents of assembly design. The windows were cellophane. Thirty pieces of the new 3/4" to 1' scale of furniture came with the house.

Three different designs of Strombecker houses were being offered by the company in the 1938 catalog. Although the houses were all similar, they differed in construction and size. The houses had an unusual design, featuring a kind of stair arrangement for the two-story dollhouses. From the outside of the house, the top floor extended out over the walls of the first floor. The other side was open for play. Again, the floors were not square. The first floor extended out from the second floor. The house came in both a four room and a six room version. The smaller four-room model was 17 3/4" wide, 16 1/2" high and 14 1/2" deep. The six-room house was 26" wide, 16 1/2" high by 14 1/2" deep. The two models were both the same house, but put together differently to make either a four- or six-room model. The houses were made of "Fibo-board" and had to be assembled. Both the insides and the outsides were printed with curtains, windows, shutters, etc. There was no roof, it was an extension of the upstairs front wall. The houses are marked with the Strombecker company name so they are easy to identify. Another different model of the house was also marketed by Strombecker. It was only 19" wide, by 13" high and was 10 1/2" deep. It was made of a lighter weight cardboard and used the same basic design of construction. Both the outside and the inside of the house were quite different, however. The printing for the windows, walls, decoration, etc. was all of a different design. Instead of the house having the look of stone, it had

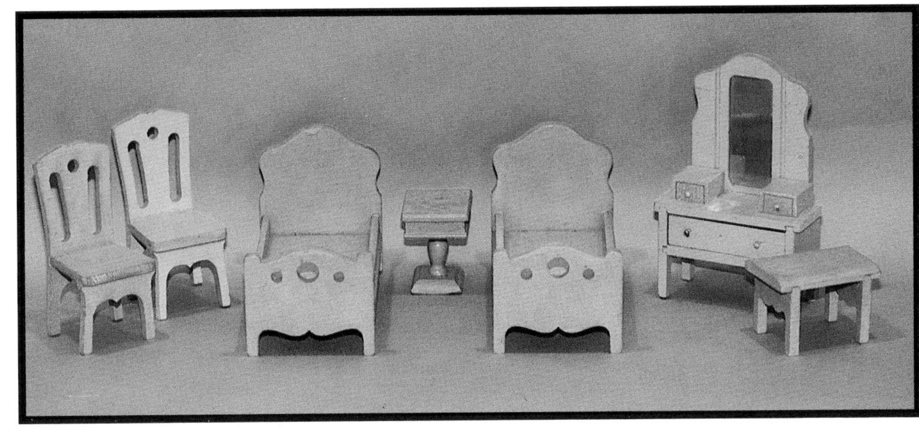

The first Strombecker dollhouse furniture was made in the 1" to 1' scale. Pictured is the bedroom furniture. The large drawer on the vanity opens. The beds were also produced without the decorative holes.

white clapboard siding. The house included a red roof, chimney, and green shutters. The houses were sold complete with furniture. The most expensive model came with fifty pieces of furniture to furnish the six rooms. The four-room house came with thirty-three pieces of furniture. The smaller house came with thirty pieces of furniture and accessories. These houses are furnished with several pieces of furniture that are seen nowhere else in the catalog. These include a three piece sofa and chair with curved arms and a yellow vanity with round mirror unlike the one pictured in the "Modern" furniture set.

By the mid 1940s, the Strombecker houses were being manufactured by Warren Paper Products who made the Built-Rite houses. In 1946 Strombecker was advertising a more conventional four-room dollhouse. The house was much like other Built-Rite houses. It came unassembed and was to be put together by the customer. The house came furnished with eighteen pieces of furniture and a car for a total cost of $2.50. The house was made of fiberboard and had an attached garage. It measured 23" long, by 11" wide by 13" high. The second floor could be lifted off the first floor if needed. The back of the house was open for play. Outside and interior decorations were printed on the fiberboard. The house had a green roof, white clapboard and brick siding.

Another house has been found furnished with Strombecker furniture that dates from around 1950. The house is very modern for its time and is complete with printed draperies and decorations both inside and out. The Strombecker furniture found with the house is the set pictured in the 1950 catalog featuring the kitchen table and chairs with wire legs. It is possible that this house was another product marketed by Strombecker furnished with their furniture (see chapter on Miscellaneous Dollhouses).

By the mid to late 1950s, as the popularity of the plastic dollhouse furniture increased, the interest in wood dollhouse furniture virtually vanished. In order to tap a different market, Strombecker designed new larger furniture to compete with the wooden furniture being marketed for the 8" - 10" dolls like the Ginny and Jill dolls made by Vogue, Inc. The larger furniture was well made and has survived the years very well. The sturdy furniture was just the right size for Ginny, Jill, and their friends but when Mattel's Barbie arrived on the scene in 1959, the Strombecker furniture was too small for her. Perhaps that is why the line was discontinued. The furniture included a living room sofa and chair with the flocked look of upholstery on the cushion area, a corner lamp table, coffee table, a dining room table and chairs, twin beds (could be bunked), wardrobe, four poster bed with canopy, chest of drawers, rocking chair, patio set, lawn swing, baby tender, baby crib, play pen, potty chair, cradle, baby bath, and a high chair.

When the 8" Betsy McCall dolls (made by American Character) became so popular, several of the same furniture designs were also used for Betsy. The difference in the furniture is that the McCall furniture was painted white and decorated with decal designs while the regular Strombecker furniture was finished in a light wood finish. The white furniture promoted by McCalls included a rocking chair, twin/bunk beds, canopy bed, table and chairs, wardrobe, and three drawer chest. McCalls magazine advertised the Betsy McCall furniture in their November, 1958 issue. The pieces listed in the ad include a rocker for $2.00, a wardrobe for $3.50, and a four poster bed for $4.50. The ad states the furniture is made by Strombeck-Becker and that other pieces are also available. The furniture is marked on the bottom "© McCalls/Made by Strombecker." The larger regular Strombecker furniture is marked "Strombecker/Moline, Ill. U.S.A."

Although dollhouse collectors think the Strombecker dollhouse furniture was a very important product, the company, itself, relied on their toy model kits to stay profitable in the toy business. By 1955, plastic products had taken over this line and Strombecker had lost 75% of the model business because they made only wood kits. After Strombecker's unsuccessful try at producing plastic kits, the Dowst Manufacturing Co. (Tootsietoy) purchased the Strombecker model kit line in 1961. The Strombeck-Becker Manufacturing Co. name was also sold to Dowst at that time and the original Strombecker Company became known as the Strombeck Manufacturing Co.

Several sets of dollhouse furniture have been found in original bubble packs with the identification "Strombeck Manufacturing Co." printed on the packages. These items were issued after the company's name change in 1961. In a directors' meeting that was held in April, 1962, it was announced that all the toy and kit lines had been eliminated so the furniture may date from this short period of time in the early 1960s. The items contained in the packages are a combination of furniture products issued in the early 1950s and new designs.

The Strombeck Manufacturing Co. continued to produce custom wood-working products, mainly handles, for many more years. The complete firm was sold in 1985 to Stuhr Enterprises Inc. The new owners moved the company from the Moline factory in 1989 and the firm is now located in Wilton, Illinois where they continue the "handle" business started so long ago by J. F. Strombeck.

There are so many different pieces of Strombecker furniture that a very nice collection can be assembled which includes only the products from this one firm. The furniture was made for a longer period of time than that of any other company so the Strombecker furniture is also still plentiful. What is surprising is that so many complete mint-in-box rooms of furniture still turn up at antique shows, shops, and flea markets. As in any collectible, these are the best and most expensive examples of the Strombecker line of dollhouse furniture.

The early Strombecker bathroom furniture in the 1" scale was produced in both plain green enamel and in green enamel with gold decorations added. The bathtub was produced both with a shower and without. A vanity, bench, opening hamper, and a medicine cabinet were all interesting additions to the basic bathroom pieces. Photograph and furniture from the collection of Patty Cooper.

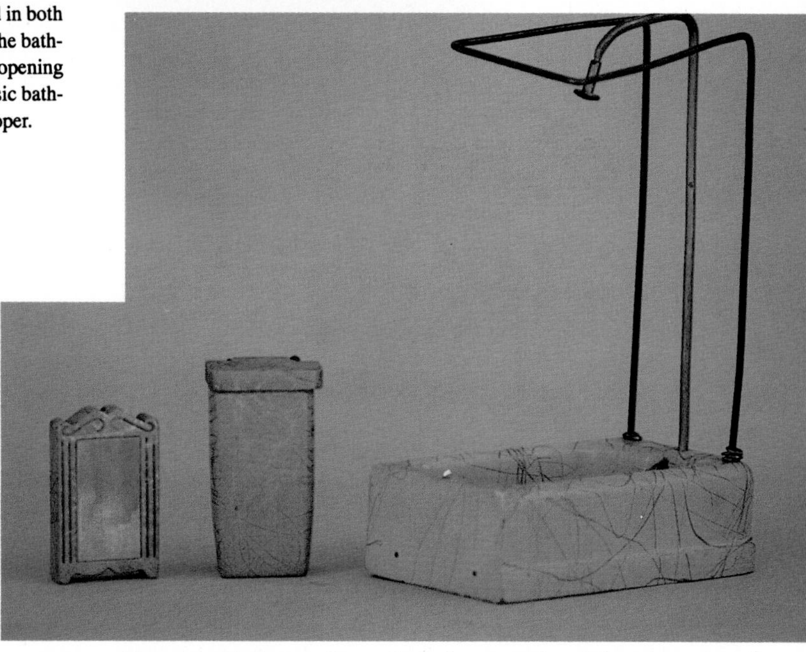

Pictured is the shower, along with the hamper and medicine cabinet, from the early 1" scale bathroom set.

This 1931 1" scale living room furniture has often been confused with Schoenhut. The Strombecker set includes a sofa, chair, footstool, end table, radio, grandfather clock, and library table. The Strombecker library table has a stretcher between the legs while the Schoenhut table has no stretcher (see Schoenhut chapter).

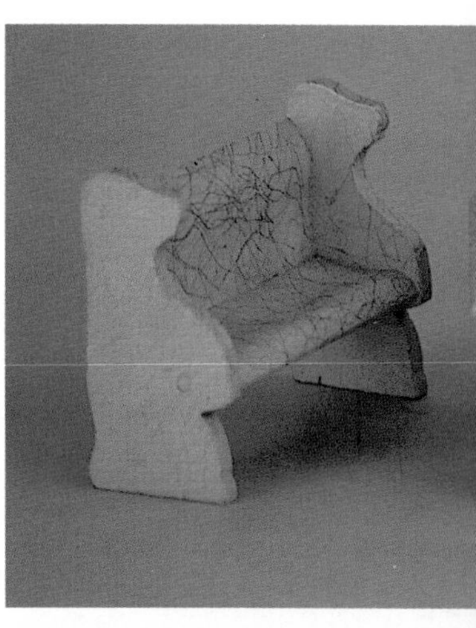

The 1" scale grand pianos were produced in both plain enamel and in red with the added gold designs. Both pianos are circa early 1930s as are the jardiniere and floor lamp.

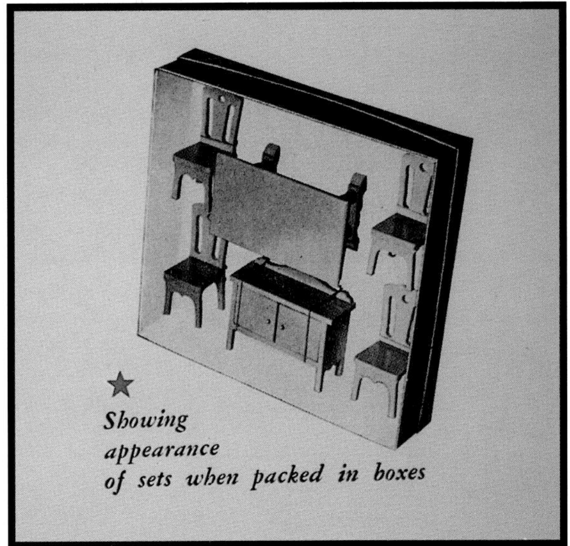

Showing
appearance
of sets when packed in boxes

Pictured is a boxed set of the 1" scale dining room furniture as shown in the Strombecker catalog for 1934.

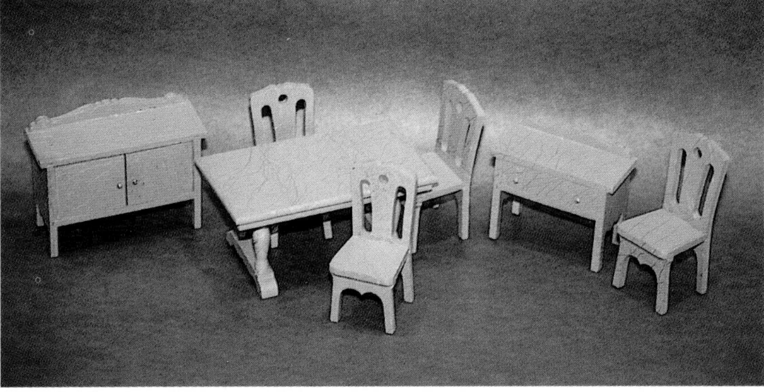

This 1" scale dining room furniture also dates from the early 1930s. This furniture seems to be just a little larger than 1" scale but it was part of the Strombecker first line of dollhouse furniture which was made in the 1" scale. The doors on the server open and the drawer in the buffet is also functional. The set was also produced in green.

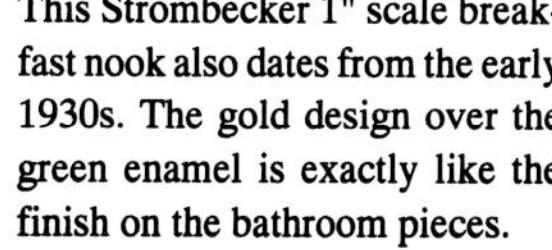

This Strombecker 1" scale breakfast nook also dates from the early 1930s. The gold design over the green enamel is exactly like the finish on the bathroom pieces.

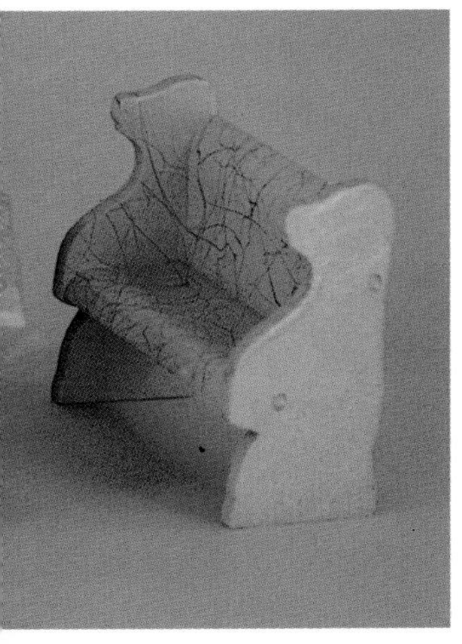

These kitchen pieces of furniture were also part of the first line of Strombecker furniture in the 1" scale offered in 1931. The kitchen cabinet contained functional doors and the door on the refrigerator also opened. The chairs are the same design as the dining room chairs without the added cutouts. Furniture from the collection of Gail and Ray Carey. Photograph by Gail Carey.

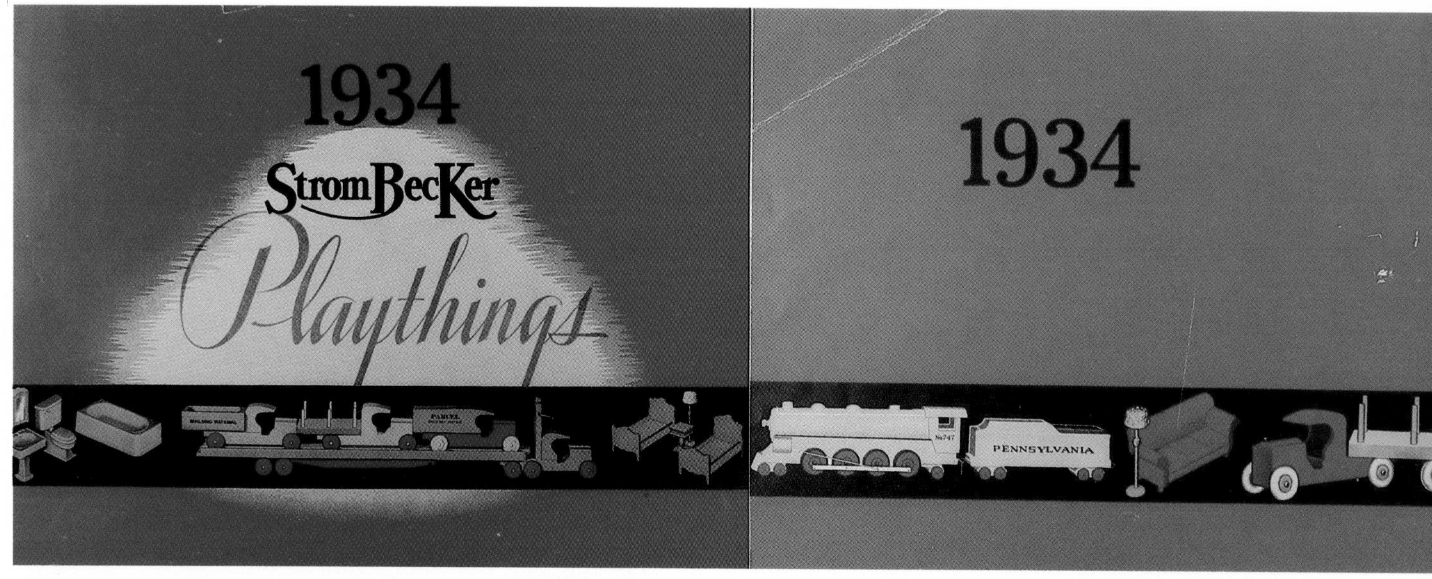

The front of the 1934 Strombecker catalog pictures several pieces of their 1" to 1' scale of dollhouse furniture. From the collection of C. S. Olson.

The back of the catalog for 1934 also shows some 1" scale furniture including the red sofa. From the collection of C. S. Olson.

Pictured is an early 1" scale Strombecker kitchen stove circa 1933 along with the refrigerator with a coil on top.

The Strombecker 3/4" to 1' scale of dollhouse furniture was first offered for sale in the company catalog for 1934. The catalog pictured this bedroom furniture in walnut but the same set was also produced for many years finished in pink enamel.

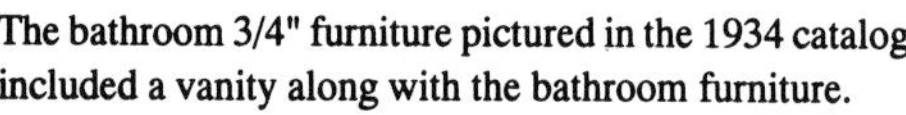

The bathroom 3/4" furniture pictured in the 1934 catalog included a vanity along with the bathroom furniture.

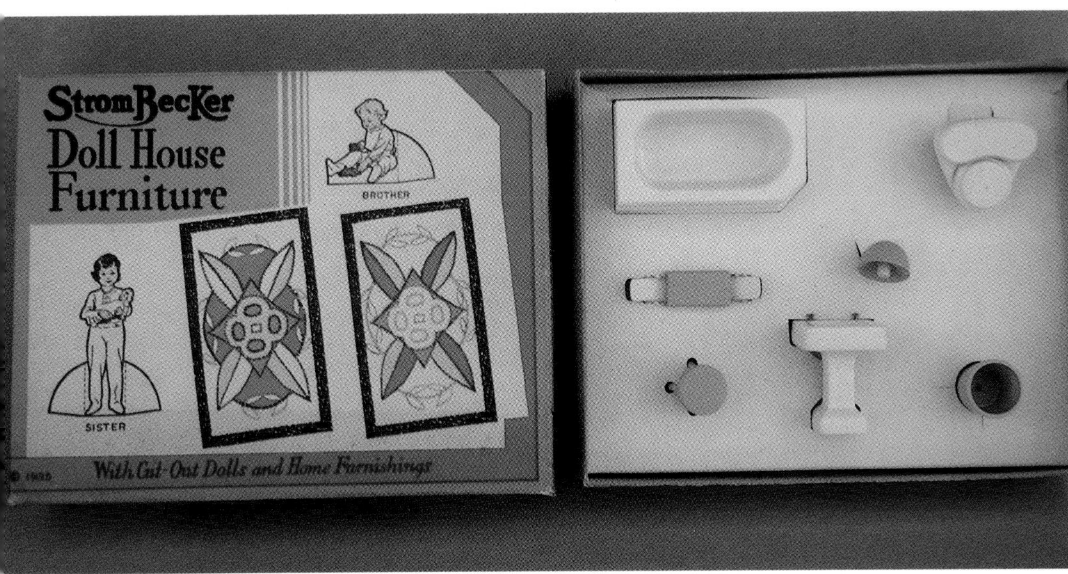

This boxed bathroom set of 3/4" scale furniture is dated 1935. The box also included several cut-outs to accompany the furniture. Photograph and furniture from the collection of Betty Nichols.

Pictured are 3/4" scale kitchen pieces first produced in 1934. This furniture continued to be sold through the 1930s.

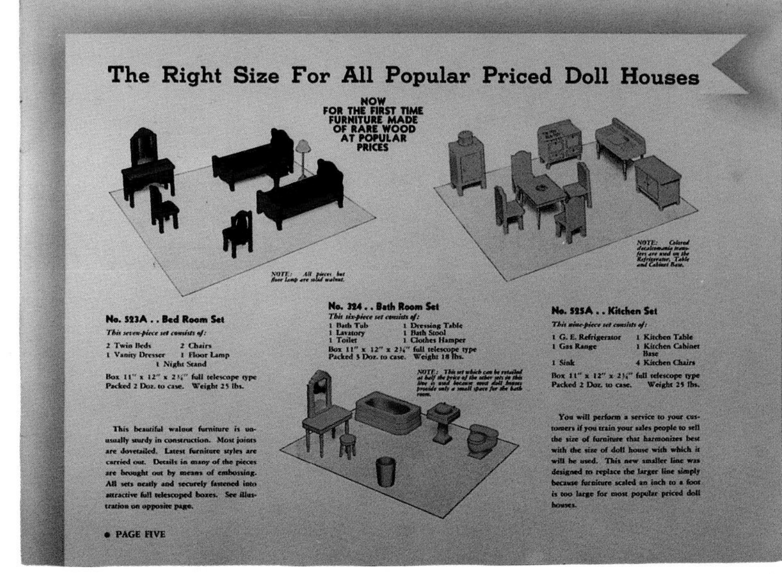

The Right Size For All Popular Priced Doll Houses

NOW FOR THE FIRST TIME FURNITURE MADE OF RARE WOOD AT POPULAR PRICES

NOTE: All pieces but floor lamp are solid walnut.

NOTE: Colored decalcomania transfers are used on the Refrigerator, Table and Cabinet Base.

No. 523A . . Bed Room Set

This seven-piece set consists of:

2 Twin Beds, 2 Chairs, 1 Vanity Dresser, 1 Floor Lamp, 1 Night Stand

Box 11" x 12" x 2¼" full telescope type
Packed 2 Doz. to case. Weight 25 lbs.

No. 324 . . Bath Room Set

This six-piece set consists of:

1 Bath Tub, 1 Dressing Table, 1 Lavatory, 1 Bath Stool, 1 Toilet, 1 Clothes Hamper

Box 11" x 12" x 2¼" full telescope type
Packed 3 Doz. to case. Weight 18 lbs.

NOTE: This set which can be retailed at half the price of the other sets in this line is used because most doll houses provide only a small space for the bath room.

No. 525A . . Kitchen Set

This nine-piece set consists of:

1 G. E. Refrigerator, 1 Kitchen Table, 1 Gas Range, 1 Kitchen Cabinet Base, 1 Sink, 4 Kitchen Chairs

Box 11" x 12" x 2¼" full telescope type
Packed 2 Doz. to case. Weight 25 lbs.

This beautiful walnut furniture is unusually sturdy in construction. Most joints are dovetailed. Latest furniture styles are carried out. Details in many of the pieces are brought out by means of embossing. All sets neatly and securely fastened into attractive full telescoped boxes. See illustration on opposite page.

You will perform a service to your customers if you train your sales people to sell the size of furniture that harmonizes best with the size of doll house with which it will be used. This new smaller line was designed to replace the larger line simply because furniture scaled an inch to a foot is too large for most popular priced doll houses.

● PAGE FIVE

A Strombecker catalog page from 1934 pictures the 3/4" scale walnut bedroom furniture, the bathroom pieces in pink, and the kitchen done in green enamel. The kitchen included a refrigerator with a coil top and a low cabinet as well as the more usually found items. From the collection of C. S. Olson.

These 3/4" scale living room pieces were first made in 1934. The 1934 catalog pictured the sofa and chair in green. The table radio was the first radio made by Strombecker in the 3/4" scale.

Unpainted furniture sets were also offered in the company catalog for 1936. The sets included in the ad were the bedroom, dining room, living room, bathroom, and kitchen. Bathroom pieces from the unpainted set of furniture are pictured. These sets could be purchased for a cheaper price than the other furniture. Furniture from the collection of Gail and Ray Carey. Photograph by Gail Carey.

Although the 3/4" dining room was pictured in walnut in the 1934 catalog, the blue enamel pieces were made in the same design and are easier to find. Photograph and furniture from the collection of Patty Cooper.

Pictured are several Strombecker kitchen pieces. The small sink and stove are 3/4" scale from the early 1934 line. The smaller coil top refrigerator is from the early 1931 1" line. The larger coil top refrigerator and the larger stove are Strombecker pieces from the 1936 line of furniture. The stove originally came with a top that could be closed to cover the burners. Photograph and furniture from the collection of Patty Cooper.

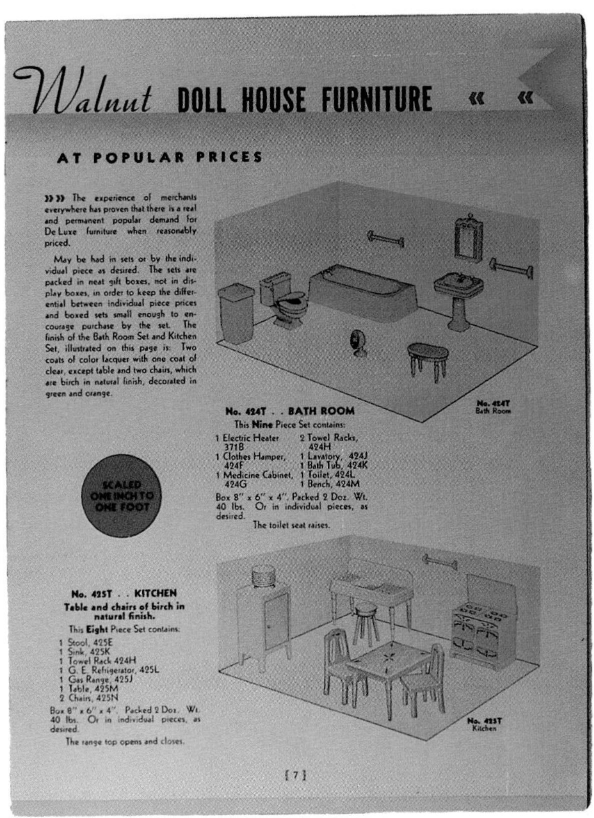

Walnut DOLL HOUSE FURNITURE « «

AT POPULAR PRICES

»» The experience of merchants everywhere has proven that there is a real and permanent popular demand for De Luxe furniture when reasonably priced.

May be had in sets or by the individual piece as desired. The sets are packed in neat gift boxes, not in display boxes, in order to keep the differential between individual piece prices and boxed sets small enough to encourage purchase by the set. The finish of the Bath Room Set and Kitchen Set, illustrated on this page is: Two coats of color lacquer with one coat of clear, except table and two chairs, which are birch in natural finish, decorated in green and orange.

SCALED ONE INCH TO ONE FOOT

No. 424T . . BATH ROOM

This **Nine** Piece Set contains:

1 Electric Heater 371B
1 Clothes Hamper, 424F
1 Medicine Cabinet, 424G
2 Towel Racks, 424H
1 Lavatory, 424J
1 Bath Tub, 424K
1 Toilet, 424L
1 Bench, 424M

Box 8" x 6" x 4". Packed 2 Doz. Wt. 40 lbs. Or in individual pieces, as desired.

The toilet seat raises.

No. 424T Bath Room

No. 425T . . KITCHEN

Table and chairs of birch in natural finish.

This **Eight** Piece Set contains:

1 Stool, 425E
1 Sink, 425K
1 Towel Rack 424H
1 G. E. Refrigerator, 425L
1 Gas Range, 425J
1 Table, 425M
2 Chairs, 425N

Box 8" x 6" x 4". Packed 2 Doz. Wt. 40 lbs. Or in individual pieces, as desired.

The range top opens and closes.

No. 425T Kitchen

[7]

Strombecker catalog page from 1936 showing the "1" scale" kitchen and bathroom pieces. The legs on the kitchen table are made from the same design as the ones on the sink. From the collection of C. S. Olson.

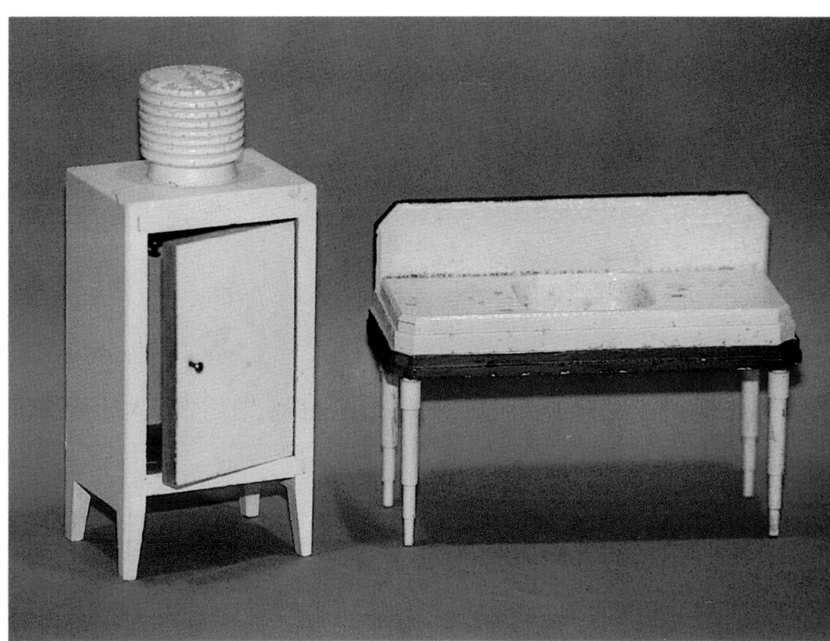

This sink on legs was also one of the 1" scale kitchen pieces pictured in the Strombecker catalog for 1936. Pictured with it is the refrigerator from the same year.

The 1" scale living room furniture shown here is also pictured in the 1936 catalog. Missing is the sofa that matches the chair.

This walnut dining room furniture was also part of the 1" line in 1936. The drawer opens in the buffet. A pair of candlesticks also came with the set of furniture.

This walnut 1" scale bedroom furniture was pictured in the 1936 catalog. Also included with the set were a pair of candlesticks and a lamp.

NEW MODERN DOLL HOUSE FURNITURE SETS

THE CORRECT SCALE FOR ALL POPULAR PRICED DOLL HOUSES

FREE MAT

No. 551 . . MODERN LIVING ROOM

This **Ten** Piece Set contains:

1 Davenport
1 Lounge Chair
1 Foot Stool
1 Coffee Table
1 Fireplace
1 Radio
1 End Table
2 Floor Lamps
1 Table Lamp

Packed 1 set to box. 2 Doz. assorted to case. Box 14" x 10" x 2½". Weight 25 lbs. to case.

The Davenport, Lounge Chair, and Foot Stool are upholstered in Izarine fabric type finish. The unbreakable round mirror is of highly polished nickeloid.

No. 552 . . MODERN COLONIAL DINING ROOM

This **Ten** Piece Set contains:

1 Dining Table
1 High Boy Dutch Cabinet
6 Chairs
1 Buffet
1 Server

Packed 1 set to box. 2 Doz. assorted to case. Box 14" x 10" x 2½". Weight 22 lbs. to case.

No. 701X . . RUGS

Simulated Broadloom type in 9" x 12" size. Available in brown, green, or blue. May be cut with scissors to room size and smaller pieces used as throw rugs. Packed 1 Dozen. Weight 1 lb.

No. 553 . . MODERN BEDROOM

This **Ten** Piece Set contains:

2 Beds
1 Night Stand
2 Chairs
1 Table Lamp
1 Vanity Dresser
1 Vanity Bench
1 Table Clock
1 Costumer

Packed 1 set to box. 2 Doz. assorted to case. Box 14" x 10" x 2½". Weight 21 lbs. to case. Unbreakable mirror is of highly polished nickeloid.

No. 554 . . 20TH CENTURY BATHROOM

This **Ten** Piece Set contains:

1 Bath Tub
1 Toilet
1 Lavatory
1 Medicine Cabinet
1 Clothes Hamper
1 Dressing Table
1 Dresser Bench
1 Bath Scale
1 Electric Heater
1 Bath Stool

Packed 1 set to box. 2 Doz. assorted to case. Box 14" x 10" x 2½". Weight 23 lbs. to case. Unbreakable mirrors are of highly polished nickeloid.

[14]

No. 555 . . MODEL KITCHEN

This **Ten** Piece Set contains:

1 Cabinet Sink
1 Electric Refrigerator
1 Gas Range
1 Kitchen Table
4 Kitchen Chairs
1 Table Clock
1 Kitchen Stool

Packed 1 set to box. 2 Doz. assorted to case. Box 14" x 10" x 2½". Weight 22 lbs. to case.

No. 556 . . MODERN SUN ROOM

This **Eight** Piece Set contains:

1 Davenport
1 Table
2 Arm Chairs
1 Radio
1 End Table
1 Floor Lamp
1 Table Lamp

Packed 1 set to box. 2 Doz. assorted to case. Box 14" x 10" x 2½". Weight 23 lbs. to case.

No. 547 . . SCHOOL ROOM

Every little girl loves to play teacher. This set has exceptional play value. We introduced it last year, and it clicked far beyond our expectations. It is sold both for doll houses, and for imaginative play anywhere. Your trade will like this good item which excels for table, floor, or doll house play.

This **Six** Piece Set contains:

1 Teacher's Desk with 2 real drawers.
1 Teacher's Swivel Chair that turns.
4 Students' Seat-Desks with tilting seats.

Also, a cut-out blackboard, waste basket, and set of books.

Packed 1 set to box. 2 Doz. assorted to case. Box 14" x 10" x 2½". Weight 23 lbs. to case.

SCALE 3/4-INCH TO ONE FOOT

16-PIECE, 4-ROOM COMBINATION FURNITURE SET

No. 548—A special 4-room, 16-piece set packed in one box, for promotion in conjunction with a 4-room doll house and for the many people who want a flashy gift at a price but do not want to give a single room of furniture—or are limited in their willingness to spend—a StromBecKer super value item.

Packed 1 complete 4-room set to box. 2 Dozen assorted to case. Box 14" x 10" x 2½". Weight 28 lbs. to case.

FREE MAT

[15]

In 1938 Strombecker offered a whole new design of the 3/4" scale of dollhouse furniture. The company called it the "Modern" design. Shown are the catalog pages from that year picturing the new line of furniture. From the collection of C. S. Olson.

The new "Modern" dining room furniture included ten pieces. The legs on this set of furniture are very unusual.

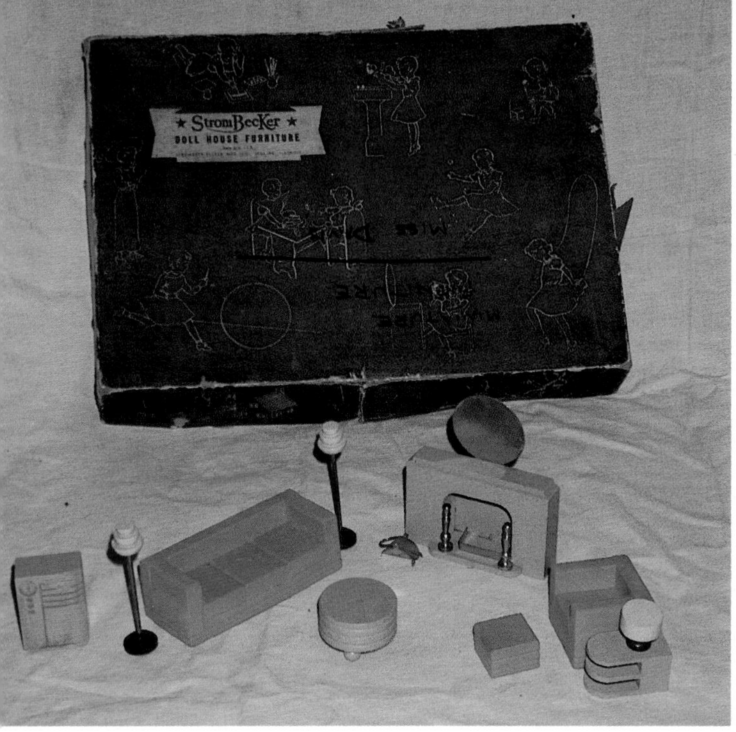

The new living room design featured a fireplace with a round mirror and very unusual lamps. This boxed set of furniture is very hard to find complete with lamps. Photograph and furniture from the collection of George Mundorf.

Two different vanities were produced that match this set of "Modern" bedroom furniture. The vanity on the left was used in the furnishings of the Strombecker dollhouses on the market in 1938. Photograph and furniture from the collection of Patty Cooper.

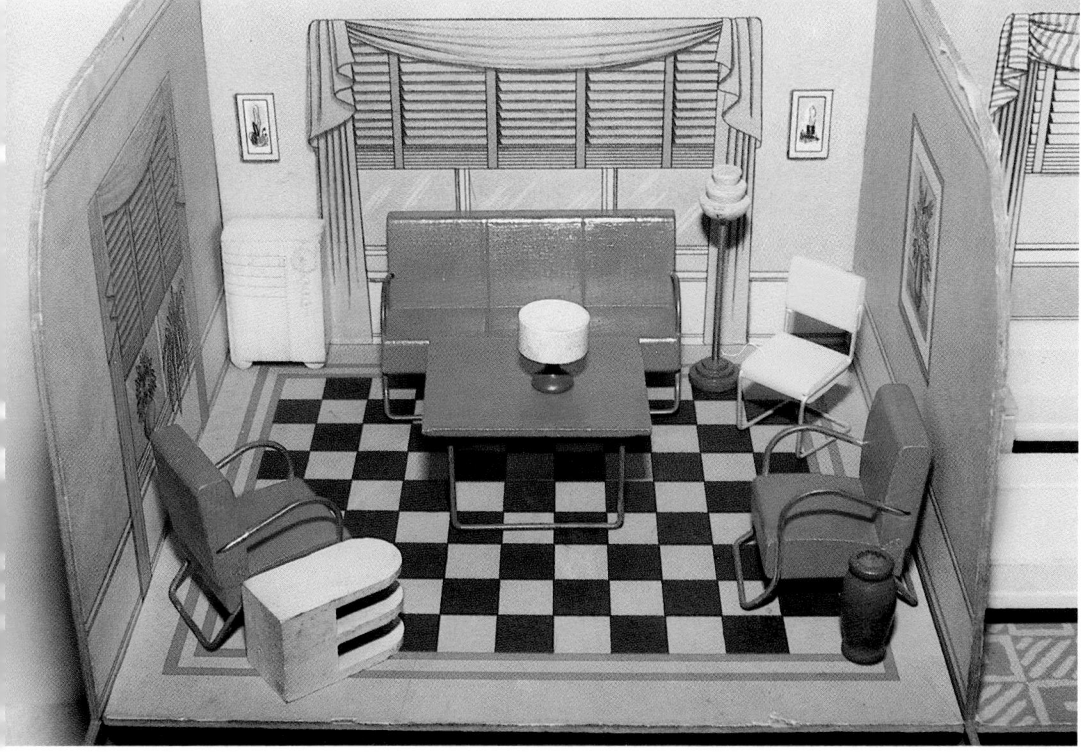

The furniture provided for the sun room was especially attractive and these wire and wood pieces are very collectible. The yellow chair probably dates from the late 1940s and is part of the kitchen set from that period. Photograph and furniture from the collection of Roy Specht.

Strombecker also brought out a new line of 1" scale furniture in 1938. All of the furniture had been updated and had a new lighter look. The living room set shown is missing the new style radio, a "Philco" design that had a slant top and was flush with the floor.

The 1" scale dining room pieces included a buffet with opening drawer and a serving cart. The accessories included a wood bowl and a pair of candlesticks.

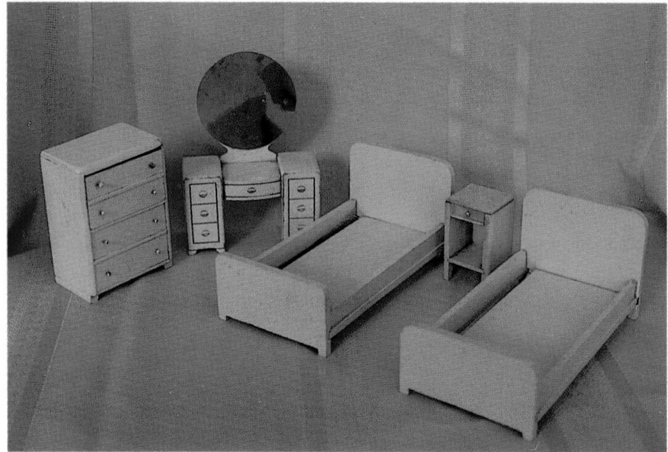

The 1" scale bedroom was pictured in the catalog in a white enameled finish. Accessories included a clock and two lamps. Photograph and furniture from the collection of Patty Cooper.

Furniture for a child's room in the 1" scale was also offered in 1938 for the first time. This set is a favorite for collectors. Missing is the shoofly in the shape of ducks. Accessories included a lamp and clock. The color of the first set was grey with orange and yellow trim.

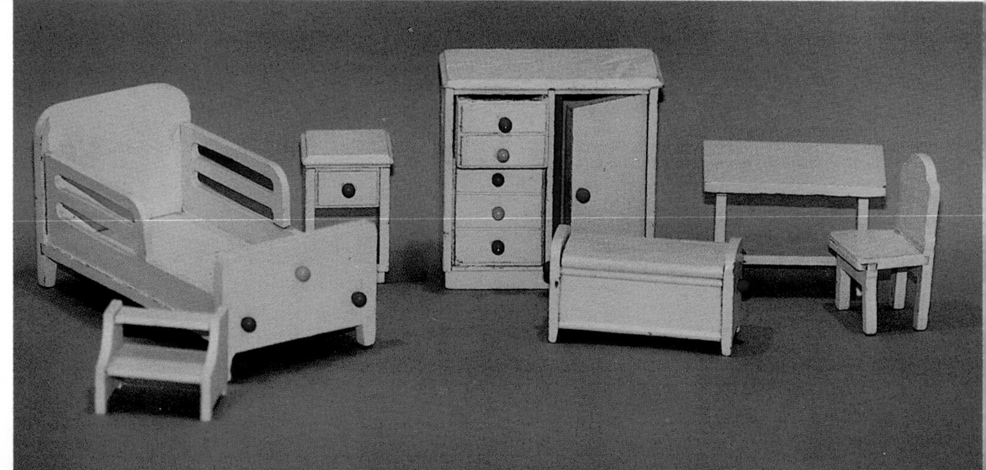

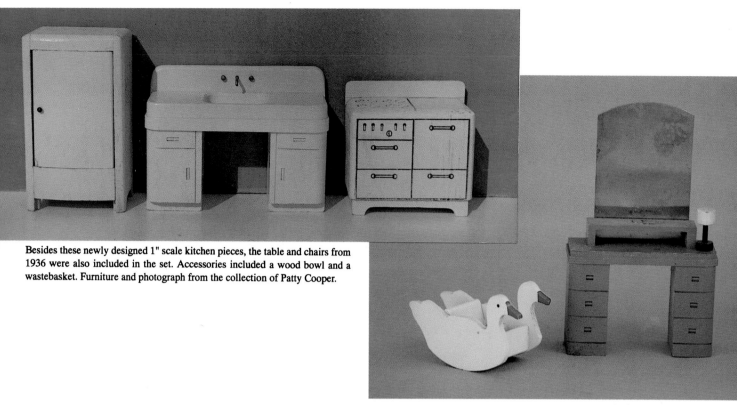

Besides these newly designed 1" scale kitchen pieces, the table and chairs from 1936 were also included in the set. Accessories included a wood bowl and a wastebasket. Furniture and photograph from the collection of Patty Cooper.

The new bathroom set in the 1" scale included an unusual vanity that is hard for collectors to find. Pictured with the vanity are the shoofly rocker from the child's room, and one of the "modern" lamps.

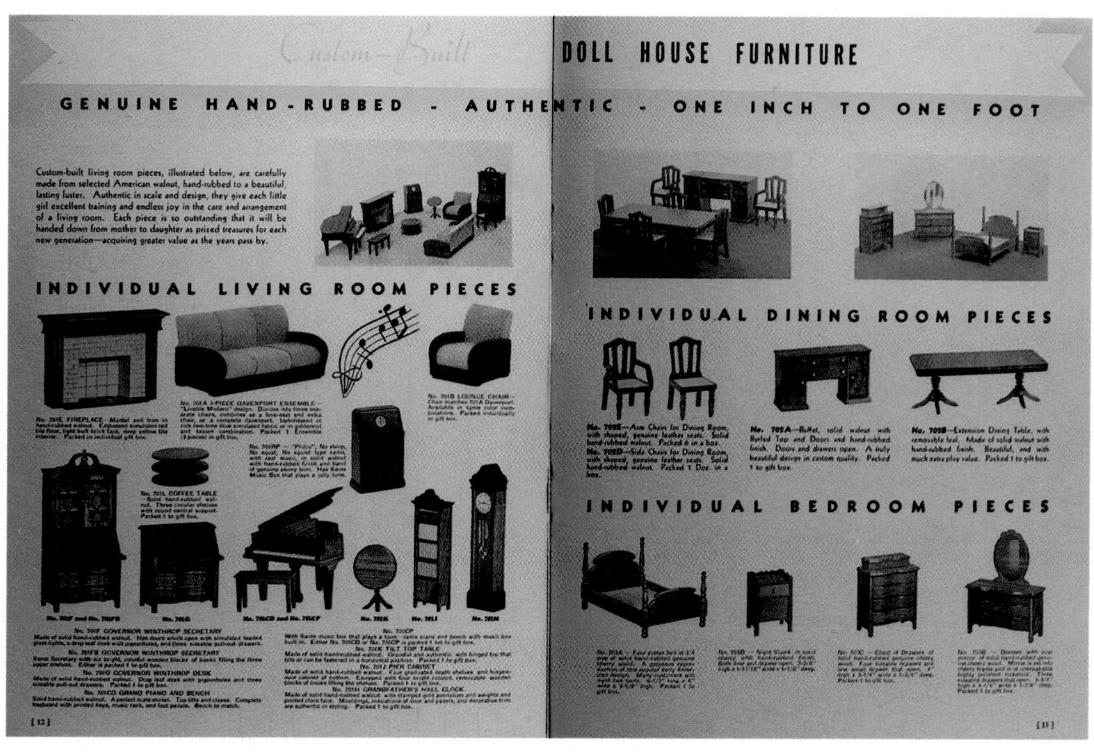

Custom-Built

DOLL HOUSE FURNITURE

GENUINE HAND-RUBBED - AUTHENTIC - ONE INCH TO ONE FOOT

Custom-built living room pieces, illustrated below, are carefully made from selected American walnut, hand-rubbed to a beautiful, lasting luster. Authentic in scale and design, they give each little girl excellent training and endless joy in the care and arrangement of a living room. Each piece is so outstanding that it will be handed down from mother to daughter as prized treasures for each new generation—acquiring greater value as the years pass by.

INDIVIDUAL LIVING ROOM PIECES

INDIVIDUAL DINING ROOM PIECES

INDIVIDUAL BEDROOM PIECES

Besides the newly designed 1" scale furniture, Strombecker also offered a more expensive line of "Custom-Built Dollhouse Furniture" at this time. The line was also in the 1" scale and has become a favorite of collectors. Pictured are pages of the 1938 Strombecker catalog showing many of the furniture items. From the collection of C. S. Olson.

The "Custom-Built" Fireplace, Grandfather's Clock, and Pier Cabinet are pictured. The cabinet was complete with books. All three items are made of walnut.

The Governor Winthrop secretary from the "Custom" line. The drawers and doors on the secretary are functional and the front drops down. The secretary is missing the piece that was originally mounted above the doors. The furniture is made of walnut.

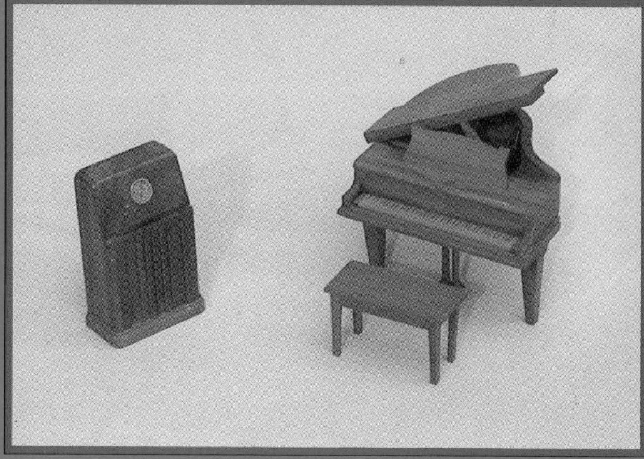

Pictured are the "Custom-Built" musical grand piano and radio. The walnut pieces contain music boxes with wind-up keys located in their backs. This radio, without the music box, was part of the living room furniture in 1938. From the collection of Kathy Garner. Photograph by Bill Garner.

Pictured are the "Custom-Built" tilt top table, Governor Winthrop desk and the regular 1" scaled radio from the Strombecker line. The drawers and the pull down desk are functional. All the pieces are walnut. The top of the table can be tilted to the front. From the collection of Kathy Garner. Photograph by Bill Garner.

This three-piece sectional sofa was featured in the 1938 Strombecker catalog. It was part of the "Custom-Built" line. The same design was also used in a matching lounge chair. From the collection of Kathy Garner. Photograph by Bill Garner.

The bedroom furniture included with the cheaper houses was not so elaborate and included a 3/4" scale of chest of drawers instead of the mirrored vanity.

Several pieces of the "Custom-Built" furniture were made of cherry. Pictured are the nightstand, dresser, poster bed, and chest of drawers. The drawers are functional in the furniture. From the collection of Kathy Garner. Photograph by Bill Garner.

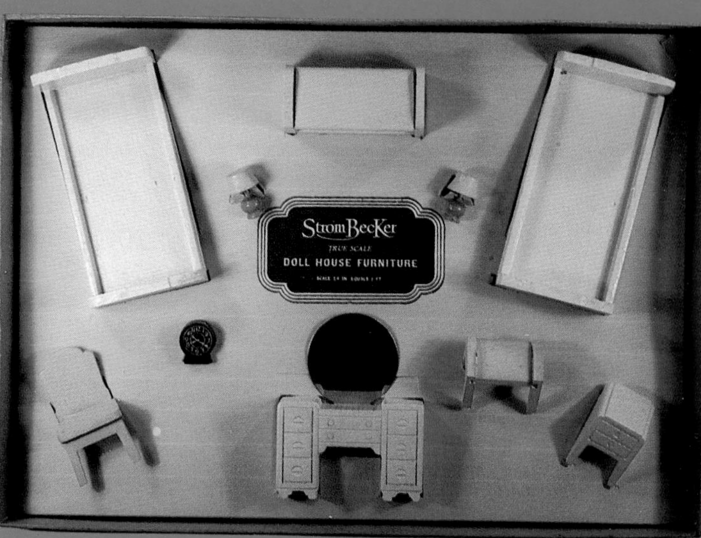

By the early 1940s, Strombecker again made changes in their 3/4" line of dollhouse furniture. The 1942 Montgomery Ward Christmas catalog pictures this bedroom furniture as part of the furnishings for a large Rich six-room house.

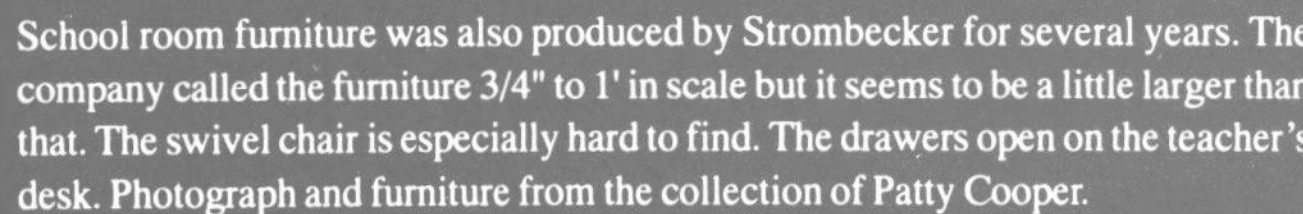

School room furniture was also produced by Strombecker for several years. The company called the furniture 3/4" to 1' in scale but it seems to be a little larger than that. The swivel chair is especially hard to find. The drawers open on the teacher's desk. Photograph and furniture from the collection of Patty Cooper.

The large Rich dollhouse in the Montgomery Ward Christmas catalog from 1942 also included the walnut 3/4" set of Strombecker dining room furniture.

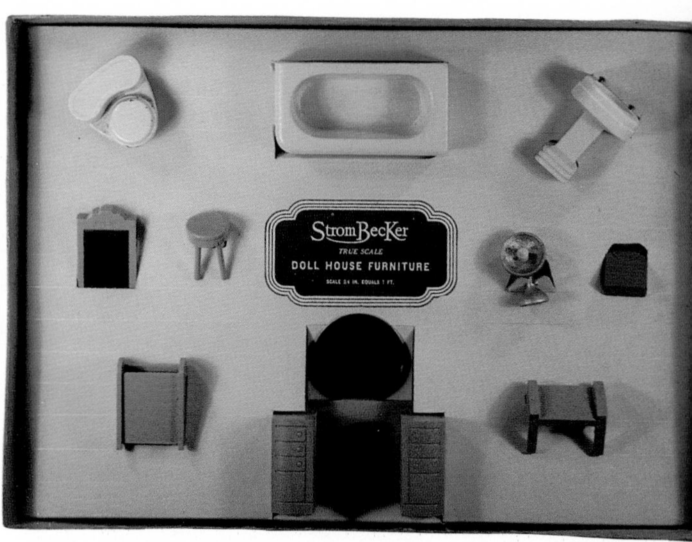

The same style Strombecker box housed this set of bathroom furniture. The vanity is especially unusual as the lavatory could fit between the drawers for a "built-in" look.

The cheaper Rich four-room house included only a Strombecker dining table and four chairs in a wood finish.

This set of 3/4" kitchen furniture also dates from the mid-1940s. It included newly designed appliances but used the trestle table from years past.

By the mid-1940s, Strombecker had made several more changes to the designs of their 3/4" scale furniture. The living room grand piano had been replaced by an upright and two different designs of floor standing radios were being used in the line of furniture. Photograph and furniture from the collection of Patty Cooper.

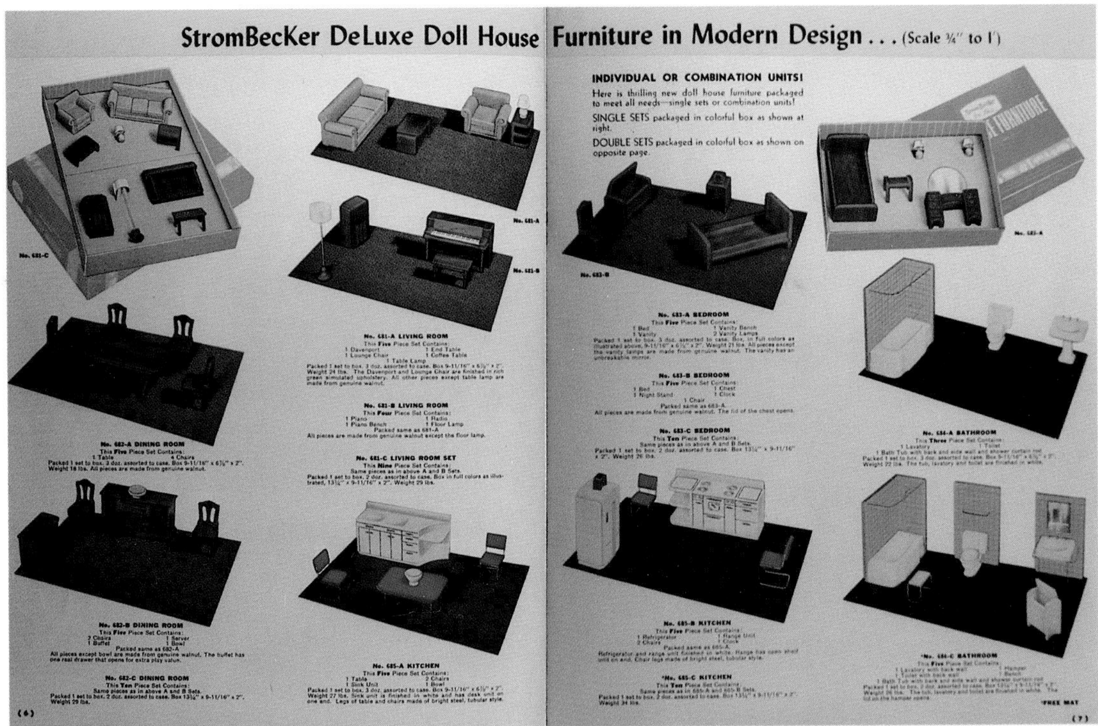
StromBecKer DeLuxe Doll House Furniture in Modern Design . . . (Scale ¾" to 1')

INDIVIDUAL OR COMBINATION UNITS!
Here is thrilling new doll house furniture packaged to meet all needs—single sets or combination units!
SINGLE SETS packaged in colorful box as shown at right.
DOUBLE SETS packaged in colorful box as shown on opposite page.

No. 681-A LIVING ROOM

No. 681-B LIVING ROOM

No. 681-C LIVING ROOM SET

No. 682-A DINING ROOM

No. 682-B DINING ROOM

No. 682-C DINING ROOM

No. 685-A KITCHEN

No. 683-A BEDROOM

No. 683-B BEDROOM

No. 683-C BEDROOM

No. 684-A BATHROOM

No. 685-B KITCHEN

*No. 685-C KITCHEN

*No. 684-C BATHROOM

By the late 1940s, Strombecker once again produced 3/4" scaled dollhouse furniture with several new changes. The company catalog for 1950 pictured the new sets of furniture. The kitchen pieces with open shelves were a new innovation along with the wire and wood kitchen table and chairs. The dining room set included all new furniture which featured a Duncan Phyfe type table. This set is very hard to locate. Catalog from the collection of C. S. Olson.

This 3/4" scaled Duncan Phyfe style dining room table, chairs, buffet, and server were produced circa 1950. This set of walnut furniture is one of the hardest sets of Strombecker furniture to locate. From the collection of Kathy Garner. Photograph by Bill Garner.

The Strombecker bathroom furniture circa 1950 which included imitation tile backgrounds. The bathtub piece originally contained a shower curtain and rod. A mirror was also originally a part of the lavatory piece. Photograph and furniture from the collection of Patty Cooper.

This box for the 1" to 1' scale Strombecker furniture (circa 1953) pictures the regular line of this furniture then offered by the company. The only changes to the line appear to be the addition of a television set for the living room and a different design of sofa, chair, and footstool.

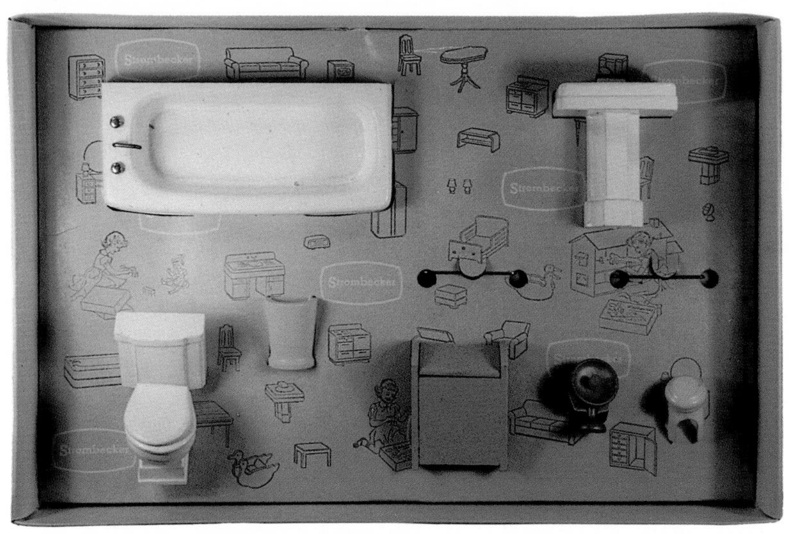

The 1" scale bathroom furniture circa 1953. No changes appear to have been made since the set was pictured in the company's catalog for 1938. The vanity was no longer a part of the set.

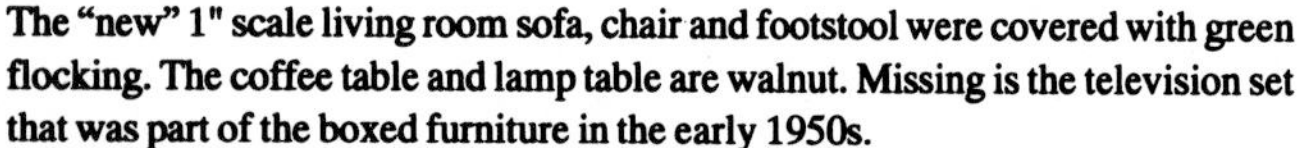

The "new" 1" scale living room sofa, chair and footstool were covered with green flocking. The coffee table and lamp table are walnut. Missing is the television set that was part of the boxed furniture in the early 1950s.

The living room for the 3/4' scale of furniture also included a television. Missing from this set of furniture is a lamp table. This sofa and chair are finished in aqua enamel as are most of these later versions of Strombecker living room pieces.

The box for the 3/4" scale Strombecker furniture from the same period also pictured the sets of furniture in the company's other line.

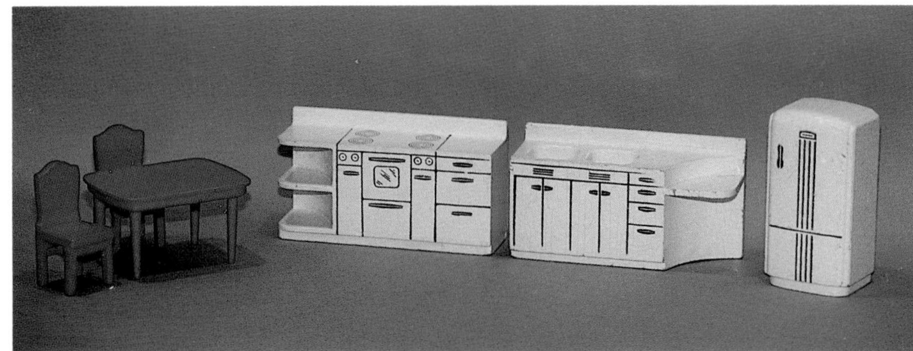

The kitchen furniture circa 1953 included the pieces from the late 1940s with the shelves removed from the sink and stove appliances. A newly designed wood table and chairs replaced the earlier wood and metal model. The kitchen appliances pictured were part of the earlier kitchen circa 1949. The bedroom, dining room and bathroom pieces were the same as those used in earlier years.

. . . No. 320F DREAM HOME DOLL HOUSE . . .

Complete with 30 pieces of furniture as illustrated on pages 4 and 5 of this catalog.
PACKING:—Doll house is packed in an individual carton, knocked down. Furniture is packed in a separate box—not displayed.

FEATURES OF DOLL HOUSE
1. Cellophane windows.
2. Yellow curtains.
3. Dormer windows.
4. Unusually rigid when assembled.
5. Licensed under Purdy Patents of assembly design.
6. Every room easily accessible.

View with front open and furniture in place.

Illustrating how front and back of house opens, making every room easily accessible.

View with back open and furniture in place.

The StromBecKer Dream Home, of authentic Dutch Colonial Architecture, measures 19⅜" in width (across front), 14⅜" in depth and 16¼" in height. It contains five rooms and is made from the finest grade cylinder kraft corrugated board attractively printed . . . a very substantial, rigid and durable construction.

• PAGE TEN

The 1934 Strombecker catalog pictures this "Dream Home Dollhouse" that was made of corrugated board. The house contained five rooms and was sold complete with thirty pieces of furniture. The house measured 19 3/8" wide, 14 3/8" deep, and 16 1/4" tall. Catalog from the collection of C. S. Olson.

Three Strombecker houses were pictured in the company catalog for 1938. The most expensive house contained six rooms and came completely furnished. The house was designed in a "step" fashion. The house was made of 3/16" Fibo-Board and was sent ready to assemble. The house measured 26" wide, 16 1/2" tall and 14 1/2" deep. Photograph and house from the collection of Roy Specht.

The house is shown with its original furnishings. There was no roof to the house. Photograph and house from the collection of Roy Specht.

The original living room furniture included this unusual three-piece sofa and chair that is seldom seen. Photograph and furniture from the collection of Roy Specht.

The pictured kitchen furniture is a combination of both early and new designs of furniture. The small kitchen cabinet was first included in sets of furniture in 1934 while the trestle table, chairs, sink, and stove date from 1938. The original catalog pictures a refrigerator and apartment size stove with this house in place of the cabinet and the larger stove. Photograph and furniture from the collection of Roy Specht.

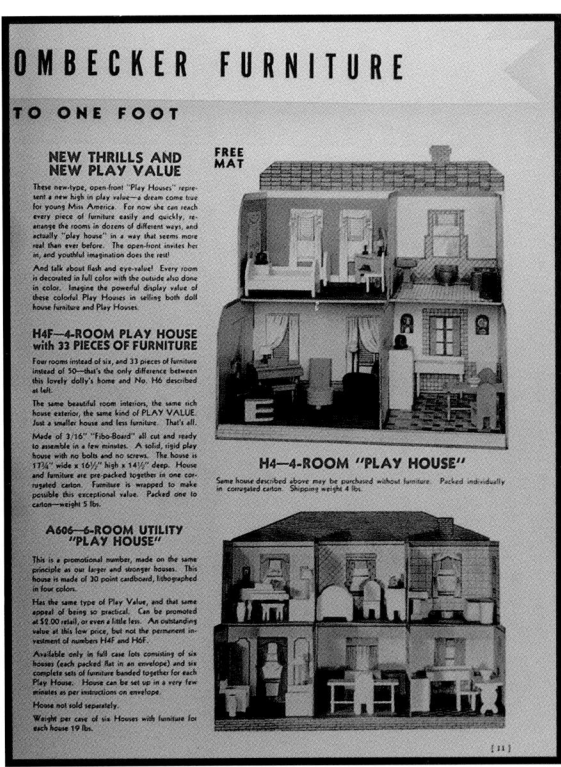

OMBECKER FURNITURE

TO ONE FOOT

NEW THRILLS AND NEW PLAY VALUE

These new-type, open-front "Play Houses" represent a new high in play value—a dream come true for young Miss America. For now she can reach every piece of furniture easily and quickly, rearrange the rooms in dozens of different ways, and actually "play house" in a way that seems more real than ever before. The open-front invites her in, and youthful imagination does the rest!

And talk about flash and eye-value! Every room is decorated in full color with the outside also done in color. Imagine the powerful display value of these colorful Play Houses in selling both doll house furniture and Play Houses.

H4F—4-ROOM PLAY HOUSE with 33 PIECES OF FURNITURE

Four rooms instead of six, and 33 pieces of furniture instead of 50—that's the only difference between this lovely dolly's home and No. H6 described at left.

The same beautiful room interiors, the same rich house exterior, the same kind of PLAY VALUE. Just a smaller house and less furniture. That's all.

Made of 3/16" "Fibo-Board" all cut and ready to assemble in a few minutes. A solid, rigid play house with no bolts and no screws. The house is 17¾" wide x 16½" high x 14½" deep. House and furniture are pre-packed together in one corrugated carton. Furniture is wrapped to make possible this exceptional value. Packed one to carton—weight 5 lbs.

A606—6-ROOM UTILITY "PLAY HOUSE"

This is a promotional number, made on the same principle as our larger and stronger houses. This house is made of 30 point cardboard, lithographed in four colors.

Has the same type of Play Value, and that same appeal of being so practical. Can be promoted at $2.00 retail, or even a little less. An outstanding value at this low price, but not the permanent investment of numbers H4F and H6F.

Available only in full case lots consisting of six houses (each packed flat in an envelope) and six complete sets of furniture banded together for each Play House. House can be set up in a very few minutes as per instructions on envelope.

House not sold separately.

Weight per case of six Houses with furniture for each house 19 lbs.

FREE MAT

H4—4-ROOM "PLAY HOUSE"

Same house described above may be purchased without furniture. Packed individually in corrugated carton. Shipping weight 4 lbs.

Two other designs of the house were also advertised in the Strombecker catalog from 1938. One of the houses was exactly like the six-room model but was produced in a four-room unit. The other house was also smaller and was made of a lighter weight cardboard. Both of these houses came furnished with fewer pieces of furniture. Catalog from the collection of C. S. Olson.

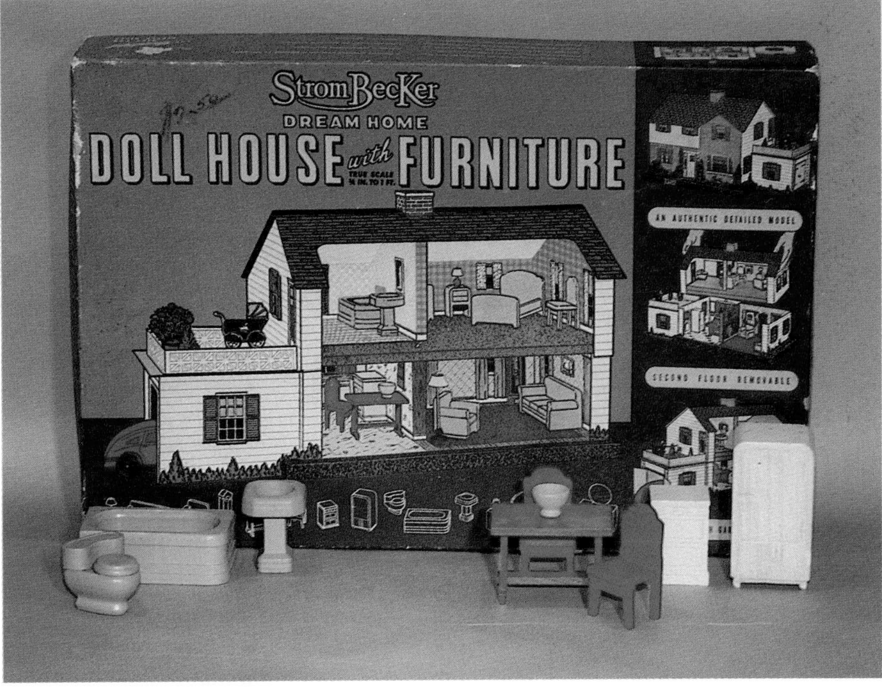

In the mid-1940s Strombecker advertised another model of a cardboard house. This house was sold in this Strombecker box. Pictured with the box are the bathroom and kitchen pieces of furniture that came with the house. The kitchen included the apartment size stove.

The bathroom furniture is also original to the house. See earlier illustration for the sun room furniture. Photograph and furniture from the collection of Roy Specht.

STROMBECKER 4-ROOM DOLL HOUSE Strombeck Becker have produced a new "Dream Home" doll house which we predict will prove to be extremely popular. It is a very attractive low priced 4-Room fiberboard house with attached garage. The illustration shows the open back of the house with the 4 rooms fully equipped with 18 pieces of painted wooden StromBecker furniture. Floors and walls are printed in pleasing decorations. House and furniture are built to true scale of ¾" to 1'. The front and sides of the house are tastefully decorated in colors to represent brick, white clapboards, and green roof. There is a toy automobile with the set and a cardboard baby carriage and Scottie dog. An exclusive idea is the "lift-off" feature, whereby the second story may be lifted off to provide easier access to the first floor. Overall dimensions are 23" long, 11" wide, and 13" high. House comes knocked down but is easily put together. **Price, postpaid, $2.50.**

The house, itself, was produced by Warren Paper Products (Built-Rite). The house contained four rooms and a garage complete with a Strombecker wooden car. This house was advertised in the *Children's Activities* magazine in March, 1946. The price was $2.50 including eighteen pieces of furniture.

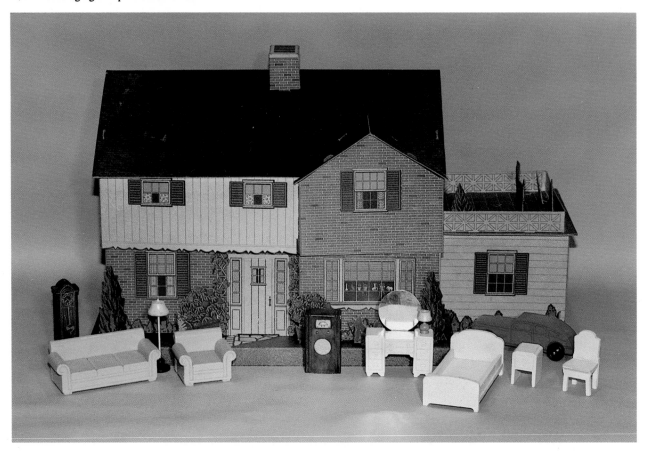

The house came with eighteen pieces of furniture. Pictured with the house is the furniture for the living room and bedroom. The radio is a different model than the walnut one that came with the sets of furniture. A wood car was also included with this Strombecker set.

Strombecker began production of wood furniture suitable for 8" dolls in the late 1950s. Pictured is a box, along with the living room chair, from this line of furniture. Furniture from the collection of Gail and Ray Carey. Photograph by Gail Carey.

Strombecker also produced several pieces of baby furniture suitable for use with 8" dolls. Included were the high chair, potty chair, crib, baby tender, cradle, playpen, and baby bath. Other items like the wardrobe could be used with either the baby or little girl dolls. Furniture from the collection of Gail and Ray Carey. Photograph by Gail Carey.

The living room furniture suitable for 8" dolls also included a sofa bed, corner lamp table, and coffee table (not pictured). Photograph and furniture from the collection of Kathleen Neff-Drexler.

In 1958 some of the regular Strombecker designs that were used for furniture for the 8" dolls were also issued under the Betsy McCall label to be used with the new Betsy McCall doll. The doll and her furniture were promoted by McCalls magazine during this time period. Pictured is a set of Strombecker bunk beds changed to the Betsy McCall design by the addition of white paint and decals. The original box is also shown. Photograph and furniture from the collection of Leslie Robinson.

This Strombecker wood table and chairs was also made to be used by the popular 8" dolls from the 1950s. It is marked on the bottom "Strombecker/Moline, Ill. U.S.A."

The Betsy McCall wardrobe and chest of drawers are pictured along with an 8" Betsy McCall doll which was made by American Character. These pieces were also based on regular Strombecker designs. Photograph and furniture from the collection of Leslie Robinson.

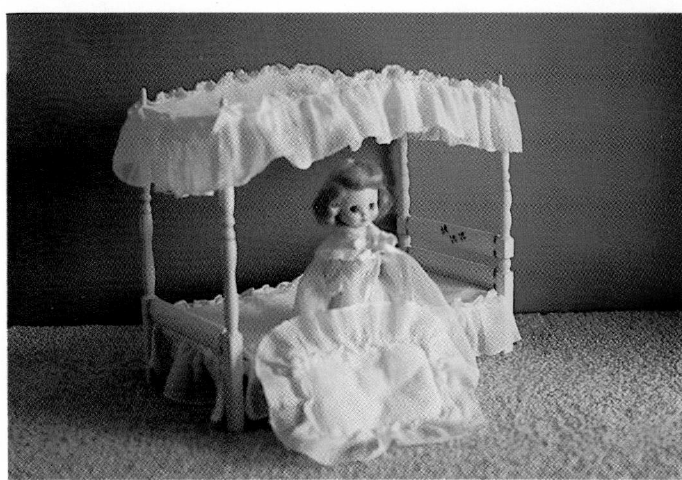

This bed with canopy is one of the most desirable of the Betsy McCall collectibles. This one includes all its original bedding. The bed was also from one of the regular Strombecker designs. An American Character Betsy McCall doll, wearing her night clothes, is pictured with the bed. Photograph and items from the collection of Leslie Robinson.

The Betsy McCall table and chairs set made by Strombecker is pictured along with an American Character Betsy McCall doll. The Betsy McCall furniture is marked on the bottom "© McCalls/Made by Strombecker." Photograph and furniture from the collection of Leslie Robinson.

A dressed up 8" American Character Betsy McCall doll is pictured beside her rocker and bed which were the regular Strombecker designs from their 8" doll line of furniture. All of the McCall furniture is marked "© McCalls/Made by Strombecker." Photograph and collectibles from the collection of Leslie Robinson.

This Strombecker umbrella table and two chairs could be used by Betsy McCall or other 8" dolls. Strombecker also made a lawn swing in its regular line. Photograph and furniture from the collection of Leslie Robinson.

Bubble pack of kitchen furniture labeled "Strombeck Manufacturing Co.". The company name changed from the Strombeck-Becker Manufacturing Co. to Strombeck Manufacturing Co. in 1961. This furniture must have been marketed in the early 1960s. Furniture and photograph from the collection of Patty Cooper.

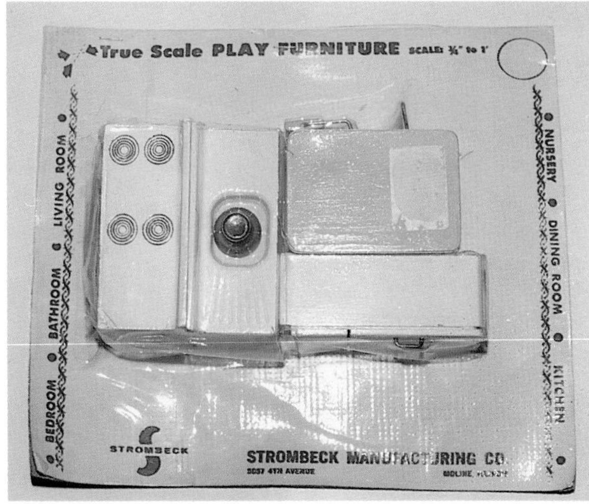

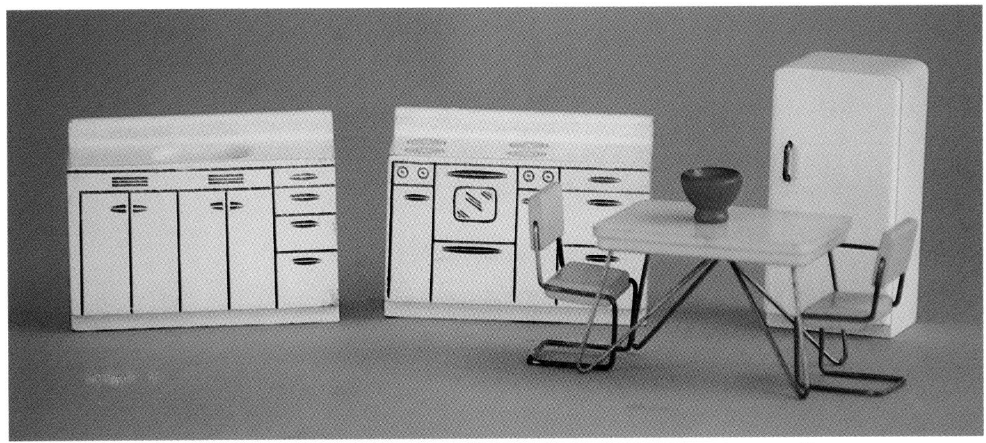

The kitchen furniture included in the bubble pack featured the table and chairs made of wood and wire.

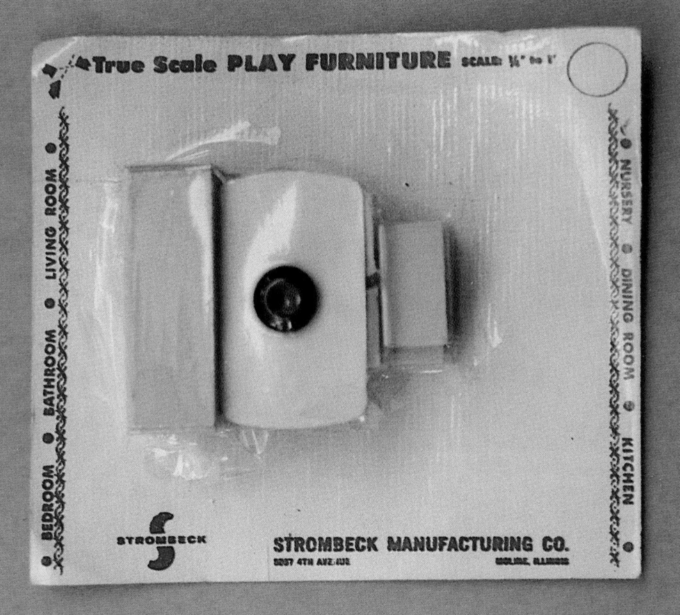

The dining room bubble pack included a table, two chairs, buffet, bowl, and serving cart. The wheels on the cart are only painted and were not functional. The legs on the table are black.

The bedroom furniture included a bed with a straight headboard and footboard, two lamps, chair, blanket chest, nightstand, and bench.

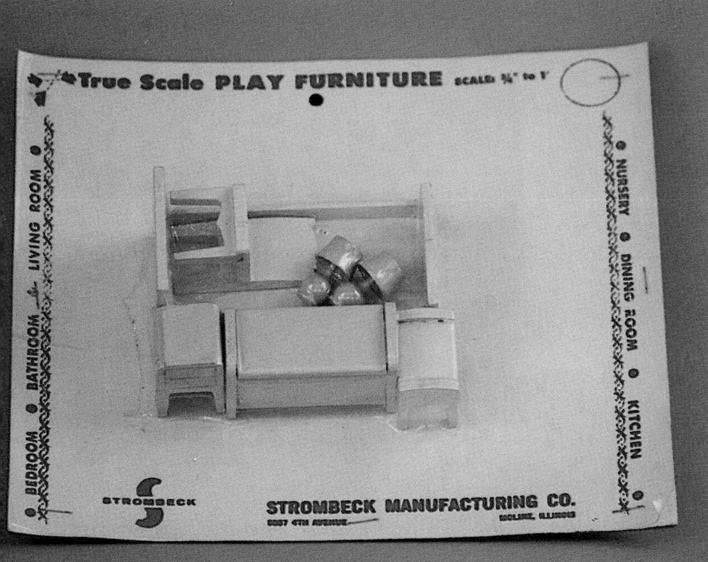

The bubble pack living room set featured a television, sofa, chair, floor lamp, coffee table, and end table.

The bathroom pieces include a bathtub, toilet, sink, scale, hamper, wastebasket, and stool. The packages state that a nursery set was also available.

Tynietoy

The houses and furniture labeled Tynietoy have appealed to collectors since their beginning shortly after World War I.

Marion I. Perkins began making miniature furniture around 1917. Because of the war, the dollhouse furniture that had formerly come from Germany was no longer available and Perkins saw a market that needed to be filled. In partnership with Amey Vernon, Perkins opened the Toy Furniture Shop in Providence, Rhode Island around 1920. The operation became successful and at the height of the furniture's popularity, fifty people were working to produce the small furniture, accessories, and dollhouses.

The furniture was supposed to be 1" to 1' in scale but it is actually just a little larger. The pieces were all based on antique designs and included Chippendale, Colonial, Empire, Windsor, Hepplewhite, Sheraton, and Victorian styles. In 1930, an "Empire Sleigh Bedroom" complete with bed, bureau, and chair cost $6.35. Furniture could also be purchased individually with bookcases listed at $2.75 each. Although the prices seem reasonable for the high quality of furniture, it was depression time in the United States and not many families could afford that much money for toys. In a time when a complete cardboard dollhouse could be purchased furnished with Strombecker furniture for under $2.00, it is unlikely that the Tynietoy products were owned by the general public. Most likely, the homes and furniture were purchased by adults as collectibles or by wealthy families as toys for their children.

The furniture was crafted with functional doors and drawers and was designed in several different finishes. Included were mahogany, walnut, pine, maple, oak, and enamel. There were over a hundred different pieces of furniture. Most of the furniture was marked with the trademark of a two-story house with a pine tree on the left and a ladder back chair on the right. This trademark was stamped into the wood on most of the furniture.

Some of the furniture pieces produced by Tynietoy included the following items: Nursery: Crib, bureau, rocker, high chair, table, chairs, and screen. Bedroom: Victorian spool bed, bureau, and nightstand; Empire sleigh bed and bureau; Pineapple bed, four-poster bed, and swell front bureau. Living room: Victorian table, arm chair, side chair; Hepplewhite love seat, side chairs, arm chairs; Sheraton love seat, arm chairs; Camel back sofa, book case, and Chippendale secretary. Kitchen: Breakfast nook and kitchen cupboard. Dining room: Victorian table and chairs; Sheraton table, chairs and sideboard; and swell front sideboard. Many other pieces of Tynietoy furniture were also produced during their many years of business.

The Tynietoy firm also offered over one hundred accessories to accompany their furniture. Many of these items probably came from other sources and some of the metal pieces were made by Tootsietoy. The ice box provided for the Tynietoy kitchen was produced by Hubley and other accessories may have been imported from Germany.

Dollhouses were also crafted by Tynietoy. Even by today's standards, they were expensive. The houses came in several different designs, from a two-room cottage to an eleven-room manor house. The larger houses were made in a Georgian style painted white with green shutters. The larger two-story styles were over 6' wide. A mansion with nine rooms, three halls, and a pantry was 6'2" wide, 2' 8" high and 1' 5 1/2" deep.

In 1930 the house sold for $145 unfurnished. If a garden was included, the total cost was $170. Furniture could be obtained for another $100. According to the advertisement for this house, it was wired for electricity, had non-breakable windows, contained a staircase, and a real fountain. The front of the house was removable.

Other house models made by Tynietoy included a Nantucket house (five rooms) with a captain's lookout, a two-room cottage, a New England town house, and a four-room village house.

Tynietoy also offered their own line of dollhouse dolls including cloth wired and wooden peg dolls.

After Amey Vernon's death, Marion Perkins sold out her interest in Tynietoy in 1942. The Tynietoy furniture was carried by Louise Fales Specialities for a few more years. By the early 1950s, the Tynietoy products were no longer on the market.

The Tynietoy furniture, houses, and accessories are still very collectible today but their high cost make it difficult for most collectors to pursue an interest in this field.

Large Tynietoy Georgian house containing eight rooms and attic including the kitchen wing on the left. The front of the house has been removed for easy access to the rooms. The house contains electric lights and a stairway as well as a music room. The house is furnished with Tynietoy furniture. From the collection of The Toy and Miniature Museum of Kansas City.

One of the rooms in the Tynietoy dollhouse. The wonderful craftsmanship used in the making of the furniture can be seen. The furniture is all in a scale of slightly over 1" to 1'. From the collection of The Toy and Miniature Museum of Kansas City.

Dining room furniture made by Tynietoy included this Sheraton sideboard, round table, and four chairs. Several other styles of dining room furnishings were also produced. Furniture from the collection of Gail and Ray Carey. Photograph by Gail Carey.

Tynietoy spool bed, dresser, and Windsor-type chair all made by Tynietoy. The drawers open on the Tynietoy furniture. Furniture from the collection of Gail and Ray Carey. Photograph by Gail Carey.

Sheraton sofa, swell front bureau, and shelf clock all produced by Tynietoy. Furniture from the collection of Gail and Ray Carey. Photograph by Gail Carey.

Tynietoy nursery crib, bureau, and rocker. These pieces were made in pink and yellow enamel as well as blue. Also pictured is the New England Cradle. Photograph and furniture from the collection of Patty Cooper.

Chippendale chair and Sheraton half moon table, also produced by Tynietoy.

Tynietoy ladder back chairs and mirror enameled pink. Furniture from the collection of Gail and Ray Carey. Photograph by Gail Carey.

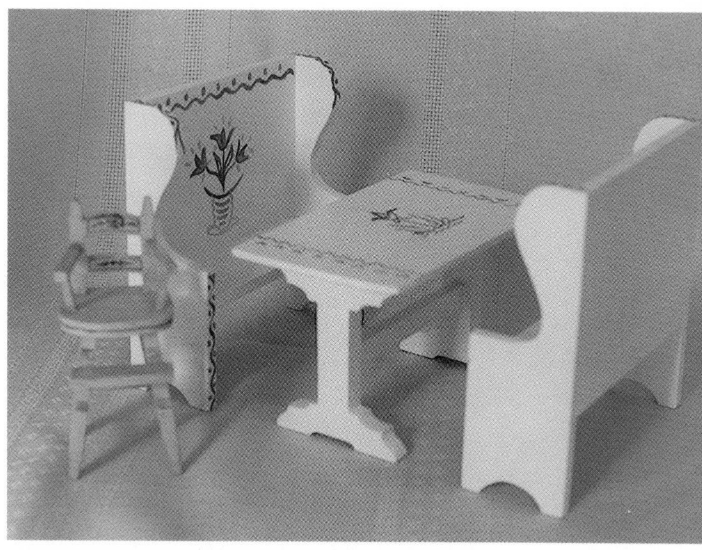

The Tynietoy baby high chair is pictured with the kitchen breakfast nook. Photograph and furniture from the collection of Patty Cooper.

Shown are the ladder back settee and rocker made by Tynietoy. Photograph and furniture from the collection of Patty Cooper.

The Tynietoy Victorian chair, grandfather clock, and wing chair are pictured. Photograph and furniture from the collection of Patty Cooper.

Rich

Maurice Rich Sr. and Edward M. Rich founded the Rich Company in 1921 in Sterling, Illinois. In 1923 the firm switched from making their original product of tops for automobiles and began the manufacture of toys. In 1935 the company moved to Clinton, Iowa and became the Rich Toy Manufacturing Co. During the 1950s the firm moved again, eventually locating in Tupelo, Mississippi where labor was cheaper. Dollhouses were made from approximately 1935 to the early 1960s.

Although most of the dollhouses are not marked, several characteristics can be used to identify the Rich products. Most of the houses were made of U. S. Gypsum hardboard and are very similar to the Keystone houses. The inside walls of the Rich houses, however, are not decorated with pictures and other decor as are the Keystone houses. The floors of the kitchens and bathrooms in the Rich houses are usually decorated with a diamond shaped pattern to indicate tile. Many of the other floors were originally flocked but this flocking has sometimes worn off. Another pattern used by Rich was an evergreen tree printed on many of the shutters on the houses. The earliest Rich houses had opening metal casement windows. Later houses had acetate windows silk screened with black or white window panes. The larger two-story houses included staircases while the smaller ones did not. Some of the fancier houses featured doorbells, and a porch light as well as electric lights.

The Rich Toy Co. made many different models of their dollhouses. The houses contained from two rooms to six rooms and varied in size from 17 1/2" by 8" by 14" tall to 35" by 14" by 25" high. The houses were made in many different architectural styles including Colonial, Tudor, Art Deco (modern), Cape Cod, and one of the last Rich designs, a ranch style house. Besides these more elaborate houses, Rich also produced several models of simple cottages.

The Rich cottages were made in various designs in the 1930s. Each house was made of pressed hardboard and contained from two to three rooms. Like all Rich houses, the cottages were open in the back to provide easy access to the rooms. The inside walls of many of these cottages were left in their natural brown color.

The "Art Deco" house produced by Rich in the 1930s is an especially attractive model. The company called it a "Modernistic Dollhouse." The house measured 25" by 11 1/2" by 15". It was made of U. S. Gypsum hardboard and included four rooms in a two-story design. The second floor featured an open porch. The front of the house included a large curved bay window and a matching curved overhang above the door. The roof was flat. The house is similar in design to the "Art Deco" model made by Built-Rite (see chapter on Built-Rite).

The most often found Rich dollhouses are those made in a Colonial or Tudor style of architecture. Many different models of these houses were made during the decades of Rich dollhouse production. Several of the Colonial and Tudor Rich dollhouses were featured in the Montgomery Ward Christmas catalogs during the 1940s. In 1942, a four-room, two-story "Colonial Mansion" was pictured that sold for $4.25. It was 27" long, 14" wide and 19 1/2" tall. The kitchen had built-in cabinets. The house could be purchased furnished with thirty-seven pieces of Strombecker furniture for $5.79. Also featured in the same catalog was a larger six-room Rich Tudor styled house. This model was 33" long, 15 1/4" wide, and 22 1/2" to the top of the chimney. It came furnished with forty-six pieces of Strombecker furniture for $9.50. A similar Tudor six-room house (35" by 15" by 23 1/4") was pictured in the Montgomery Ward Christmas catalog for 1948. This model was priced at $8.25 unfurnished. Judging by Rich advertising material, this same house was still being made by the company during its last years in business.

The ranch house was quite different from the other Rich designs. The house was 26" long by 12 1/4" high by 15 1/2" deep. It was a one-story house containing four rooms. The model also included a covered patio. The house had a hinged roof as well as an open back. This house was produced at the end of the Rich production of dollhouses.

In addition to its regular dollhouses, the Rich Toy Co. also produced the unusual Colleen Moore Castle dollhouses during the mid-1930s. Colleen Moore was one of the most popular film stars of the silent screen. She retired from movies shortly after the coming of sound but she did not retire from the public spotlight.

Besides her films, Colleen Moore was very well known for her famous castle dollhouse. The castle took nine years to complete and was finished in the mid-1930s. It measures 9' by 9' by 14' tall. The castle contains eleven rooms and more than 700 artisans and craftsman worked on the structure to make it the marvel it became. Colleen Moore's father, Charles Morrison, supervised the project while it was being built in Hollywood, California.

When the Colleen Moore Castle was completed, it was taken on a tour of the country in order to raise money for charity. The castle was displayed in many of the big cities of America during this tour. It is possible that the Rich castles were sold by department stores in these cities during this time. After the national tour ended, the castle was given a permanent home at the Museum of Science and Industry in Chicago, Illinois.

Because of the popularity of the dollhouse and all the publicity it generated, several products were produced to tie-in with the castle publicity. The EFFanBEE Doll Co. marketed several editions of their six-inch Wee Patsy Dolls as "Fairy Princess The Colleen Moore Dollhouse Doll." The boxed composition dolls came in sets of either one or two dolls. The larger sets included Wee Patsy dressed in both girl and boy costumes. The inside cover of the doll box described the Colleen Moore Castle as well as the Fairy Princess Doll.

Of most interest to dollhouse collectors, however, are the commercially made dollhouses produced in the image of the Colleen Moore Castle. The Rich Toy Co. produced two different sizes of the castle. The houses were made of Masonite and the larger model contained eight rooms while the smaller one had only five rooms. The larger castle included a courtyard, two-story hall, chapel, drawing room, kitchen, dining room, bathroom, and two bedrooms. The house measured 20" high, 30" wide, and 16" deep. The smaller five-room castle was 24" wide, by 17" high by 13" deep. Both of these castles were authorized Colleen Moore products.

The outsides of the castles were painted cream with a grey trim. The roof and turret tops were red. The inside walls were left in the natural brown of the Masonite product while the floors were painted. The windows were covered with isinglass which featured black individual pane divisions. The scale of the Rich castles was 1/2" to 1'.

The original Colleen Moore Castle still draws visitors and the wonderful items it contains continue to bring a feeling of wonder to young and old alike. Although Colleen Moore died several years ago, her legacy of interesting films and her fascinating castle will keep her memory alive for years to come.

According to the Clinton County Historical Society in Clinton, Iowa, the Rich Company met hard times when they moved South and the firm was forced to discontinue business.

Because the Rich houses are sturdy and were sold through major mail order catalogs, many of these products can still be found by today's collector. Although not very many collectors are lucky enough to own a Rich Castle, most collectors can locate a Rich Colonial Mansion to add to a dollhouse collection.

Rich Toy Co. made several different models of cottage dollhouses beginning in the 1930s. Each house was made of pressed hardboard and contained from two to three rooms. The early cottages featured metal casement windows. Pictured is an early cottage attributed to Rich. It contains the early metal windows, the evergreen tree designs on the shutters, and a front door similar to other larger Rich dollhouses. All of the Rich houses were decorated on the outside with shurbs, trees, vines, and other landscaping additions. These designs were probably added through a silk screening process. Photograph and house from the collection of Marilyn Pittman.

This cottage, attributed to Rich, includes only two rooms. The house is 17 1/2" wide, 8" deep, and 14" tall (not including chimney). The house has metal windows on the second floor and plastic on the lower floor. The insides of the upstairs windows have been replaced. The interior of the cottage contains no wall or floor decorations but has been left in the natural brown of the hardboard.

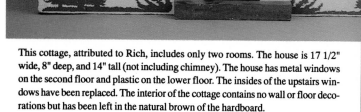

This cottage was probably also made by the Rich Toy Co. The house has a similar chimney to the "Birchwood" house advertised by Rich in 1939. The two wood pieces on the chimney top were also used by Rich on several of their houses. The evergreen design on the shutters also indicates the cottage was made by Rich. Photograph and house from the collection of Marilyn Pittman.

This Rich advertisement dates from 1938. Pictured are three of the company's hardboard houses from the period. The cottage was priced at only $1.47. The ad appeared in the All-American Products Corp. 1938 Christmas catalog. The company was located in Chicago. Catalog from the collection of Marge Meisinger.

Two more Rich dollhouses were pictured in an ad for Massey's Drug Store located in Shirley, Indiana. The 1940 Christmas catalog was from the Famous Funn Family Service. A cottage was priced at $2.25 while a four-room Colonial house was listed for a cost of $3.25. The Colonial house also featured an electric light. Catalog from the collection of Marge Meisinger.

Rich used several different Colonial dollhouse designs through the years. Pictured is a four-room model. The house measures 27" long by 14" deep by 19 1/2" to the chimney top. It was made of heavy pressed wood and contained metal framed transparent windows upstairs. The house featured a porch light and the familiar diamond pattern floors in the kitchen and the bathrooms.

The Rich Company also produced Colonial six-room houses. Pictured is a six-room model that measures 28" wide, 19" high (not including chimney) and 12" deep. The house features the diamond shaped pattern on the kitchen and bathroom floors as well as the familiar evergreen tree design on the shutters. The house was made of U. S. Gypsum board.

The inside of the six-room Colonial house included a stairway, as did most of the six-room, two-story Rich houses.

Like Keystone, the Rich company also produced several houses designed in the Tudor manner. These houses were made in both four-and six-room models. Pictured is one of the four-room designs. The house had two rooms upstairs and two rooms downstairs. The inside was finished in white and the floors included the diamond pattern in the bathroom and kitchen, a green "rug" in the living room, and a blue "rug" in the bedroom. The windows were clear plastic with white "panes." The outside of the house included a front step. Photograph and house from the collection of Marilyn Pittman.

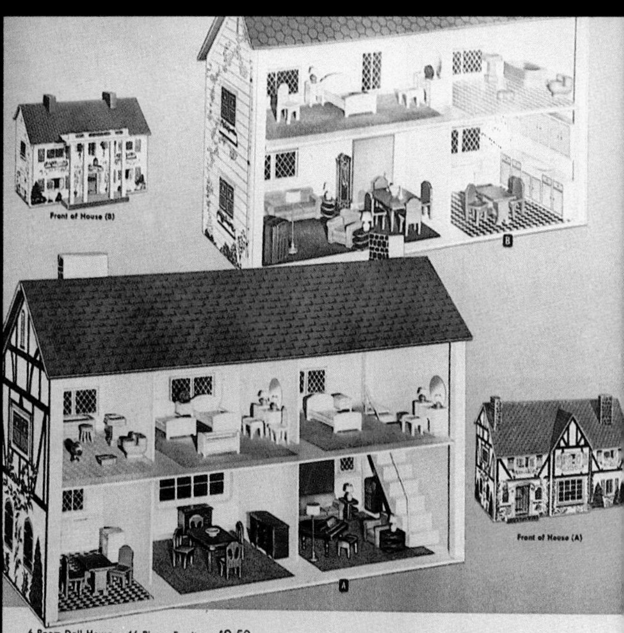
Front of House (B)

Front of House (A)

6-Room Doll House—46 Pieces Furniture $9.50

oorbell rings . . . front door opens! Wards largest and finest. English-style—ade of ⅛ in. smooth pressed-wood; will not warp, split or crack. Windows clear plastic panes—3 metal casement windows on 2nd floor hinged to . House entirely open in back for "playing dolls." 33 in. long; 15¼ in. 22½ in. to chimney top. First floor painted to resemble stone—2nd floor er and stucco. Interior staircase. Kitchen and bathroom floors painted tile-. Two bedrooms, living and dining rooms have sprayed floss rugs!

ks. sturdy Wood furniture included; piano, radio, davenport, end tables, , beds, chests, chairs, complete bathroom set—and others shown; many Walnut. Davenport and lounge chair finished to look like real upholstery; pieces painted. Shipped flat, easily assembled.

408—Doll House and 46 Pieces Furniture. Ship. wt. 28 lbs. Mailable **$9.50**

4-Room Doll House—37 Pieces Furniture $5.79

[B] **Open the front door**—"walk right in!" You'll love this Colonial Doll Mansion—so sturdy it will last to be used by several children. Size 27 in. long; 14 in. wide; 19½ in. to chimney top. Made of heavy ⅛ in. pressed-wood; won't crack, split or warp. Four metal-frame transparent windows—2 hinged double windows downstairs. Back of house is open to permit placing of furniture. Painted White with Green roof and shutters. Swinging wood porch lantern. Four big rooms—Kitchen has built-in "play" cabinets and sink. Tile-effect floor in kitchen and bathroom. Living room and bedroom "carpeted" with felt-like fabric sprayed on floors. Shipped flat; easily assembled.

448 T 407—Doll House only (Furniture not included). Mailable. Wt. 14 lbs. **$4.25**

449 T 409—Doll House Complete with 37 pieces Wood Furniture (shown and described under 48 T 405 at bottom of page). Ship. wt. 19 lbs. Mailable **$5.79**

BELOW—37 PIECES OF WOOD DOLL HOUSE FURNITURE—WILL FURNISH 5 ROOMS

. Substantially made Wood Furniture, brightly painted. Living Room: and chair "upholstered" to look and feel like real material. Sofa 4¼ in. other pieces in proportion. Radio, 2 end tables, "grandfather" clock

sticks, bowl. Bedroom: Bed, chair, vanity table with metal mirror, vanity bench, night stand, lamp, 2 candlesticks, clock. Kitchen: Range, sink, table, 2

Rich also manufactured several six-room Tudor houses. Although the houses were quite similar, there were differences in the design of the windows and the doors. Montgomery Ward featured a large six-room model in their Christmas catalog in 1942. The house was 33" long, 15 1/4" deep, and 22 1/2" to the chimney top. The house sold for $9.50 furnished with Strombecker wood furniture. The other Rich house advertised, along with the Tudor model, was a smaller Colonial house. It was priced for $5.79 furnished with thirty-seven pieces of Strombecker furniture. Catalog from the collection of Betty Nichols.

Another Rich Tudor dollhouse was advertised in the Montgomery Ward Christmas catalog for 1948. This house was listed as 35" long by 15 1/4" deep by 23 1/4" to the chimney top. The house was built of Duron Fiberboard (made by U. S. Gypsum Co). Although the house pictured appears to be the same house shown in the Christmas catalog it measures only 34" wide.

"Duron" Fiberboard Doll House

6-ROOM...RIGIDLY BUILT...FURNITURE SEPARATE

Rear View of Doll House shown above, showing room locations and suggested furniture placement. Packed flat for shipment; easily assembled.

Buy Christmas Gifts on Wards Monthly Payment Plan—See Credit Terms on Page 176.

EXTRA LARGE DOLL HOUSE FOR LITTLE GIRLS TO PLAY AT KEEPING HOUSE. Neat and sturdy ... built of Duron Fiberboard made by U. S. Gypsum Co. Finished in white with red, brown and green stencil designs; red blinds, green roof, stone chimneys. Door opens. Clear casement stenciled plastic windows. Six large rooms—Living room from which stairway leads to upper floor. Dining room and Kitchen; Bedroom, Nursery and Bath on second floor. Open at back. Kitchen and Bath have stenciled tile pattern floors, other rooms colorful napped rugs. Shipped flat. Walls and roof hinged for quick assembly with screws included.

Furniture, shown in small picture of house, is n cluded with house—it may be bought separately listing directly below (48 T 4520). House: 35 by 23¼ in. high to chimney top. Ship. wt. 21 lbs.
48 T 4476M—Doll House only. Mailable........

49-Pc. PLASTIC FURNITURE SET. Selected from bottom of this and opposite page; designed for house above and other Doll Houses of approx size. 49-pc. set includes: 12-pc. Living Room (a place), 9-pc. Dining Room, 8-pc. Bedroom, Kitchen, 6-pc. Bathroom, 6-pc. Nursery—Se from (C) Page 104; other sets described below and opposite pages. Ship. wt. 2 lbs. 8 oz.
48 T 4520—49-Pc. Furniture Set............

PLASTIC DOLL HOUSE FURNITURE—MINIATURE REPRODUCTIONS OF FULL SIZE PIECES

THIS MOLDED PLASTIC FURNITURE is beautifully made—copied from fine adult furniture. Well proportioned with careful reproduction of details. Bright lustrous colors on some pieces—woodlike finish on others. Most Doors and drawers open. Fits houses above.

(A) 13-Pc. LIVING ROOM. Fireplace, Grand Piano and Bench, 2 End Tables, 2 Table Lamps, Coffee Table, Radio, Floor Lamp, Barrel and Club Chairs, Sofa 5 in. long; others proportional.
48 T 4513—Boxed. Ship. wt. 1 lb. 4 oz......1.95

(B) 10-Pc. DINING ROOM. Sheraton Ta Chairs, top drawer opens on Server a let, door opens on China Cabinet. Two lamps. Table 4 in. long. Other pieces in pro
48 T 4514—Boxed. Ship. wt. 1 lb. 2 oz....

The inside of the large Rich Tudor house featured the stairway. The floors were flocked except for the kitchen and bathroom floors which were printed with the usual diamond tile pattern. The one inch scale Strombecker wood furniture fits perfectly in these large Rich houses. The windows have been replaced.

Many other designs of Rich dollhouses were produced over the years, including houses that fit no architectural catagory. Pictured is a four-room house 27" long by 9" deep by 16 1/2" tall. The front door does not open and there are two round holes on either side near the roof line to make moving the dollhouse an easier task.

The inside of the house features the diamond shaped flooring in the kitchen and the bathroom. The other two rooms have white floors with a colored line used for decoration around the space where a rug would be. The plastic windows have been replaced and were probably originally made with white panes.

Another similar Rich dollhouse which came complete with Rich labeled instructions to help in assembling the house. The roof of this house is two-tone green, more like the roof of a Keystone house. Photograph and house from the collection of Kathleen Neff-Drexler.

The floors of this Rich house are very different from the floors in most Rich houses. Besides the usual diamond shaped pattern for the kitchen and bathroom floors, the other two rooms have green floors with designs in the middle. Photograph and house from the collection of Kathleen Neff-Drexler.

Another similar Rich four-room house used the same door arrangement but different window styles and exterior decoration. The two-story house contained two rooms upstairs and two rooms downstairs. House from the collection of Susan Jenkins Lacerte. Photograph by Norman R. Lacerte.

A more recent Rich house is this six-room design which features pink shutters. The house is circa late 1950s. The house also has two rounded picture windows on the front.

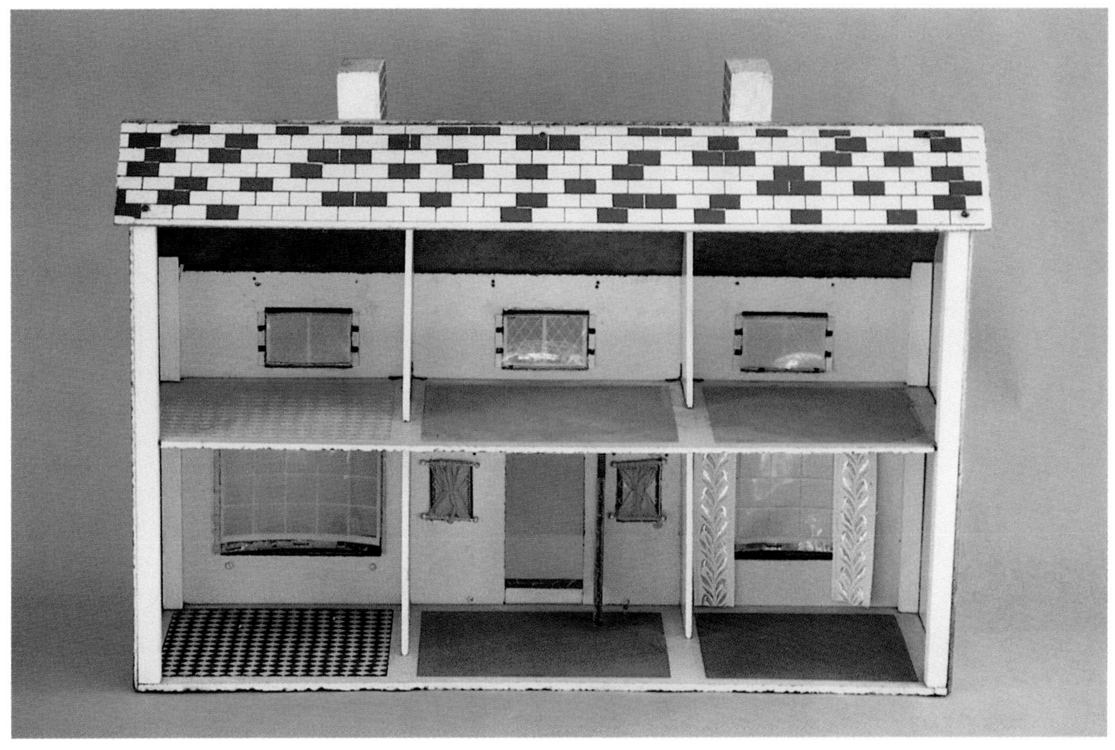

This six-room house measures 27" long by 9" deep by 19 1/2" to the top of the chimney. The house features the usual Rich floor coverings.

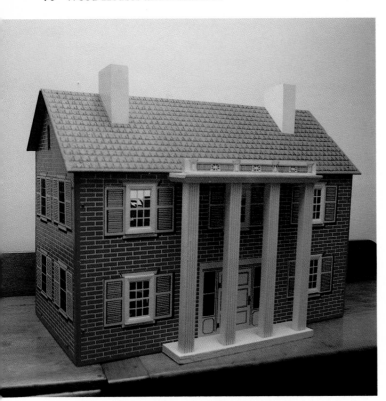

One of the last designs of dollhouses, made by the Rich company, was this six-room Williamsburg dollhouse. The house measures 28" long by 15 1/2" wide by 20 1'2" high. The design includes red brick siding with white, green, and black stencil designs. It is made of U. S. Gypsum hardboard. The inside floors have diamond tile designs on the kitchen and the bathroom and green or blue floors in the other rooms. The company sold the house complete with plastic furniture. House and photograph from the collection of Marilyn Pittman.

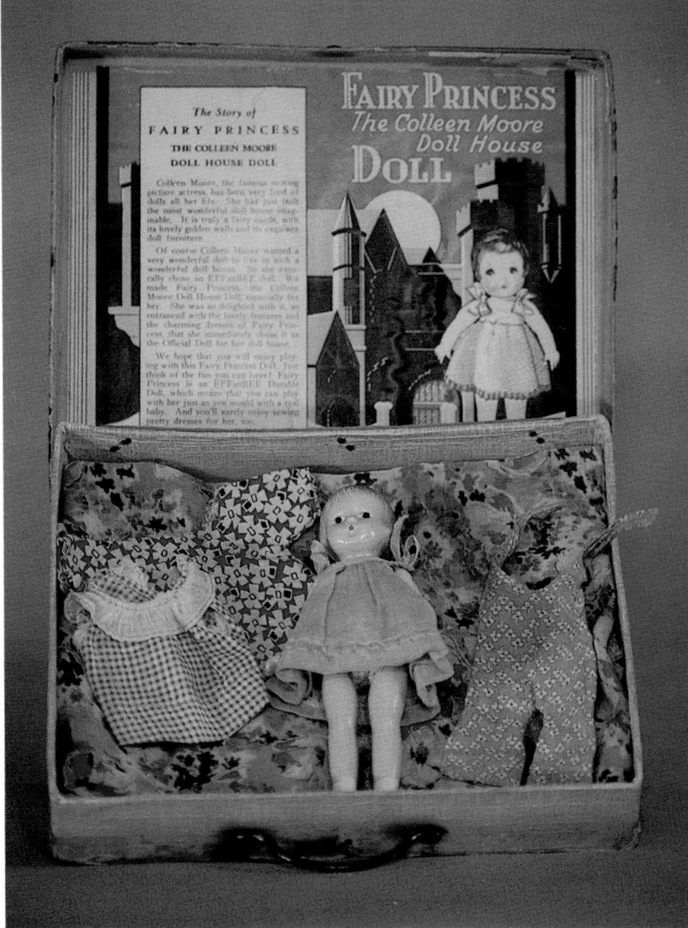

Fairy Princess Colleen Moore Dollhouse Doll produced by the EFFanBEE Doll Co. during the mid-1930s. The 6" tall composition doll is really a special version of the company's Wee Patsy Doll manufactured to tie in with the publicity generated by the national tour of the Colleen Moore Dollhouse.

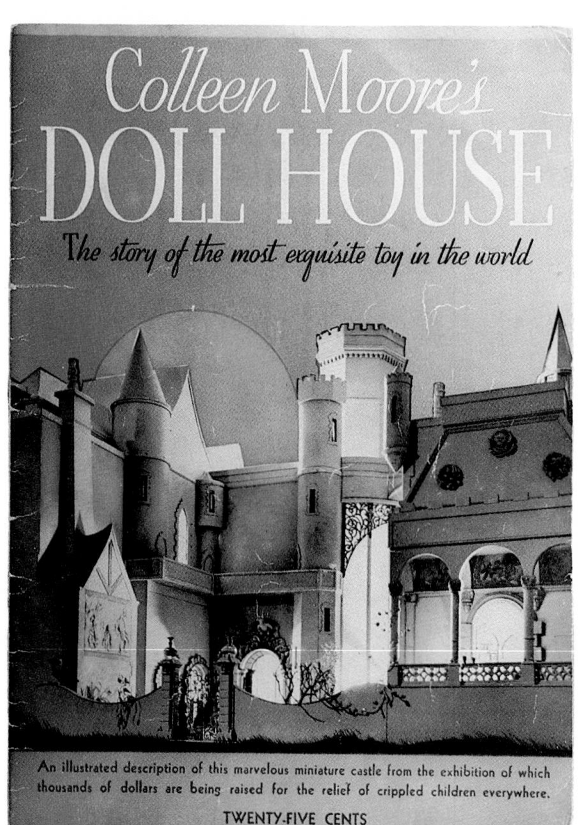

Colleen Moore's Dollhouse. This booklet was published by Garden City Publishing Co., Inc in Garden City, New York in 1935. It tells the story of the making of the famous Colleen Moore Dollhouse.

Rich Toy Co. made this authorized Colleen Moore Castle Dollhouse out of Masonite. Two models were produced by Rich during the mid-1930s. This smaller version measures 24" wide by 17" high by 13" deep. From the collection of Kathy Garner. Photograph by Bill Garner.

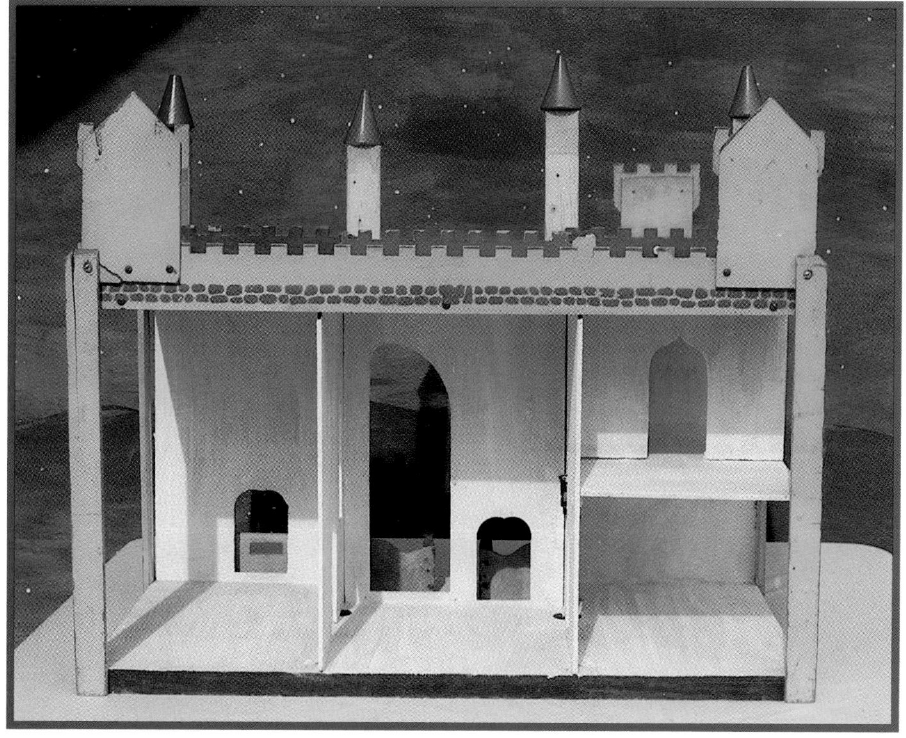

The inside of the Rich Castle contains five rooms. From the collection of Kathy Garner. Photograph by Bill Garner.

Keystone

The Keystone Manufacturing Company was founded in Boston in the early 1920s by Chester Rimmer and Arthur Jackson. It was first called Jacrim. The company produced movie projectors.

Later in its history, the Keystone Co. also manufactured dollhouses for a period of approximately ten years (circa 1940-1950). The houses were manufactured from Masonite. Most of the houses were made with either four or six rooms. The earliest houses in 1941 could be purchased either furnished or unfurnished and with inside decorations or without decorations. The houses were all two-story and ranged in size from 23" by 10" by 16" to 34" by 13" by 20". The only houses that did not fall into these specifications were the "Put-A-Way" dollhouses from 1949.

Most of the Keystone houses were decorated both inside and outside. The outside walls were printed with shutters, plants, trees, and shrubs as well as the siding of the houses. Inside, many of the houses had printed "wallpaper" and pictures on the walls. The earlier houses included metal casement windows which opened, while the later houses were fitted with plastic windows. Some houses in the 1947 line were equipped with windows that could be opened. Many of the houses, through the years, were also made with electric lights. The larger six-room houses are well suited for the Strombecker 1" scale wood furniture.

The Keystone houses are similar to the houses made by the Rich Toy Co. but there are ways to easily identify the large Keystone houses. The more elaborate six-room houses have a curving stairway, an upstairs closet, and a fireplace. Most of the Keystone houses contained slots on the roof for the placement of the chimneys. Many of the houses were marked on the bottom of the outside walls but others contained no identification.

Keystone houses have similar architectural designs to those used by Rich. The company made many different Colonial and Tudor models and also put several of their dollhouses on turntables so they were easy to move for play.

One of the Keystone houses featured in the company's advertising in the early 1940s was made with an additional two-story wing. In this way, the basic four-room house was changed into a six-room model. The house contained other special features including individual room lights, kitchen cabinet, staircase, awnings, flower boxes, decorated wood floors, and carpets on floors and staircase. The interior walls were finished in separate room colors. The exterior was finished with white walls, decorated in brown and green with ivory trim, and a two-tone green roof. The size of the house was 34" by 13" by 20" tall. The company also featured similar houses with only four rooms during this same period.

The Sears 1947 Christmas catalog featured a Keystone six-room dollhouse for $8.29 unfurnished. This model was 12 1/8" by 24 1/2" by 22 1/8" high. The house included a circular staircase, and metal casement windows that opened and closed. The house was shipped already assembled, a plus for the consumer. It featured a rounded shelter over the front door.

In 1949 Keystone brought out a new concept in dollhouses called the "Put-A-Way" dollhouse. There were three different designs for these houses. Two contained one extension and the other contained two extensions. The larger one extension house became a six-room dollhouse measuring 32" long by 12 3/4" deep by 18" high when opened. The extension could be replaced inside the basic house for storage. The roof could also be removed. The house included two bedrooms, bathroom with a built-in shower, dining room, living room, and a kitchen with a terrace roof. There was also a staircase and a closet. These houses also featured built-in cabinets and bookcases. The smaller house with one extension became a five-room house. When it was open, it measured 25" by 11" by 15" high. This model included a kitchen with terrace roof, two bedrooms, bathroom, and a dinette/living room. The largest "Put-A-Way " dollhouse opened to 42" by 12 3/4" by 18" tall. This model became a seven-room house. The additional wing was to be used as a garage and utility room. Both of the extensions on this house could be placed back into the house for easy storage. These new models of dollhouses were also equipped with turntables.

The 1949 Sears Christmas catalog featured the new Keystone double extention house at a price of $9.69. The house was described as having a built-in shower, a bookcase with movable books, and a garage with a sliding door. The house was built of Masonite and Tekwood.

Another interesting house was featured in the Keystone advertising circa 1950. The two-story, six-room house featured plastic window frames with transparent windows, unlike the earlier metal casement windows. Dormer windows in the house were not just decorative but were windows for the upstairs rooms. Other features of the house included electric lights in each room, large picture windows, operating garage door, decorated interior, staircase, fireplace, and turntable. The house was 34 3/4" by 15 3/4".

In addition to making the dollhouses, Keystone also produced toys for boys during this same time period. These products included a fire department and a service station. The Fire Department was advertised in 1948 for $5.25. It was made of fiberboard and measured 17 1/2" by 8 1/2" by 7 1/2". The Keystone Service Station was produced in the same material and included two overhead doors that opened and a round plastic window. It also had a lifting grease rack.

In 1953, Keystone sold its toy division and concentrated its business in the camera and projector field. Keystone, itself, was sold to Berky Photo, Inc. in 1966 and moved to Clifton, New Jersey.

Like Rich, Keystone used the same designs for houses during different years. Sometimes the door placement was moved or the size of the house was different. This basic design was used in at least three different houses from 1941 to 1947. The orange shutters and the flower boxes are unusual characteristics of this house. Another model was also made which included blue awnings. The Masonite house contains four rooms, two on each floor. No stairway is included in this house. The windows are metal casement and the floors have printed designs. The house measures 25" by 13" by 17" tall.

The inside of the house was quite plain, unlike the Keystone houses which had decorated walls. The house is marked at the bottom of the outside wall, "Keystone-Boston/Made in U.S.A."

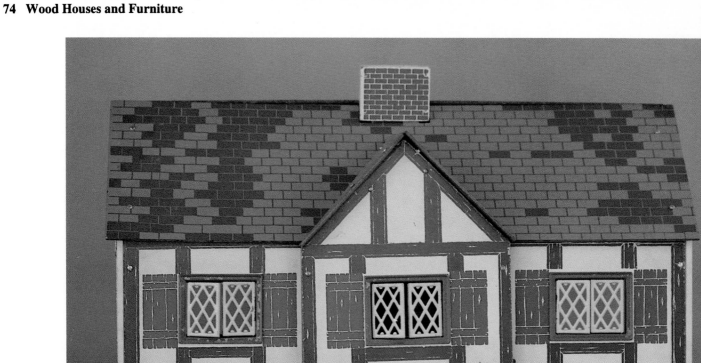

This Keystone Tudor house is one of the large Keystone models. It measures 29" by 22" tall. The six-room house was marketed in 1947.

The inside walls of the house are decorated with wallpaper designs as well as pictures. The house features a curved stairway, a built-in fireplace, and an upstairs closet.

This six-room Keystone house was featured in the Sears Christmas catalog in 1947. The Masonite house measured 12 1/8" by 22 1/2" by 24 1/2" long. It sold for $8.29.

The inside of the house contained the usual fireplace, curved staircase, and two-door closet. The house also featured the metal casement windows.

6-room Doll House . . . already assembled
Door and metal casement windows open
A cheerful roomy house for some lucky little girl, just waiting to be 'lived in.' Sturdy Masonite Presdwood, with beautiful wood doorway and steps, all enameled white. Metal casement windows open and close; have bright red painted-on shutters. The two floors are connected by a graceful circular staircase. Inside walls decorated to resemble wallpaper. Downstairs there's a living room, dining room, kitchen. Upstairs: two bedrooms and a bath. Beautifully decorated outside with painted-on trees and shrubs. Overall size: 12⅛ in. wide, 22⅛ in. high, 24½ in. long. Sturdy wood corner posts and partition supports. Shipped set-up; no assembling required. For a thrilling gift order this house and furniture described below.
$8.29
79 N 02178—Unfurnished Doll House. Shipping weight 29 lbs......$8.29

This large Keystone house dates from 1947. Because of the decorations on the front of the house, collectors call this model "The Birches." The house measures 32" wide by 22" to the roof peak by 12 1/2" deep. The tall pillars and the porch light give this house character. The house has metal casement windows. The inside of the house contains the usual curved stairway, fireplace, and closet. The original chimneys and front door are missing.

Keystone came up with a new innovation in dollhouses in 1949. The company made three models of their "Put-A-Way" dollhouse that year. One of the smaller models contained one wing that could pivot and nest into the house to make the house easier to store. The roof also could be moved to allow a child easy access to the house. The house contained plastic windows. Photograph and house from the collection of Roy Specht.

The inside of the "Put-A-Way" dollhouse featured a stairway, built-in kitchen cabinets, and storage shelves in the dining room. The two-story house contained six rooms. Photograph and house from the collection of Roy Specht.

The larger "Put-A-Way" Keystone house measured 32 1/2" by 10" by 20" high and it contained two wings. The house was made of Masonite and Tekwood. Photograph and house from the collection of Roy Specht.

The inside of the larger "Put-A-Way" house was just like the smaller model with the addition of a garage wing with opening garage door. The front of the garage contained an open breezeway with a terrace roof. This house has been redecorated. Photograph and house from the collection of Roy Specht.

Miscellaneous Wood Dollhouse Furniture

Besides major dollhouse furniture manufacturers like Strombecker, Schoenhut, and Tynietoy, there were many other minor manuracturers of wood furniture.

Perhaps the best-known of these makers was the firm of Rapaport Bros. located in Chicago. Their Nancy Forbes 3/4" scale wooden dollhouse furniture was featured in many Christmas catalogs during the 1940s. The company produced at least two different designs of the furniture. The early furniture which was being sold in 1940 included the following pieces: Bedroom: Bed, blanket chest, nightstand, chest of drawers, vanity with mirror, bench and dresser with mirror, and two lamps. Dining Room: Table, four chairs, buffet with mirror, stacked storage piece, and two lamps. Living Room: Sofa, fireplace, radio, chair, footstool, coffee table, end table, lamp table, side table, and two lamps. Kitchen: Sink, stove, refrigerator, table, four chairs, iron, and ironing board. Bathroom: Toilet, lavatory, medicine cabinet, bathtub, scale, hamper, vanity with mirror, and bench. Child's Room: Youth bed, blanket chest, steps, nightstand, chifforobe, two lamps, and perhaps a small table and chair. The youth bed looks very much like the design used by Strombecker. The Nancy Forbes furniture sold for around $1.00 for each room in 1940. The furniture is very block-like and not nearly as detailed as that made by Strombecker. The living room, dining room, and bedroom furniture sets were finished in walnut, while the kitchen, bathroom, and child's room pieces were much lighter in color.

The Nancy Forbes furniture featured in the Montgomery Ward Christmas catalog in 1945 was the new line which included many different designs. Furniture for a child's room was no longer listed. All of the kitchen pieces were of a different design and included a refrigerator, sink, stove, cabinet, table and two chairs. All of the living room furniture had also been redesigned. The room set included a sofa, two end tables, mirrored coffee table, chair, fireplace, radio, and two lamps. The dining room pieces were also different. They included a tall mirrored hutch, a low serving piece, table, four chairs and two lamps. The bedroom also had a new look. The pieces included a bed, night side tables and lamps, a vanity with a round mirror and bench, a dresser with round mirror, and a chest of drawers. The new bathroom included a medicine cabinet, toilet, lavatory, bathtub, scale, hamper, and vanity with a round mirror. The catalog price for each of the room sets of furniture was ninety-four cents. The company also made boxes of furniture available at higher prices which included a few pieces from each of the five rooms of furniture. The bathroom and kitchen furniture pieces from 1945 were painted in white enamel and the other pieces had a walnut finish.

There were at least three designs of boxes for the Nancy Forbes furniture. Each box featured the photograph of a little girl but they didn't appear to be the same child. The latest issue of the boxes shows a change in both the name of the furniture and the company who made it. The top of the box reads "Nancy Forbes 'Dream House' Furniture." It was manufactured by the American Toy and Furniture Co. in Chicago, Ill.

Dollhouse furniture that is similar in construction to the Nancy Forbes line is the Donna Lee 3/4" scale dollhouse furniture. Donna Lee items were advertised during the 1940s. The Spiegel catalog for fall and winter in 1944 also advertised a four-room Donna Lee dollhouse. The house was made of constructo board. It had a tile-like roof with a red chimney. The outside was painted with windows, doors, and shubbery. The house was 21" wide by 15 1/2" tall by 7 1/4" deep. The house came with sixteen pieces of wood furniture and sold for $2.98 complete. The furniture included living room, dining room, kitchen, and bedroom pieces.

Some of the Donna Lee furniture designs are identical to the 1940 Nancy Forbes furniture. These include the large scaled iron that was made to accompany the 3/4" scale ironing board. Perhaps both sets of furniture were produced by the same company and the Donna Lee line was a cheaper product. Boxes of Donna Lee furniture sold for as little as seventy-five cents each. The copy on the box reads "Donna Lee Doll Furniture Custom Built Playthings." The copy on the Nancy Forbes box also states that the furniture is "Custom Built." Both the Nancy Forbes and the Donna Lee furniture are made in block style with no moving parts and only crude lines used to mark divisions like drawers or doors in the furniture.

Furniture pieces included in the Donna Lee boxes are as follows: Kitchen: Stove, refrigerator, sink, table, two chairs, iron (large), and ironing board. These items are all enameled white. Bathroom: Bathtub, lavatory, toilet, scale, vanity, bench, medicine cabinet, and wastebasket. All of this furniture is finished in white enamel except the wastebasket, which is red. Bedroom: Bed, mirrored dresser, mirrored vanity and bench, nightstand, cedar chest, and chaise lounge. Dining room: Table and chairs, stacked china cabinet, and serving piece. Living room: Sofa, chairs, end tables, coffee table, and radio.

Another more interesting set of dollhouse furniture was apparently sold in the dime stores circa 1940s. The furniture is not marked and may have been imported, perhaps from Japan. Because of World War II, it would have been impossible for the furniture to have come into the country from that source until the late 1940s and the furniture seems to have been made earlier. Many of the wood pieces are "upholstered" in printed flowered material. This material varies from piece to piece and is of cheap quality similar to the old printed "feed sack" line of yard goods. The most attractive pieces are the chairs and sofas but beds, dressers, vanities, pianos, radios and other items were also produced. Some of the furniture pieces have attached round metal mirrors. The mirrors are joined to the furniture with two nails. One of the most interesting items from this set of furniture is a bookcase equipped with fake books. A vanity and stool set still carries its original store price tag of 15 cents. Some of the legs of the furniture are made of wood dowels. The furniture is a little larger than 3/4" scale.

Dining room furniture made to accompany the printed sofas included a table, chairs and buffet (all with turned legs). The buffet drawers were outlined in black. The kitchen furniture often found with this set included a table, chairs, and sink in white enamel. All these pieces have identical turned legs. The stove and refrigerator that accompanied this furniture were also white enamel with the doors outlined in black. The refrigerator seems to be smaller in scale than the other items.

Another very sturdy line of wood dollhouse furniture was produced in this country during the Depression years. Collectors sometimes refer to it as Grand Rapids furniture. The furniture is easy to identify because it was made of plywood and the edges of the furniture expose the plywood origin of the pieces. The furniture is scaled approximately 1" to 1'. All of the furniture was finished in a light honey color. The furniture may not have been made in Grand Rapids, but it was definitely produced in this country, as many of the items are marked "Made in U.S.A." The furniture was assembled with small nails which still show in the finished product. The drawers and doors do not open on this line of furniture. Furniture items known to have been made by this company include: Living room: Sofa, chair, half circle table, square table, rocking chair, and magazine rack. Dining Room: Table and chairs, hutch, and perhaps a server. Bedroom: Bed, dresser with mirror, nightstand, and cedar chest.

There were many other countries around the world who competed in the wooden dollhouse furniture market during the twentieth century, including Germany and England. Because of the two World Wars, the manufacturers in the United States were able to corner the biggest share of the market in this country and it is their furniture that is more easily available for today's collector.

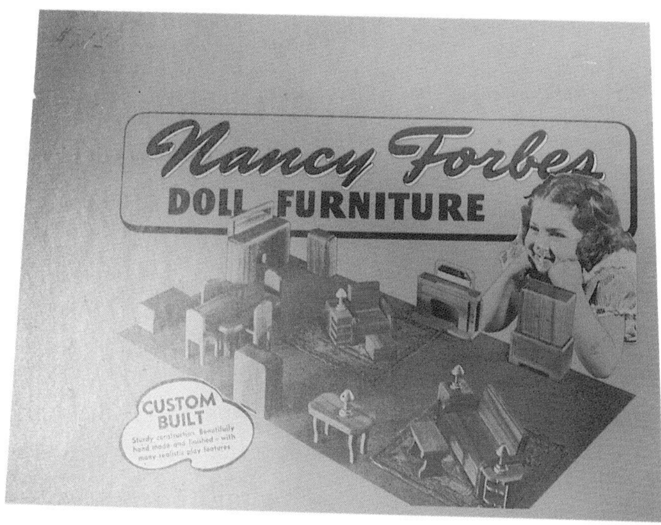

Box cover picturing the early Nancy Forbes furniture design which was advertised in 1940. Photograph and box from the collection of Patty Cooper.

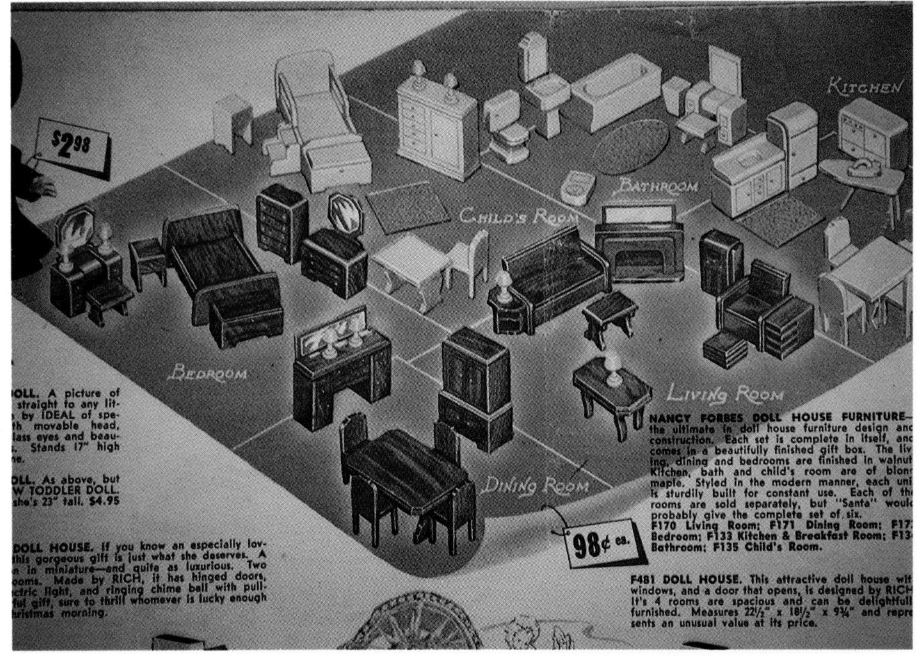

Catalog page from the Famous Funn Family Service Massey's Drug Store, Shirley, Indiana, Christmas 1940. The Nancy Forbes wooden dollhouse furniture is pictured. From the collection of Marge Meisinger.

Wood furniture attributed to Nancy Forbes circa 1940. The furniture is approximately 3/4" to 1' in scale. Included is a bed, cedar chest, dining room cabinet, sofa, and lamp.

These bathroom and kitchen pieces of wood furniture are also part of the Nancy Forbes designs circa 1940. Photograph and furniture from the collection of Patty Cooper.

Additional pieces of Nancy Forbes furniture from the early design include a dresser, radio, and nightstand. Photograph and furniture from the collection of Patty Cooper.

The Nancy Forbes bedroom furniture from the later 1945 design is pictured along with the company box from the period.

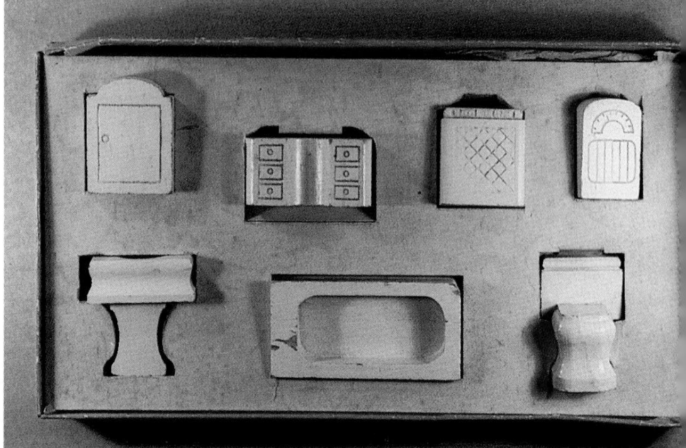

Pictured is the boxed set of Nancy Forbes 1945 bathroom furniture.

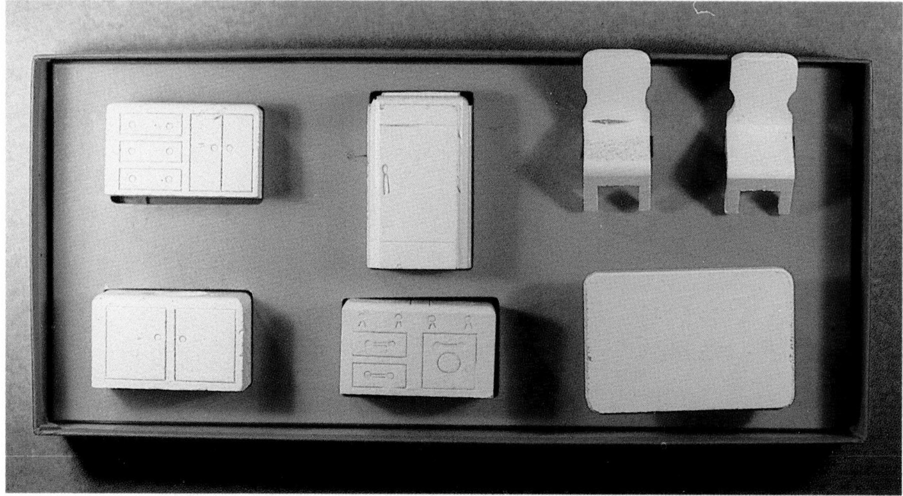

Set of Nancy Forbes kitchen furniture from the later 1945 design. In addition to the usual pieces of kitchen furniture, the Nancy Forbes set also included an additional cabinet.

The boxed living room furniture from 1945 using the later design.

The Nancy Forbes wood dining room furniture included nine pieces in the complete set and it is pictured in front of a new-style Nancy Forbes box circa late 1940s. The company listed as the manufacturer of the furniture was the American Toy and Furniture Co., Chicago.

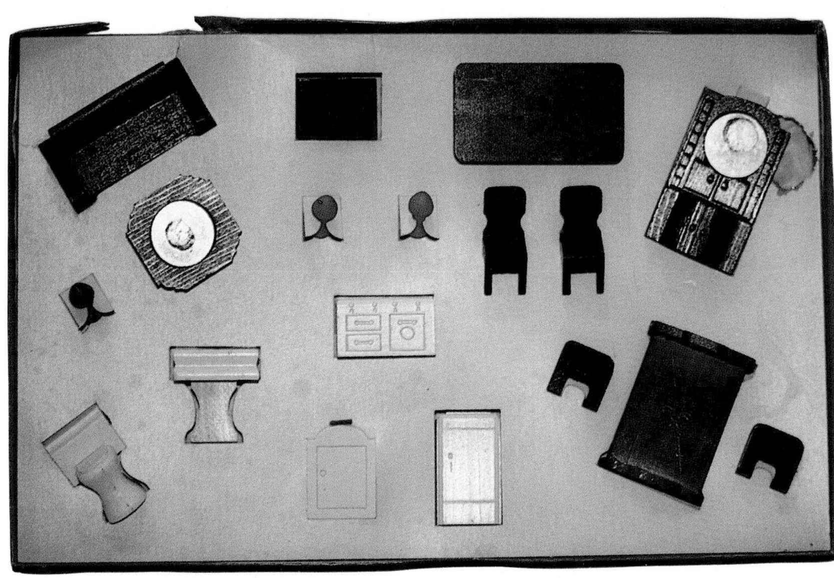

A large boxed set of Nancy Forbes furniture included pieces from each of the five sets of furniture and was sold to furnish a complete dollhouse.

Dining room pieces of Donna Lee furniture. The back of the chairs and the back legs are all one piece.

Picture from the Spiegel fall and winter catalog for 1944 which advertises a Donna Lee four-room dollhouse. The house was 21" wide by 15 1/2" tall by 7 1/4" deep. The house came with sixteen pieces of wood furniture and sold for $2.98 complete.

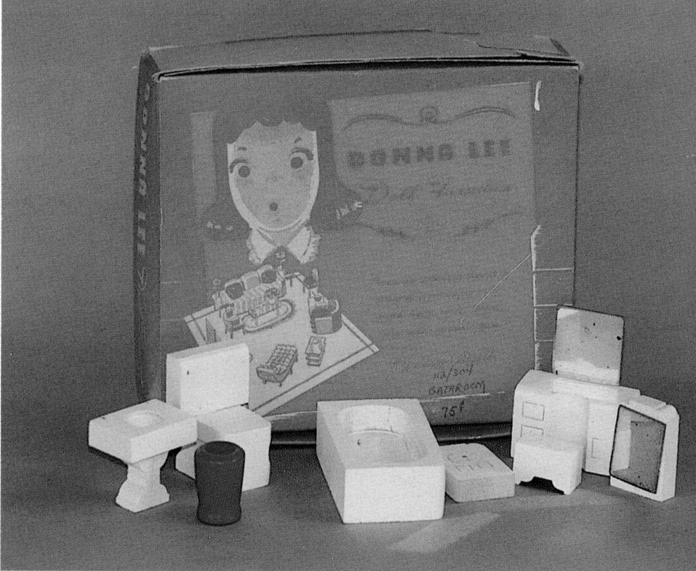

Donna Lee furniture from the boxed bathroom set of furniture.

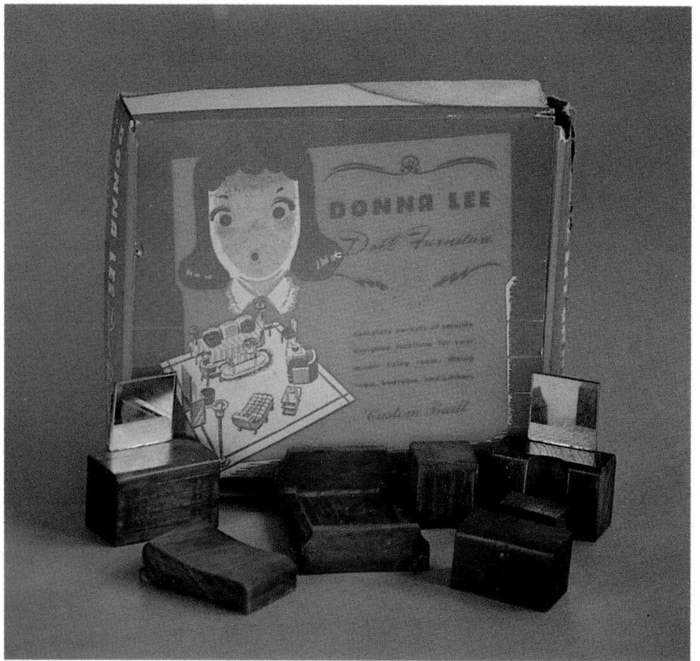

The Donna Lee boxed bedroom furniture included a crude chaise lounge as well as the usual bedroom furnishings.

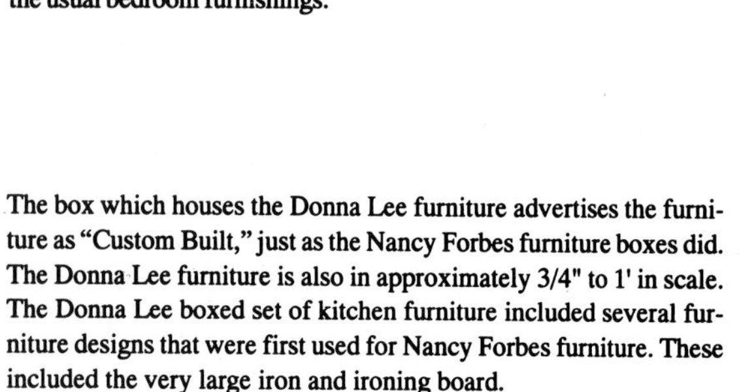

The box which houses the Donna Lee furniture advertises the furniture as "Custom Built," just as the Nancy Forbes furniture boxes did. The Donna Lee furniture is also in approximately 3/4" to 1' in scale. The Donna Lee boxed set of kitchen furniture included several furniture designs that were first used for Nancy Forbes furniture. These included the very large iron and ironing board.

Furniture of unknown origin that was "upholstered" in inexpensive material. These pieces may have been made in Japan but they appear to be circa 1940s, when the war prohibited the import of Japanese goods. This furniture is a little larger than 3/4" to the 1' scale but is not as large as the 1" to 1' scale of furniture. Photograph and furniture from the collection of Patty Cooper.

These living room pieces appear to be part of the same set of furniture and were purchased together. This furniture, also, is unidentified.

Bedroom furniture that also seems to be made by the same unknown company. The vanity and stool set still has its original price tag of 15 cents for the pair.

Plywood furniture with a scale slightly over 1" to 1' which was produced during the Depression years. This furniture is sometimes called "Grand Rapids" and may have been produced there. None of the drawers open in this line of furniture.

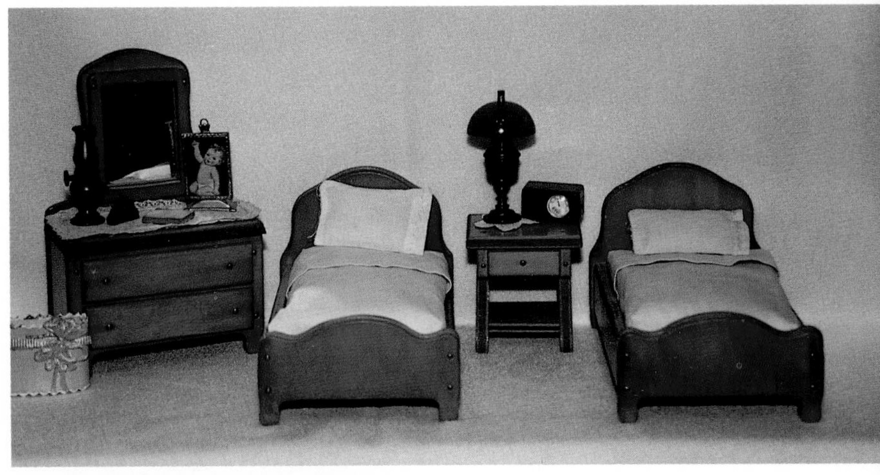

"Grand Rapids" bedroom furniture included beds, nightstand, and a dresser with mirror. Furniture from the collection of Gail and Ray Carey. Photograph by Gail Carey.

Magazine rack, rocker, and cedar chest from the same line of furniture. The pieces are made of plywood. Furniture from the collection of Gail and Ray Carey. Photograph by Gail Carey.

Dining room pieces of furniture called "Grand Rapids" by collectors. Furniture from the collection of Gail and Ray Carey. Photograph by Gail Carey.

These 1" to 1' scaled wood kitchen pieces were purchased from F.A.O. Schwarz in 1938. The doors function on all the furniture. The maker is not known.

Metal Furniture

Tootsietoy

Tootsietoy dollhouse furniture was made by the Dowst Brothers Company, located in Chicago, from approximately 1922 to 1937. The company had its beginning in 1878 when brothers Charles and Samuel published a trade publication called "Laundry Journal." Around 1900 the company began making metal miniatures to be used as premiums. Soon they enlarged the business by making party favors and eventually were responsible for many of the metal Cracker Jack prizes.

After the company began making dollhouse furniture with some success, the Dowst Brothers expanded, merging with the Cosmo Manufacturing Co. in 1926. The new firm was called Dowst Manufacturing Co. In 1961 as the company continued to prosper, Dowst acquired the model kit line, the trademark of "Strombecker," and the Strombeck-Becker Manufacturing Co. name from the former owners. After these changes, the company became the Strombecker Corporation.

The wonderful Tootsietoy metal dollhouse furniture and early die cast cars, trucks, and other toys have become quite collectible today. Sears, Roebuck and Co. catalogs featured the Tootsietoy dollhouse furniture yearly beginning in 1923 and ending in 1934. The furniture was made of metal alloy in a scale of approximately 1/2" to 1'. Other firms' catalogs continued to carry the furniture for another year or so but the once-strong demand for the small metal furniture was over. Consumers were more interested in the wood furniture that did not break so easily.

The Tootsietoy pieces were so small and delicate that the furniture could easily lose a leg or a mirror from a child's rough handling. The paint also scarred easily so it is difficult today to find the furniture in excellent condition unless it is still mint-in-the-box.

The first models of the Tootsietoy furniture were based on the furniture of the early 1920s. The dining room table was round, the icebox was the kind that needed a chunk of ice inside, and the bathtubs had "feet."

The boxed sets shown in the Sears, Roebuck catalog in 1923 sold for eighty-three cents for each room. The following pieces of furniture were contained in each room setting: Kitchen: cabinet, sink, stove, ice box, table, and two chairs. Bathroom: toilet, bathtub, lavatory, stool, chair, two towel racks, and medicine cabinet. Bedroom: two beds, mirrored dresser, mirrored vanity, bedside table, rocker, and chair. Dining Room: round table, four chairs, buffet, server, and cart on wheels. Living Room: sofa, rocker, chair, table, floor lamp, table lamp, and victrola. Most of these early pieces of furniture continued to be made until the end of the Tootsietoy dollhouse furniture production.

By the mid-1930s this line of furniture was often marketed under the name "My Dolly's Furniture" and one Sears catalog used the name "Daisy" to identify the furniture. In the later years, this line of furniture was sold in smaller sets and at cheaper prices than the newer line of furniture.

In 1930 the Dowst firm brought out an entirely new design of dollhouse furniture. Although it was still made of metal, it was designed to reflect the changes that had taken place in the real furniture then being sold to consumers.

Sears also featured the new Tootsietoy line in the 1930 catalog. Each box sold for eighty-seven cents. Items in each room are as follows: Bedroom: two twin beds, mirrored dresser, rocker, table, and lamp. Living Room: sofa (overstuffed type), matching chair and ottoman, long table, floor lamp, and end table. Dining Room: table (oblong), four chairs, buffet, and china closet. Kitchen: stove, refrigerator (with coil on top), cabinet, sink, table, two chairs. Bathroom: modern "built in" tub, toilet, lavatory, clothes hamper, medicine cabinet, stool, and two towel racks. Many more of the new pieces had moving parts than had the earlier furniture. The stove, refrigerator, and cabinet had doors that opened and the dresser drawers also were functional.

Additional pieces of furniture were produced by Dowst that were sometimes sold in other boxed sets or alone. These included a grand piano and bench, a radio (with opening doors), a desk with a drop front, candlelabras, a vacuum sweeper, iron, chaise lounge, bathroom scale, and a vanity dresser and bench. Some of the 1930 designs were also produced later with a covering of flocking to simulate upholstery. These included a sofa, chair, chaise lounge, and beds. A few miniature pieces of Tootsietoy furniture were also manufactured by the company. These included: sofa, chairs, piano, lamp/table, library table, beds, dresser, and vanity. These items make nice additions to a Tootsietoy dollhouse furniture collection.

Besides making dollhouse furniture, the Dowst Brothers also marketed dollhouses to be used with their furniture pieces. The Sears catalog for 1924 pictures what appears to be the first dollhouse for the new metal furniture. It sold for $3.79 and was colored red, green, and white. It was a two-story colonial home of red brick with a green roof and white window frames with shutters. The house contained four rooms and measured 26 3/8" by 17 3/4" by 20 1/4". The inside was cream color and the front opened for play. The house was made of heavy cardboard and had a framework of wood. This higher priced house was featured only the one year.

By the following year, 1925, the more familar house is shown. That is the house pictured in the 1925 Tootsietoy catalog that folds up easily and is held together by brads. The Sears version is only four rooms and measures 18" by 16" by 12" deep. It sold for $1.98. The Tootsietoy catalog house is the same size but it had extra partitions to make the house six rooms. The inside of the new house was decorated with printed curtains, rugs, pictures, linoleum, and tile. The six-room version consisted of a living room, kitchen, and dining room downstairs, and two bedrooms and a bathroom upstairs. The house was made of heavy cardboard. In 1926 Sears carried a six-room version of the house as well as the earlier four-room model. These houses featured a rounded design printed with bricks over the front door. Although this house has been found marked "Wayne Paper Co. Ft. Wayne, Indiana", it appears to be the same house advertised in the Tootsietoy catalog as a Dowst product. It may be that the company contracted with other firms to produce their dollhouses just as Strombecker did in later years.

Another Tootsietoy house was made with outside walls printed to resemble stucco, with dormer windows on the roof. This two-story, six-room house had green shutters and a red roof. The front of the house opened with large doors. A similar house marked with the "Daisy" trademark contained opening doors in both the front and the back for even easier access to the inside rooms. This innovation was important because the bathroom and kitchen were located behind the dining room and bedroom and it would have been difficult for a child to reach those rooms without a back opening.

In 1930, besides bringing out the new furniture designs, the Dowst Company also marketed a newly designed dollhouse. It was called the Tootsietoy Mansion but is now known to collectors as the Spanish Mansion because of its architecture. According to the company's advertising, the house was designed by a real architect who specialized in Spanish architecture. The influence of California and the mansions of the Hollywood movie stars must have made an impact on the Dowst Co. to encourage them to market a house totally foreign looking to most of their potential customers. It should be noted that Sears continued to carry the old brick model instead of featuring the new Mansion. The new higher price of $5.00 also could have influenced the buyers at Sears. The old model was still selling for under $2.00. The rooms contained in the new mansion were as follows: vestibule 3" by 4", hall, living room 9" by 12", dining room 9" by 14", kitchen 10" by 6 1/2", upstairs bedrooms (12" by 6" and 8" by 9"), and bathroom 8" by 6". The company also brought out a boxed set of their newly designed furniture to be sold to furnish the mansion. The box contained furniture for five rooms. The Spanish Mansion is currently a very desirable house for a dollhouse collector and it will bring top dollar when one is found in excellent condition.

Since all of the Tootsietoy houses were made of cardboard, the chance of a collector finding any of these houses in mint condition is very remote. Most collectors have to settle for a house in "played with" condition, if they can find one at all. With the new color copying machines now available, inside partitions can be duplicated to replace missing ones if an original can be located.

The slogan of the Dowst Brothers in describing their metal dollhouse furniture was "All the Strength of Metal, All the Beauty of Wood." The furniture did not live up to its advertising. Because the pieces were so small, the metal had little strength and the paint did not adhere to the metal. Most pieces now found are less than perfect specimens. The company did meet a need in the toy field. No longer would only wealthy children be privileged to own a furnished dollhouse. With the coming of the inexpensive Tootsietoy furniture and their reasonably priced cardboard houses, little girls from the middle class could also be proud "home owners." Even with the price of a room of furniture at less than $1.00 and the cost of a house at less than $2, the families of most little girls could still not afford such a luxury. The best these children could hope for was perhaps one or two of the cheaper boxes of furniture containing only four Tootsietoy pieces. These could be placed in a house made from a cardboard or wood box. The "playing house" was just as much fun in either case, and Dowst Brothers still receive credit for making commercial dollhouse furniture available to the "common folk."

Tootsietoy metal furniture was marketed in several different style boxes. These packages contained from four to thirty-seven items of furniture in each box. The furniture was made by the Chicago-based Dowst Brothers Company. Photograph and box from the collection of Patty Cooper.

Another style Tootsietoy box contained metal bedroom furniture. Pictured is the early bedroom furniture consisting of a bed, dresser, rocker, and a chair.

Pictured is the early Tootsietoy bathroom furniture which featured the bathtub with "legs." The scale is from the later bathroom set. The Sears, Roebuck and Company catalog sold these Tootsietoy rooms of furniture for seventy-nine cents each in 1924.

The earliest Tootsietoy dining room set represented the "Golden Oak" furniture by featuring a round table that was then popular.

The early Tootsietoy living room also dates from the 1920s. The lid of the phonograph opened. Although this furniture was recently purchased with the box shown, the box features illustrations of the later Tootsietoy furniture and may not be original to this set.

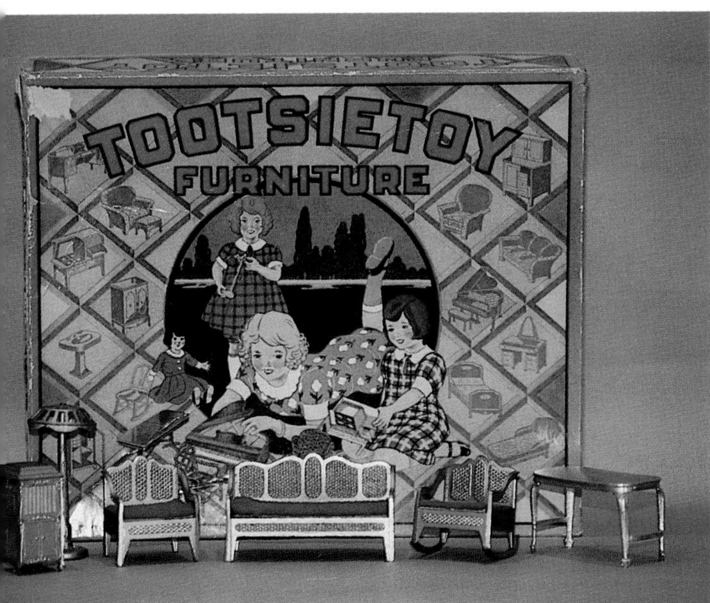

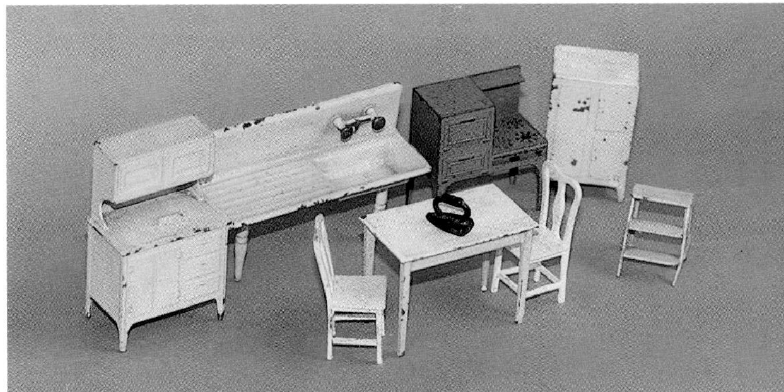

The first Tootsietoy kitchen designs included an ice box that needed a chunk of ice to operate. None of the doors or drawers opened in this furniture.

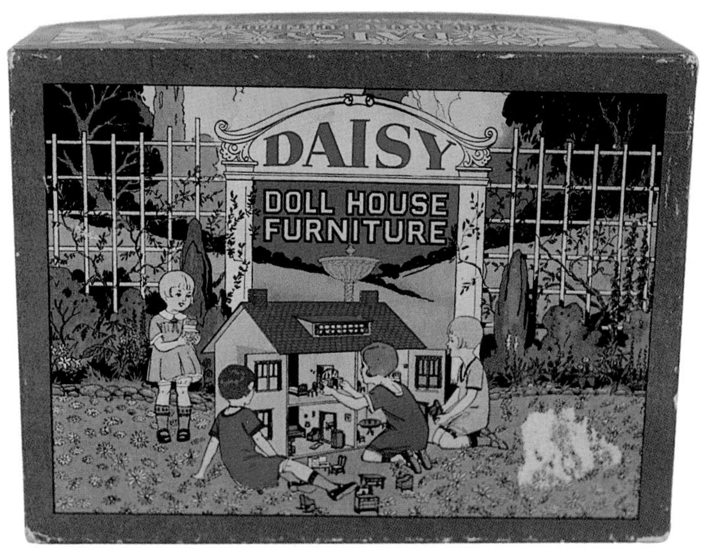

Dowst Brothers marketed their metal furniture under several different names. The Sears catalog carried a furniture line called "Daisy" in 1928 that was identical to the Tootsietoy furniture. Pictured is a box with the "Daisy" trademark. Photograph and box from the collection of Patty Cooper.

The four kitchen pieces of furniture contained in the "Daisy" box are the early Tootsietoy designs. Photograph and furniture from the collection of Patty Cooper.

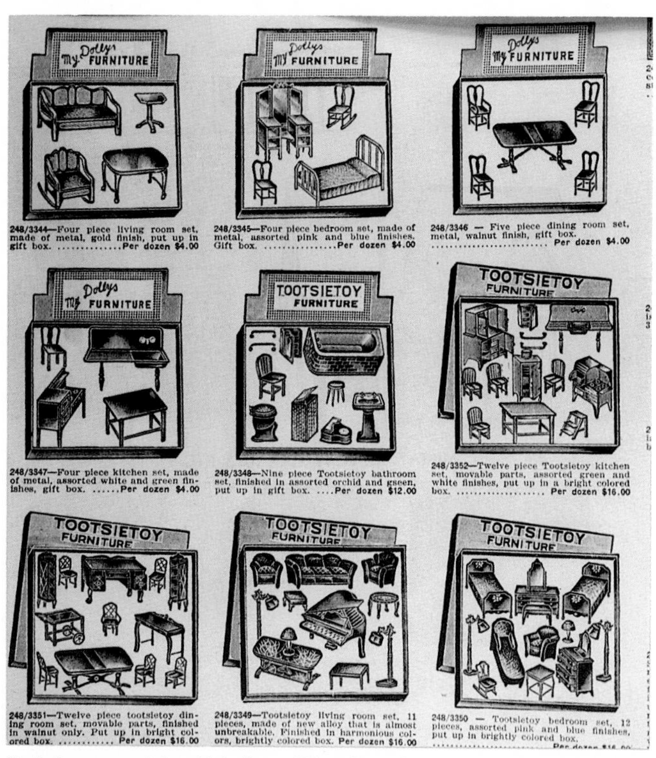

248/3344—Four piece living room set, made of metal, gold finish, put up in gift box. ....Per dozen $4.00

248/3345—Four piece bedroom set, made of metal, assorted pink and blue finishes. Gift box. ....Per dozen $4.00

248/3346 — Five piece dining room set, metal, walnut finish, gift box. .... Per dozen $4.00

248/3347—Four piece kitchen set, made of metal, assorted white and green finishes, gift box. ......Per dozen $4.00

248/3348—Nine piece Tootsietoy bathroom set, finished in assorted orchid and gseen, put up in gift box. ....Per dozen $12.00

248/3352—Twelve piece Tootsietoy kitchen set, movable parts, assorted green and white finishes, put up in a bright colored box. .... Per dozen $16.00

248/3351—Twelve piece tootsietoy dining room set, movable parts, finished in walnut only. Put up in bright colored box. .... Per dozen $16.00

248/3349—Tootsietoy living room set, 11 pieces, made of new alloy that is almost unbreakable. Finished in harmonious colors, brightly colored box. Per dozen $16.00

248/3350 — Tootsietoy bedroom set, 12 pieces, assorted pink and blue finishes, put up in brightly colored box. ....

Both the new and the old designs of Tootsietoy furniture were advertised in the 1935-1936 Blackwell Wielandy Co. catalog. The company was based in St. Louis. The early furniture was labeled "My Dolly's Furniture" while the later design carried the Tootsietoy label.

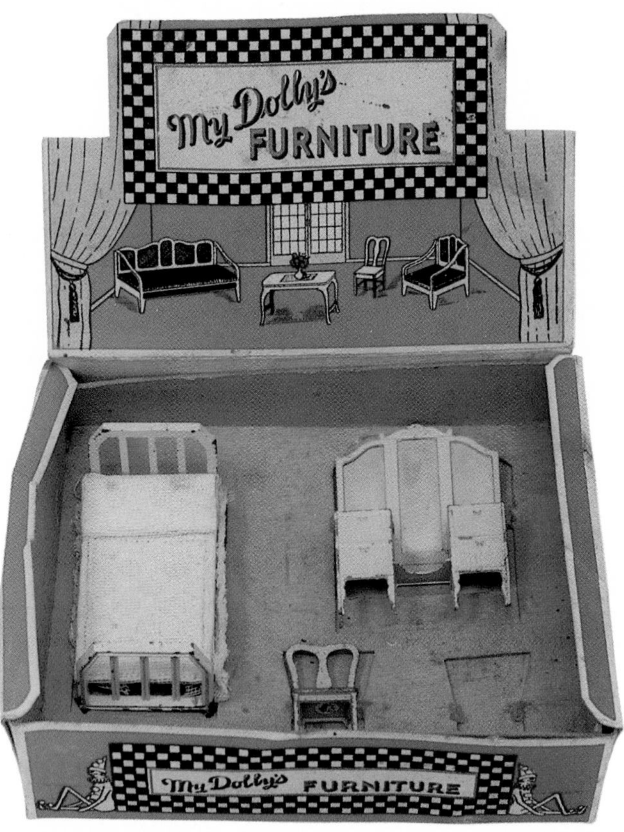

Pictured is a boxed set of "My Dolly's Furniture." The Tootsietoy pieces include a bed, dresser, and chair. A rocker was also originally part of the set. Photograph and furniture from the collection of Patty Cooper.

In 1930 Dowst Brothers marketed a new design of dollhouse furniture. Pictured are the new kitchen pieces which included doors that opened. The original box is also shown. Photograph and furniture from the collection of Patty Cooper.

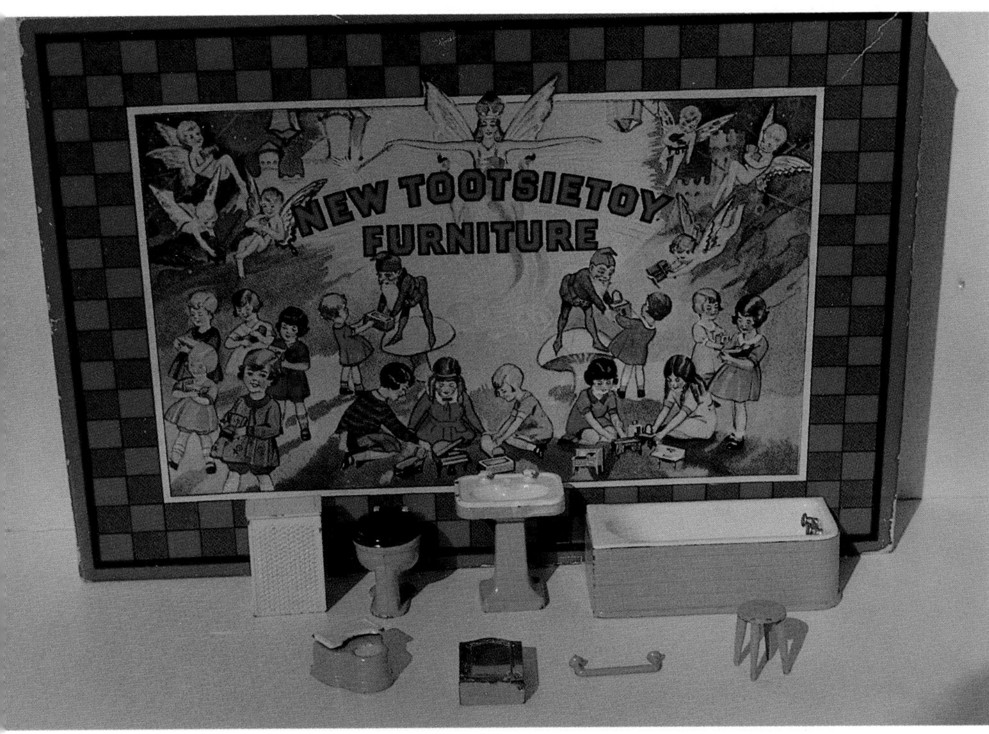

Boxed bathroom furniture from the 1930 design. The different colored hamper is the original piece that came in the box. The bathtub has a more built-in look. The medicine cabinet opens in a different direction than did the first model. Photograph and furniture from the collection of Patty Cooper.

The Sears catalog advertised the new design of Tootsietoy furniture for eighty-seven cents a box in their 1930 catalog. Pictured is a boxed bedroom set which included a chaise lounge in addition to the furniture pictured in the 1930 catalog. Photograph and furniture from the collection of Patty Cooper.

The new design of living room furniture that was sold in 1930. The sofa and chair were made to look like the overstuffed pieces of the time period. The new radio also was equipped with opening doors. Photograph and furniture from the collection of Patty Cooper.

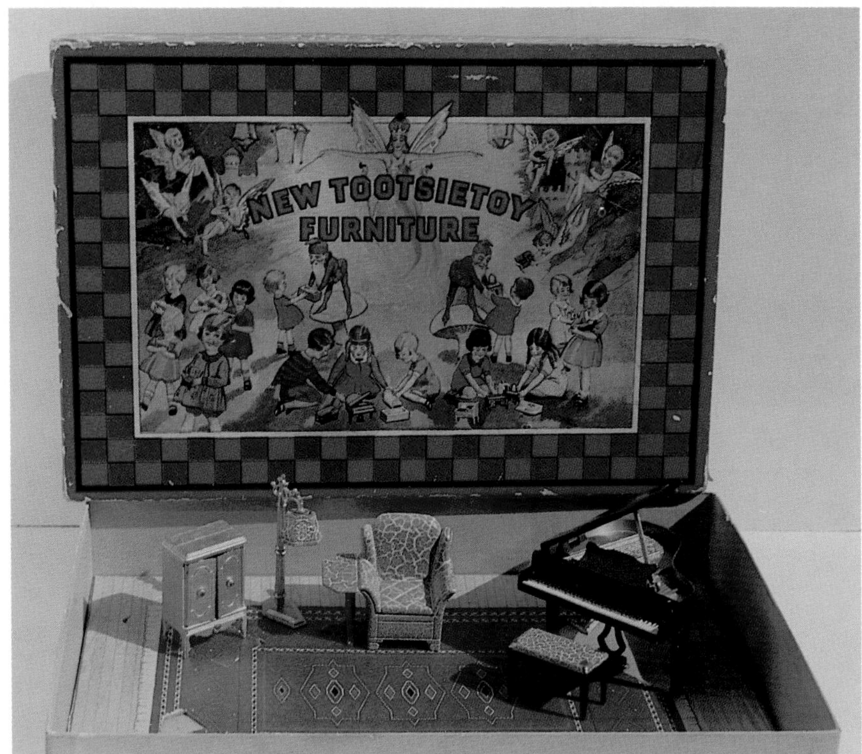

Tootsietoy boxed music room which used the new design of furniture. The lid of the piano opened as did the keyboard. Photograph and furniture from the collection of Patty Cooper.

The Tootsietoy dining room in the 1930 design was painted green. Photograph and furniture from the collection of George Mundorf.

The beds from the later Tootsietoy design included removeable "bedspreads." The table lamp sits on the vanity dresser used with some of the early Tootsietoy sets of bedroom furniture.

Additional pieces of Tootsietoy furniture. The desk was listed in the 1925 Tootsietoy catalog as a single item. The sofa is flocked in green and the coffee table was designed to accompany the sofa of a similar design.

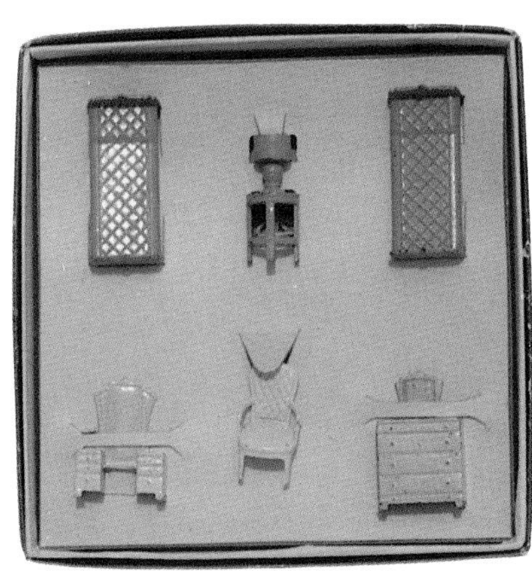

The "Midget" Tootsietoy bedroom furniture is pictured. This metal furniture is approximately 1/4" to 1' in scale. Furniture from the collection of Kathy Garner. Photograph by Bill Garner.

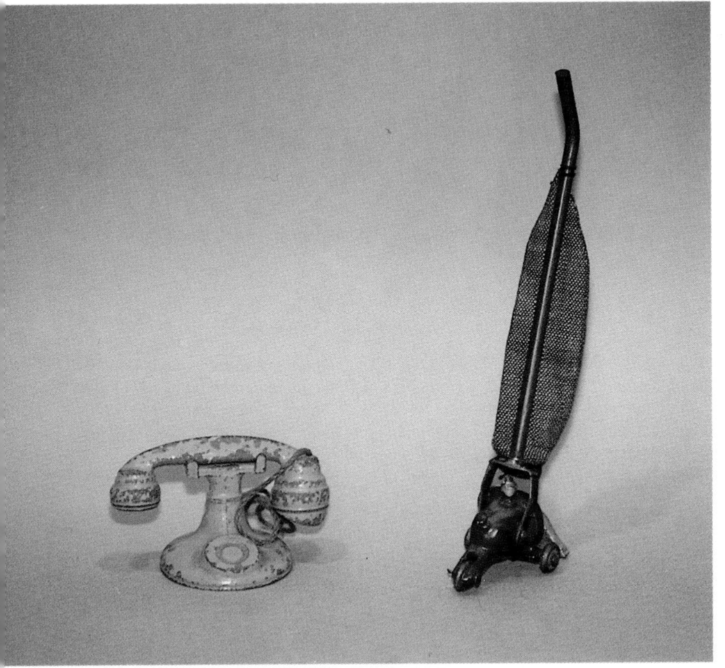

Although the Tootsietoy telephone and vacuum sweeper are too large to be used along with the Dowst furniture, they do make nice accessories for larger dollhouses. The sweeper is in the 1" to 1' scale while the telephone is approximately 1 1/2" to 1' in scale. Both items are marked with the Tootsietoy name as are all the Tootsietoy pieces.

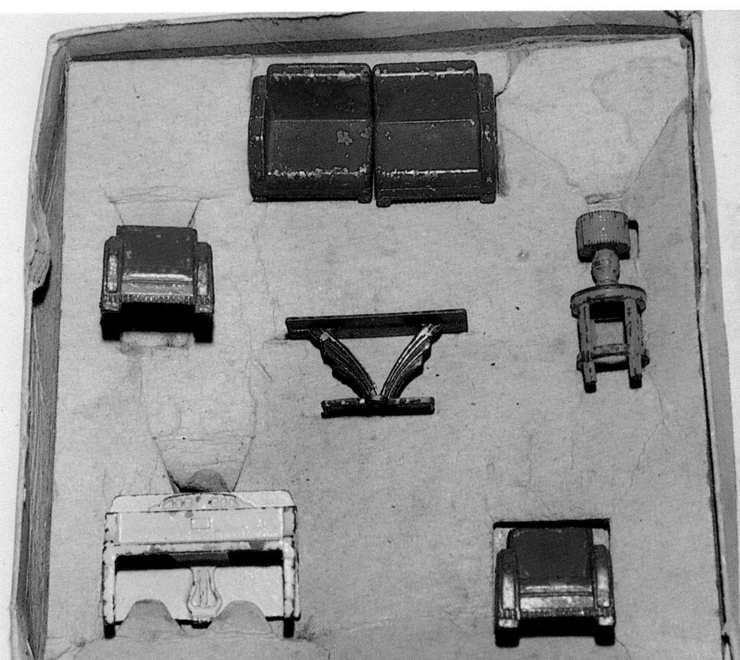

The "Midget" living room furniture is shown in its original box. Photograph and furniture from the collection of Patty Cooper.

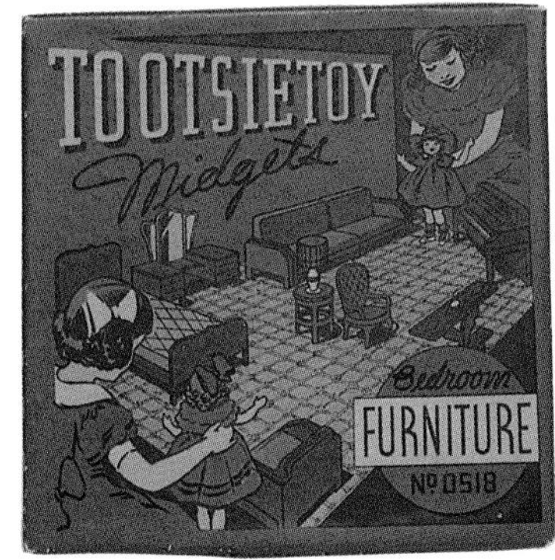

Miniature pieces of Tootsietoy furniture were also produced. The box is labeled "Tootsietoy Midgets." Box from the collection of Kathy Garner. Photograph by Bill Garner.

Tootsietoy cardboard dollhouse circa 1925. The house contains four rooms and measures 18" by 16" by 12". The front opens with two hinged doors. The house is marked "Wayne Paper Co., Ft. Wayne, Indiana." Photograph and picture from the collection of Arliss Morris.

The inside of the Tootsietoy house contains six rooms. These include a living room, dining room, kitchen, bathroom, and two bedrooms. The bathroom and kitchen are located behind the front rooms and were next to impossible to access.

Cardboard Tootsietoy house circa mid-1920s. It contains six rooms. The house opens at the front with hinged doors. The house was designed for the company's metal furniture. The house measures 17" by 11 1/2" by 17 1/2" tall including chimney.

The cardboard house marketed by Dowst Brothers under the "Daisy" name included hinged doors on both the front and the back of the house. The two hinged doors on the front were exactly like those of the original Tootsietoy house. The back of the Daisy house also featured one large hinged door that would allow access to the kitchen and bathroom. The inside decor of the two houses is identical.

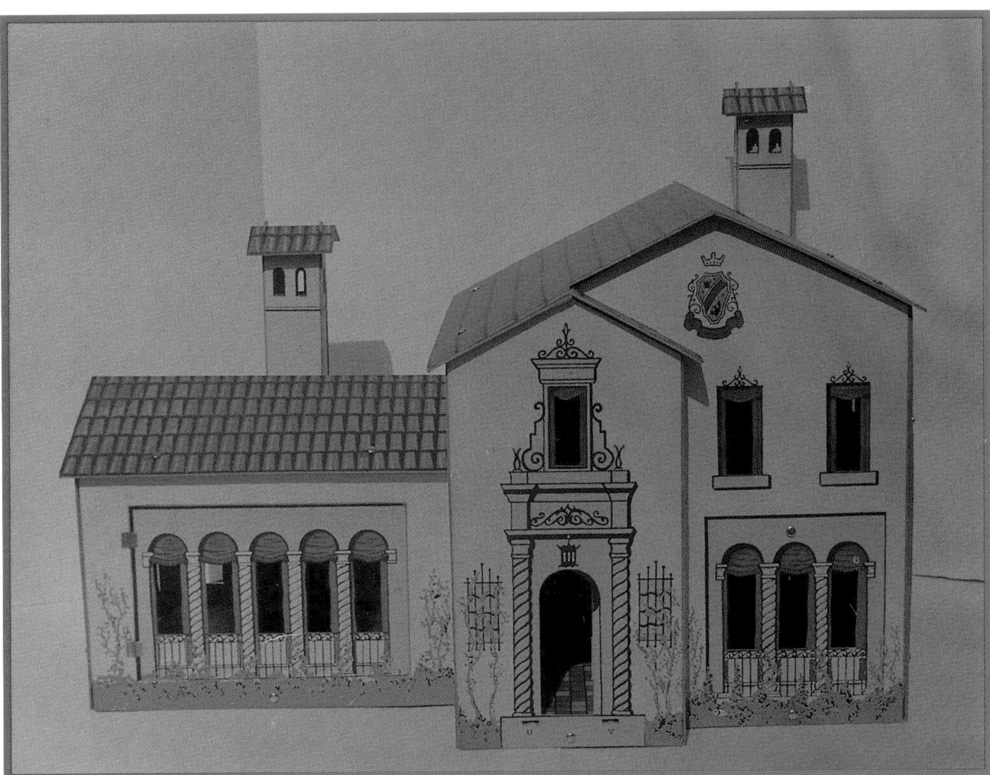

The Tootsietoy Spanish Mansion was first produced in 1930. The house contained a living room, kitchen, dining room, bathroom, two bedrooms, hall, and vestibule. The house was made of heavy book board and sold for $5.00 when new. From the collection of Mrs. David K. Large. Photograph by Patty Cooper.

The sides of the Spanish Mansion opened to allow access to the rooms inside. From the collection of Mrs. David K. Large. Photograph by Patty Cooper.

Miscellaneous Metal Dollhouse Furniture

Arcade, Kilgore, and Hubley are all companies known for their production of iron dollhouse furniture. Both the Arcade Manufacturing Co. and the Kilgore Manufacturing Co. also produced dollhouses for their furnishings.

The Arcade Company was located in Freeport, Illinois and was founded in 1885. The firm's popular toys were designed to look like their real counterparts and the company used the slogan "They Look Real" to promote their toys. The iron furniture was produced from 1925 to 1936. Most of these pieces were 1 1/2" to 1' in scale but the company did also manufacture some items that were slightly smaller than 1" to 1' in scale. Each of the furniture items was marked "Arcade" or "Arcade Mfg. Co.."

Many of the furniture pieces could be purchased along with Arcade room settings which included printed decor. These rooms included the following items: Kitchen: gas stove, sink, refrigerator, table and chair, cabinet, and dining alcove. Later an electric kitchen was added which included a Hotpoint stove and a Frigidaire refrigerator. Living room: secretary, ladder back chair, davenport, arm chair, end table, reading table, grand piano, and bench. Dining room: table, six chairs, sideboard, and china closet (Spanish style). Bedroom: bed, dresser, chair, rocker, and writing desk. Bathroom: corner bathtub, pedestal sink, toilet, stool, and shower. Laundry: wringer washer, ironer, heater, boiler, and laundry tray. Many of these pieces of furniture were marked with brand names such as Roper, Crane, Simmons, and Thor.

The Arcade Company also made several models of houses to be used with its furniture. In 1932 a house was advertised as being 2 1/2' tall, 5 1/2' wide and 19 1/2" deep. The house had six rooms and was made to look like stucco on the outside. It also had electric lights. The house had a removable glass front. The cardboard backgrounds that accompanied the furniture may have been used to decorate the individual dollhouse rooms. An advertisement for an Arcade house shows a two-story house with the bedroom and bathroom upstairs and the living room, dining room, and kitchen (kitchen behind dining room) located downstairs. Also downstairs was a one-story laundry room adjoining the kitchen.

In 1946 the Arcade Manufacturing Co. was purchased by the Rockwell Manufacturing Company in Buffalo, New York.

The Hubley Manufacturing Co. also produced iron doll house furniture. The company, located in Lancaster, Pennsylvania, was founded by John Hubley around 1894. The firm specialized in cast-iron toys. They first made trains and trolleys, but had added stoves and banks by 1909. The company began the production of iron dollhouse appliances in the 1920s. Many of their pieces look very much like the Arcade furniture, especially the ice boxes. The furniture included gas ranges marked "Eagle," ice boxes labeled "Alaska," kitchen cabinets, kitchen tables, and chairs. The items were produced in several different sizes. The stoves came in seven sizes from 4 1/2" to 12 1/2" tall. In 1940 Hubley was the largest manufacture of cast iron toys in the world. The company was bought by Gabriel Industries in 1965.

The Kilgore Manufacturing Co., located in Westerville, Ohio, was also a big producer of iron dollhouse furniture during the 1920s and 1930s. Some pieces are marked with the company name while others are unmarked. Most of the Kilgore furniture ranged in scale from 1/2" to 3/4" to 1', but there were some pieces as large as 1" to 1'. The slogan of the company was "Toys That Last." The firm produced iron toy tractors, motorcycles, trucks, cars, planes, and trains as well as dollhouse furniture.

The 1931 catalog lists the following pieces of iron furniture: bassinet, baby carriage, potty chair, high chair, stroller, cradle, bathtub, toilet, washing machine with separate wringer, ironing board and iron, refrigerator, gas stove (on tall legs), sink (also on tall legs), carpet sweeper, bed, dresser, dressing table, rocking chair, dining room table, chair, china closet, floor lamp, davenport, chair, tea cart, telephone, telephone stand, and grandfather clock. Most of the pieces were finished in either blue, green, grey, ivory or lavender enamel.

Besides the individual pieces of furniture, the Kilgore firm also produced room settings and a dollhouse for their furniture. The trade name used for these sets was "Sally Ann." The Sally Ann Playhouse consisted of five units that could be put together to form a four-room, two-story doll house. The rooms measured 7 3/4" by 7 3/4" by 6" tall. The rooms were furnished as a bedroom, dining room, kitchen, and bathroom. Twenty-five pieces of furniture were included with the house. The front lawn of the house provided room for a slide, teeter totter, and glider swing.

The Kilgore Company also marketed their furniture in room settings similar to Arcade. According to the 1931 catalog these rooms included a dining room, bedroom, kitchen, bathroom, and living room. The rooms were decorated with rugs, windows, and wall decorations. The outside of each room was designed to look like stone. Each of the rooms came complete with five pieces of furniture. These included the following: Dining Room: table, two chairs, buffet, and china cabinet. Bedroom: bed, chair, dressing table, stool, and highboy. Kitchen: table, refrigerator, sink, gas stove, and chair. Bathroom: bathtub, lavatory, toilet, stool, and hamper. Living Room: davenport, floor lamp, grandfather's clock, and two easy chairs.

In addition to the iron furniture, other companies produced furniture made of tin or other metals. The National Bellas Hess Co. catalog for 1930 featured a larger kitchen set made of metal which came with cardboard walls and floor to make a room for the furniture. The set sold for $1.00 and consisted of an icebox, cabinet, washtubs, and stove as well as the cardboard room pieces. The boxed set was called a Play House Kitchen Room No. 241. The box was marked "Katz Company, N.Y., U.S.A."

Since the metal furniture from all of these companies was so sturdy, the pieces can still be found in good condition. Today's dollhouse collectors must also compete with toy collectors in order to secure pieces of the furniture. With this added interest in these products, the prices continue to escalate.

The Arcade Kitchen, complete with room setting, circa late 1920s. The kitchen pieces included a stove, sink, ice box, cabinet, table, chair, and dining alcove. The kitchen measures 21 1/4" long by 17" deep and the dining alcove is 5" by 7 1/2". Kitchen from the collection of Gail and Ray Carey. Photograph by Gail Carey.

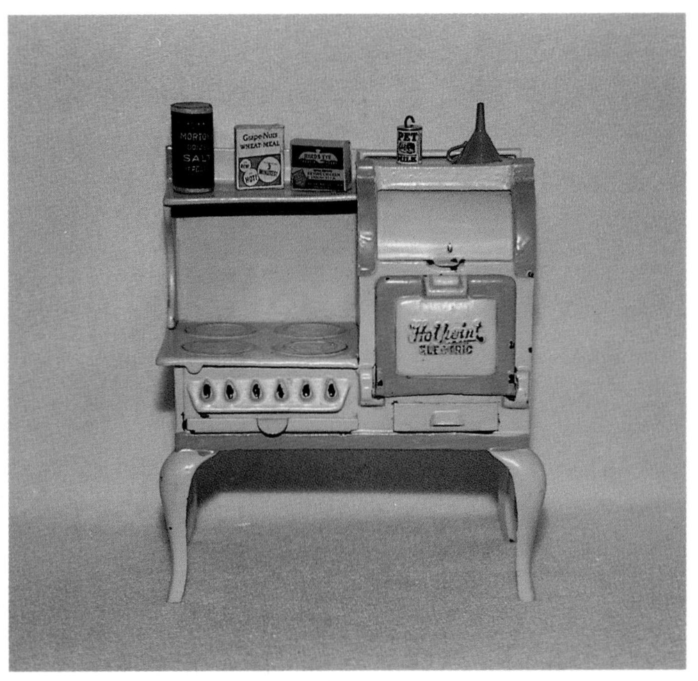

The Arcade iron furniture items followed the motto "They Look Real" and the company modeled their furniture after well known brands. This is a "Hotpoint" electric stove. The iron Arcade furniture is easy to identify as it is marked "Made by Arcade Mfg. Co., Freeport, Ill." The pieces also originally carried a seal featuring the company name. Stove from the collection of Gail and Ray Carey. Photograph by Gail Carey.

Arcade iron "Boone" kitchen cabinet and chair. The scale of these kitchen pieces is approximately 1 1/2" to 1'. Furniture from the collection of Gail and Ray Carey. Photograph by Gail Carey.

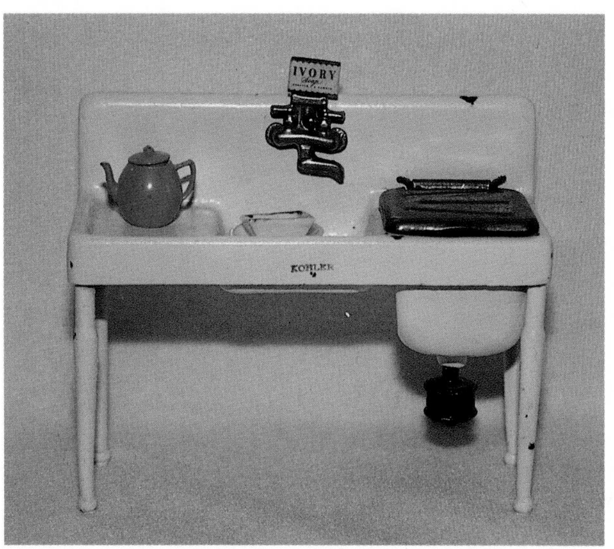

Arcade "Kohler" electric sink. The Arcade kitchen pieces are especially realistic because the iron is appropriate for these items. Sink from the collection of Gail and Ray Carey. Photograph by Gail Carey.

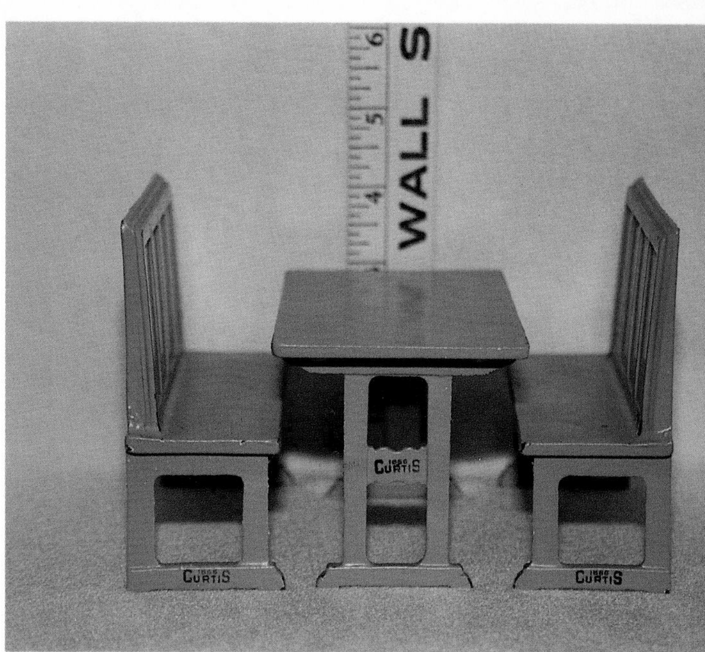

The Arcade iron dining alcove furniture reflected a trend of the 1920s for kitchens to include an eating space to supplement the formal dining room. Furniture from the collection of Gail and Ray Carey. Photograph by Gail Carey.

An Arcade kitchen table and chair pictured with a 1" to 1' Strombecker dining room table and chair to show the size difference.

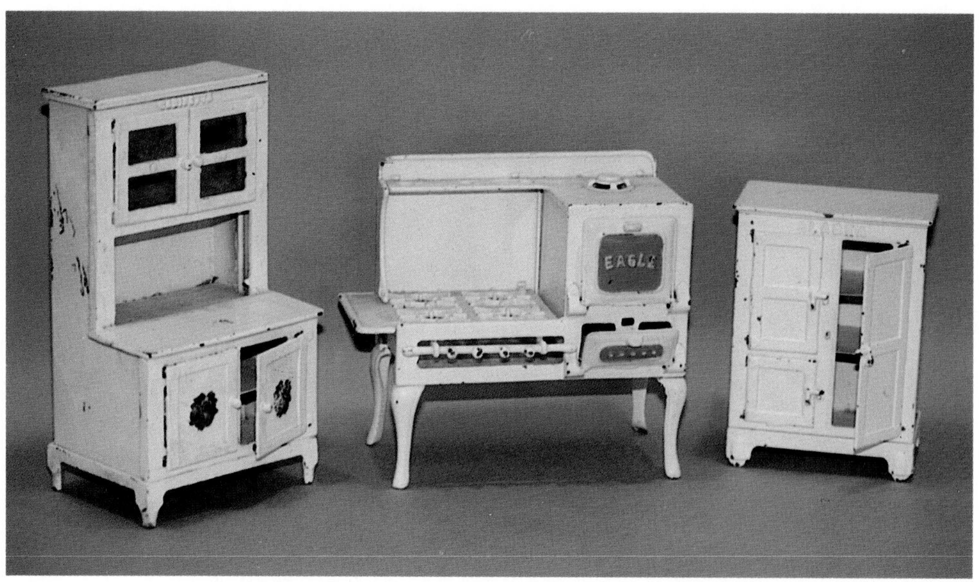

Iron kitchen furniture produced by the Hubley manufacturing Co. circa 1920s. This furniture is suitable for use with a large dollhouse scaled for 8"-9" size dolls. The doors on all three items function. The ice box measures 6" tall by 4 1/2" wide. The furniture also came in smaller sizes.

Kilgore iron kitchen pieces showing two sizes of kitchen chairs. The furniture ranges in scale from 1/2" to 1' to 3/4" to 1'. Photograph and furniture from the collection of Patty Cooper.

Arcade iron dresser and secretary. The bedroom pieces also included a bed, chair, rocker, and writing desk. The living room contained a chair, davenport, arm chair, end table, grand piano, bench, and reading table. The drawers are functional in the Arcade furniture. Furniture from the collection of Gail and Ray Carey. Photograph by Gail Carey.

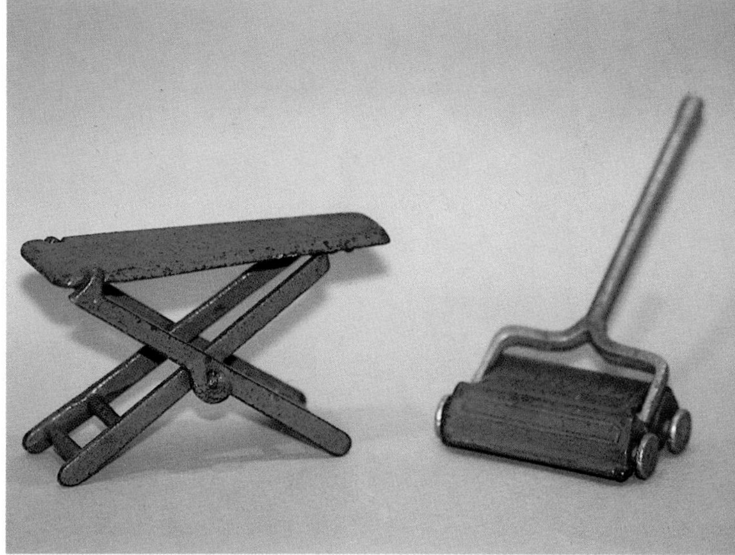

This Kilgore ironing board and carpet sweeper were made in a larger scale. The ironing board appears to be 3/4" to 1' while the carpet sweeper is 1" to 1' in scale. The sweeper is labeled "Sally Ann" which was the trademark used by Kilgore for their iron furniture sets and dollhouses.

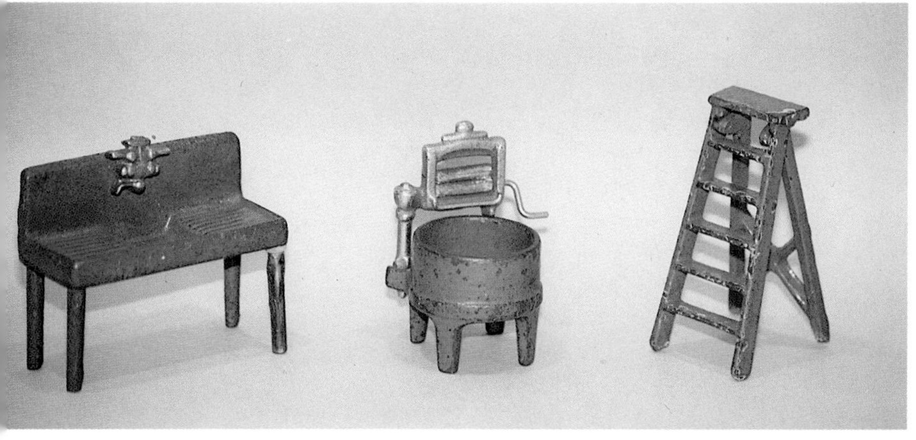

These iron Kilgore pieces were sometimes sold as part of the kitchen or household toy sets or as separate items. The furniture is in the 1/2" to 1' scale. The ladder was often included in the kitchen sets.

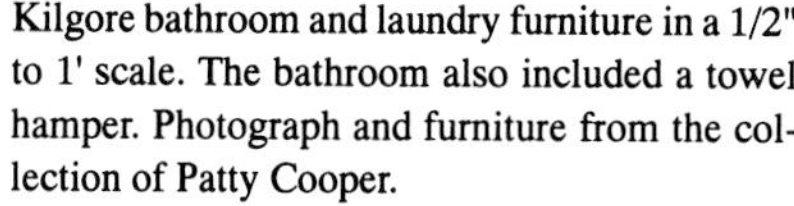

Kilgore bathroom and laundry furniture in a 1/2" to 1' scale. The bathroom also included a towel hamper. Photograph and furniture from the collection of Patty Cooper.

Kilgore produced several pieces of iron furniture that could be used in a dining room. These included an extension table, buffet, china closet, chairs, and a tea cart (not shown). Also pictured is the grandfather clock. Photograph and furniture from the collection of Patty Cooper.

Another version of the Kilgore dining room. These pieces are pictured as the dining room set in the 1931 company catalog. They include the table, two chairs, buffet, and china closet.

Kilgore bedroom pieces feature a bed, dressing table, bench, chest of drawers, and chair. The company also made a dresser with a steel mirror. Photograph and furniture from the collection of Patty Cooper.

A 1/2" to 1' scale Kilgore dressing table and bench are pictured with the high-boy. The furniture is from the Sally Ann Playhouse Bedroom set which also included a bed and chair.

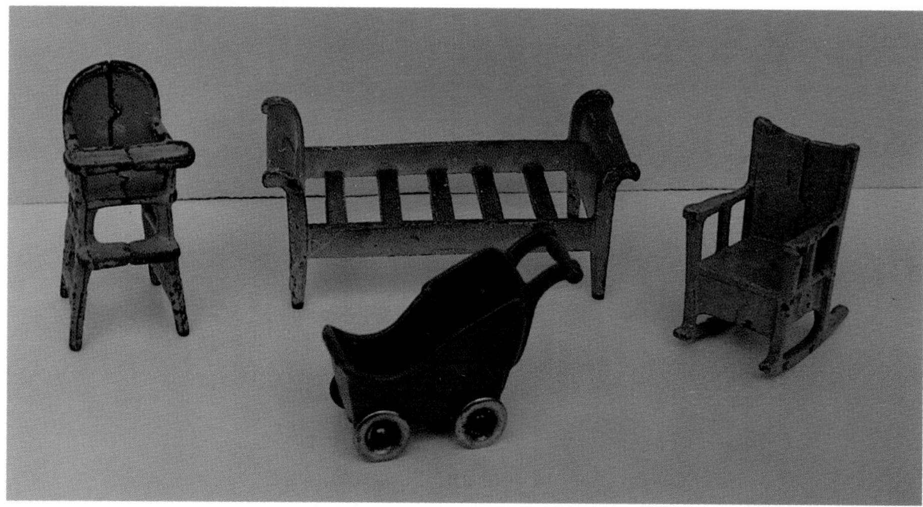

Kilgore baby pieces of furniture included the high chair, buggy, and a rocking chair. The bed is probably not a Kilgore item. This furniture is in the 3/4" to 1' scale. Photograph and furniture from the collection of Patty Cooper.

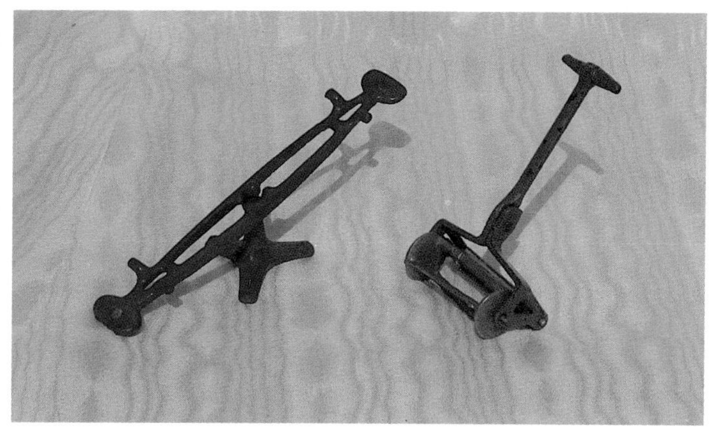

The Kilgore teeter totter and lawn mower are pictured. Both toy collectors and dollhouse collectors are interested in the iron furniture. From the collection of Kathy Garner. Photograph by Bill Garner.

The hardest Kilgore furniture to find seems to be the living room pieces. Shown is the easy chair that was sold as part of that set. The other items included a sofa, grandfather clock, floor lamp, and another chair. Furniture and photograph from the collection of Patty Cooper.

The potty chair and crib nursery pieces were produced by Kilgore both in the 3/4" to 1' scale and the 1" to 1' scale. The Kilgore nursery set included the potty chair, high chair, rocking chair, baby carriage, and crib on wheels. Photograph and furniture from the collection of Patty Cooper.

Kilgore also produced several outdoor pieces of dollhouse furniture. In the Sally Ann Playground set were the slide, lawn swing, kids car, stroller, and teeter totter. In place of the teeter totter, a Kilgore wheel barrow is pictured.

Metal furniture from the Play House Kitchen Room No. 241, Katz Company, N.Y. The furniture, circa 1930, came with cardboard walls and floor to provide a room for the furniture. The furniture pieces are suitable for use with a 7" or 8" doll. From the collection of Gordon and Judy Svoboda. Photograph by Judy Svoboda.

Cardboard Houses and Furniture

Built-Rite

The company which made Built-Rite Toys began as the Warren Paper Products Co. The firm was founded in Lafayette, Indiana in the early 1920s as a paper box manufacturer.

The company began the manufacture of paper toys in the mid-1930s. At first the products were sold under the Warren Paper Products Co. name. Later the fiberboard toy line was marketed under the Built-Rite trademark.

These toys were made for both boys and girls. Boys' products included: forts, railroad stations, farms, airports, and service stations. For girls, the company made a number of different designs of dollhouses. These houses included from one to five rooms. The toys were all manufactured in pieces and the consumer was expected to assemble the products after purchase. The boxes containing the various models were also used as the floors or the bottoms of the various buildings.

Many different dollhouse designs were manufactured by the Warren Paper Products Co. One of the early houses sold under the Warren name was labeled Set No. 11, Style No. 4. The house contained only one room and the bottom of the box, which was used as the floor of the house, measured 8" by 11". It was called "Play-time Doll House." The box lists the patent number as 1,890,269. In this early doll house, there are no wall or floor decorations and the outside of the house is quite plain. A larger house was also produced by the company in a two-story model. It was No. 14, and style No. 20. Both of these houses probably date from the mid-1930s.

Many of the later doll houses that were marketed under the Built-Rite trademark were designed in the Tudor manner. These houses came in one-, three-, or five-room models. Sometimes the furniture was also included with the houses or it could be purchased separately. The large five-room Built-Rite Country Estate model was sold by Montgomery Ward in 1944. Their Christmas catalog for that year pictured the house, complete with thirty-one pieces of Strombecker wood furniture, priced at $2.63. The house sold for only $1.79 furnished with ninety-six pieces of cardboard Built-Rite furniture. These items probably included small cardboard dishes and accessories as well as the larger pieces of furniture. This same house had been advertised in *Child Life* magazine as early as December 1939. The house also contained printed rugs, draperies, and wall decorations unlike some of the earlier houses.

A three-room house with a garage and car was also produced which bore the Built-Rite name. It came equipped with twenty-eight pieces of fiberboard furniture and cost $1.25. The house (No. 115) had a white clapboard design. It was 20" long by 13" wide by 11 1/4" tall. This house was advertised in *Children's Activities* magazine in November 1949. A four-room Built-Rite house was also advertised in *Child Life* in December 1939. The house was called a four-room cottage, and contained a living room, dining room, bedroom, and kitchen. It came complete with furniture for $1.00

One of the hardest models of Built-Rite houses for collectors to locate is the "Art Deco" house. The cardboard house has a flat roof and the outside walls are cream in color. The house is a two-story model and contains one room and a deck upstairs and two rooms downstairs. Rich Toy Co. also made a similar house of Masonite during the late 1930s. The Built-Rite house probably dates from the same period.

Besides making their own cardboard houses, Warren Paper Products also contracted with Strombeck-Becker to make houses for that company. These cardboard houses were sold under the Strombecker trade name, and were furnished with Strombecker wood dollhouse furniture. See the chapter on Strombecker for more information.

Cardboard dollhouse furniture was also produced by the company. The furniture, too, had to be assembled. There were five rooms of furniture manufactured to furnish the Built-Rite houses. The living room furniture included: sofa, radio, end tables, desk, chair, coffee table, two arm chairs, footstool, and magazine rack. The dining room pieces consisted of a table, four chairs, china closet, chest, and buffet. The bedroom set included twin beds, bureau, chifforobe, dressing table, bench, nightstand, and chair. The bathroom furniture was very modern for its time and included a fancy bathtub, lavatory with mirror, vanity, bench, toilet, hamper, rack, scale, and stool. The kitchen was complete with an ironing board. The set also included a stove, refrigerator, sink, cabinet, table, and two chairs. The furniture was made in a scale that was a little smaller than 3/4" to 1' but larger than 1/2" to 1'. Like most dollhouse furniture, some of the pieces seem to be in larger scale than others.

The Built-Rite paper toys went out of style in the 1950s with the coming of the new plastic and metal playsets and dollhouses. The firm continued to make puzzles and games and the company name was changed to the Warren Company in the mid-1970s.

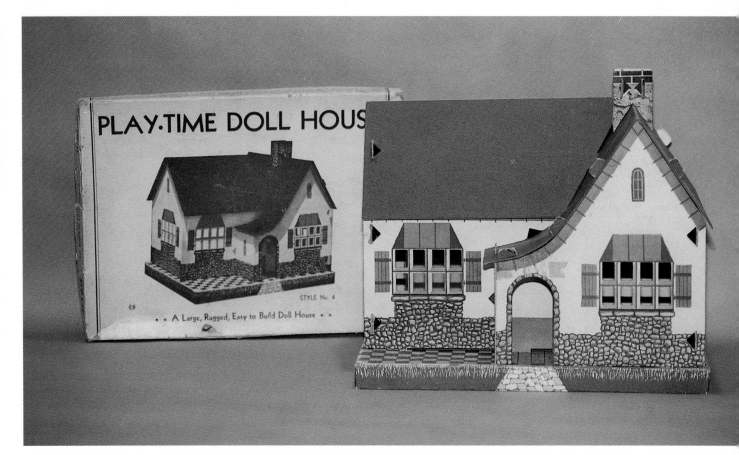

Early Warren Paper Products house dating from the mid-1930s. The house contains only one room and the box bottom measures 8" by 11". There are no wall or floor decorations. The box identifies the house as Set No. 11, Style No. 4.

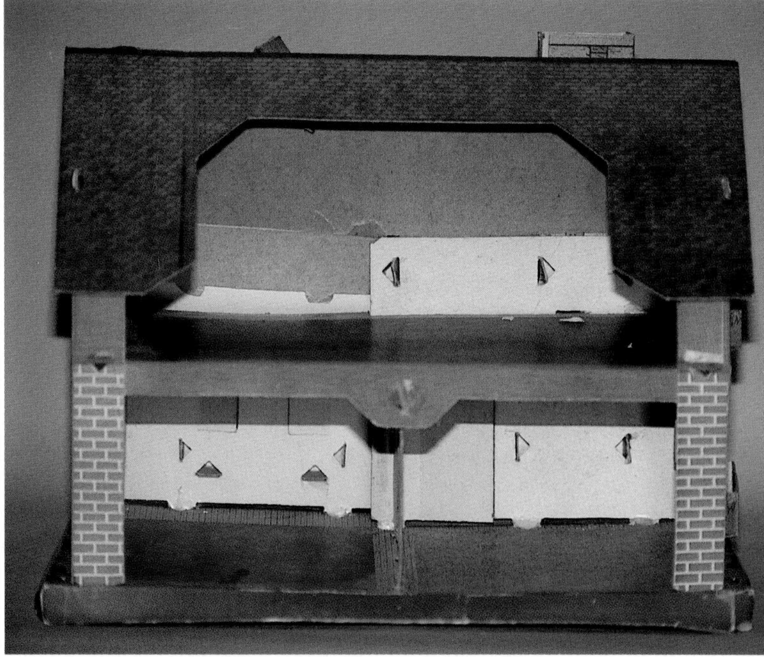

The house was made of heavy cardboard and featured no decorations on the walls.

Box and three-room Tudor Built-Rite dollhouse, circa 1940. The house measures 13" by 10" by 11" tall. From the collection of The Toy and Miniature Museum of Kansas City.

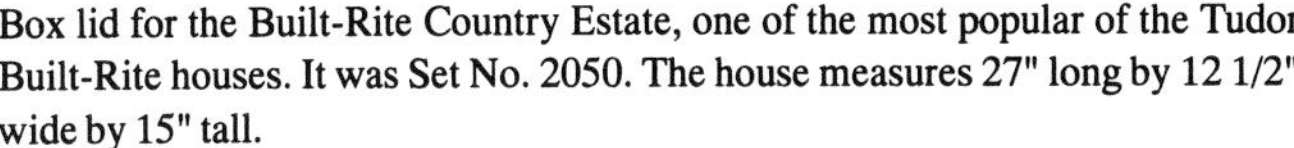
Box lid for the Built-Rite Country Estate, one of the most popular of the Tudor Built-Rite houses. It was Set No. 2050. The house measures 27" long by 12 1/2" wide by 15" tall.

This cardboard Country Estate house was featured in the Montgomery Ward Christmas catalog in 1942. The company continued to sell the house in their Christmas catalogs for several years. In 1945 the house sold for $2.63 furnished with Strombecker furniture or $1.79 furnished with Built-Rite cardboard furniture.

The two-story Country Estate house contained five rooms: living room, dining room, upstairs bedroom and bathroom, and a one-story kitchen wing. The inside of this house was printed with rugs, wall decoration, draperies etc.

Another design of a Built-Rite house was advertised in the *Children's Activities* magazine in November 1949. The house also contained five rooms but was a white clapboard model instead of Tudor. The cardboard house sold for only $1.25. The same house was also made with a garage attached to a three-room house. It measured 20" by 13" by 11 1/4" high.

One of the most sought-after Built-Rite houses is the cardboard "Art Deco" model from the late 1930s. House from the collection of the late Vilora Kergo. Photograph by Kay Houghtaling.

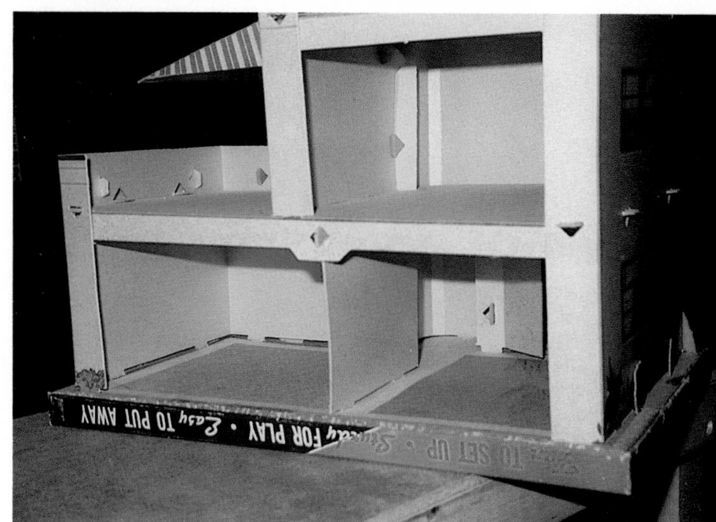

The "Art Deco" house includes three rooms and a covered deck. Although the floors are printed in color, the inside walls are plain. House from the collection of the late Vilora Kergo. Photograph by Kay Houghtaling.

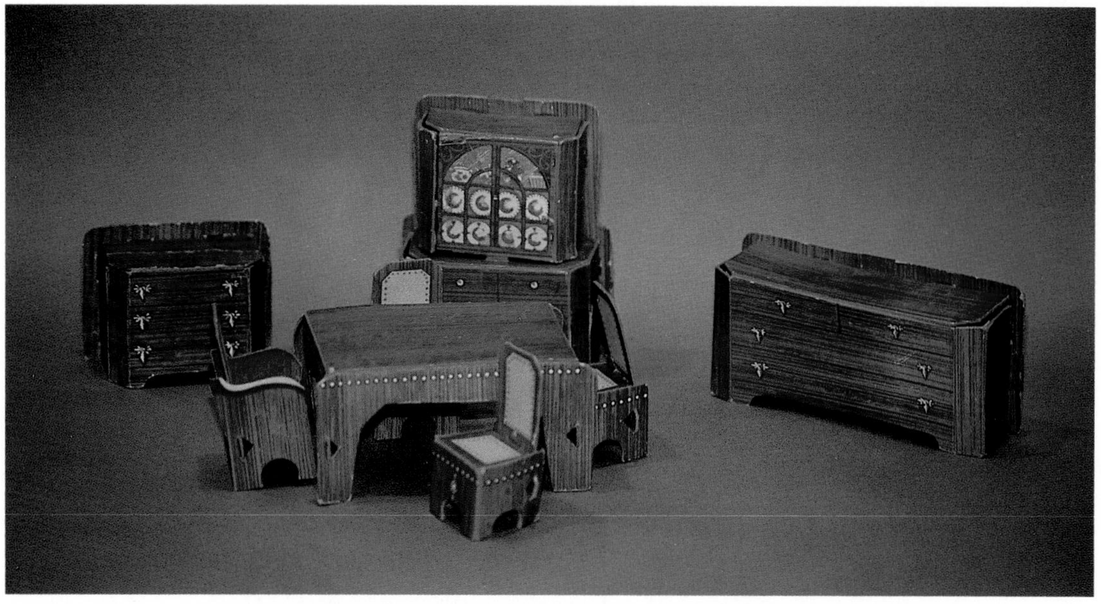

All of the Built-Rite cardboard furniture had to be assembled by the consumer. Pictured is the set of dining room furniture.

Built-Rite made a line of cardboard furniture that could be sold with their houses. The furniture was also marketed in separate packages without the houses. Pictured is the living room furniture. Missing is the book rack.

The Warren Co. followed the lead of other dollhouse manufacturers and produced twin beds to furnish their bedrooms.

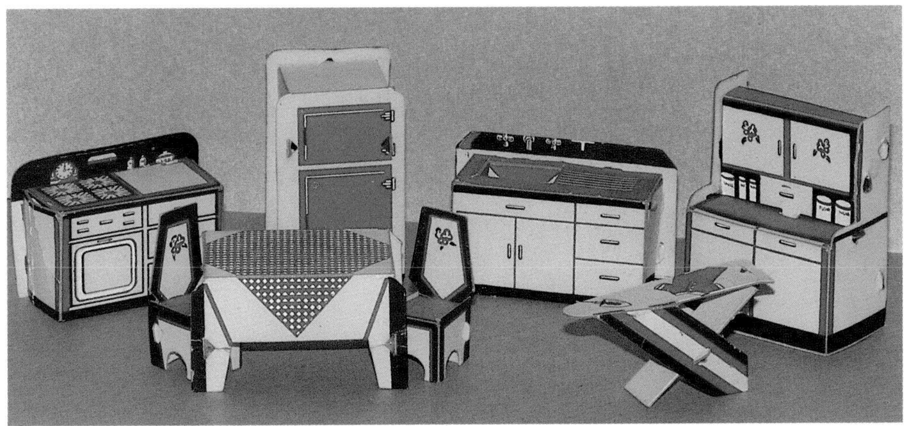

The Built-Rite kitchen furniture was unusual in its color scheme of red and yellow.

The Built-Rite bathroom furniture was quite modern for its time with the built-in look.

Cardboard Dollhouses

Dollhouses made of several different types of cardboard material have been popular throughout this century. These dollhouses could be produced cheaply and were packaged unassembled so they required little space on store shelves and were easily handled by mail order companies.

An early firm to make use of this concept of dollhouses was McLoughlin Brothers. This New York company was active in producing children's books, games, and puzzles beginning in the late 1850s. Their later products made use of color lithography and this technique was especially effective when used to make dollhouses and dollhouse furniture. McLoughlins began producing dollhouses as early as 1875. Their early two-story house was a lithographed paper on wood model. The most well known McLoughlin house is "Dolly's Play House" which was made from 1884 to 1903. The house was made of strawboard. The house and its furniture came unassembled. The house contained two rooms, one upstairs, and one down. There was no front to the house as the rooms were open for play. The house unfolded in order to be set up. The house measured 17 1/2" high by 12" wide by 9" deep. The McLoughlin Brothers catalog for 1909 featured another folding doll house. Although the house was also a two-story, two-room model, the front dropped down to reveal a lithographed garden. This house is sometimes called the "Garden House." It was listed at a price of $1.50 in 1909.

Because the cardboard dollhouses could be packaged unassembled in small boxes, they offered companies a good premium idea to promote products. An early company to take advantage of this promotion was the publisher of the *Ladies Home Journal* magazine. In 1912, after featuring the Letty Lane paper dolls for several years, the *Ladies Home Journal* offered a Letty Lane Doll House to its subscribers. In order to secure a dollhouse, a patron needed to obtain three new yearly subscriptions to the magazine. The house was called a bungalow and included a living room, dining room, bedroom, and kitchen. A small German bisque doll was included in the package.

The following year (1913), another cardboard bungalow dollhouse was produced by the Bungalow Book and Toy Co. of New York. "Betty's Bungalow Doll-House" came packaged as a book that could be made into a two-story house. The house was made of cardboard and came furnished with heavy paper furniture. There were two models of the house. One contained a living room and bedroom, and the other featured a kitchen and a dining room. The furniture was advertised as needing no glue, and was to be folded into place for assembly.

Another concept in the manufacture of the cardboard house is the use of the boxes containing the products in the construction of the house itself. The Embossing Co. from Albany, New York made a series of doll rooms using this idea. The boxes are labeled both "Tiny Town Furniture" and "Toys That Teach." Both boxes indicate "Patent Applied For." There were several different rooms in the set. Included were the living room, dining room, and bedroom. Each room was boxed separately and included the wood furniture and three walls for that particular room. The heavy cardboard walls were printed on both sides to indicate both the inside walls (including pictures) and the outside walls which pictured the windows and brick facade. Part of the box was to be used as the floor of the room (with a printed rug) and the other half of the box functioned as the roof of the room. The rooms could be stacked or played with separately. An unusual feature of these cardboard houses was that the furniture was not paper or cardboard but was made of wood. Although the boxes are not dated, they seem to be circa late 1920s. The furniture was made in pieces with friction slots so it could be put together and taken apart. The furniture was made by using a sturdy block method and seems to have survived the years very well.

This same concept was used much later in the "Add-A-Room Playhouse" series made by the National Paper Box Co. in West Springfield, Massachusetts. The rooms did not contain any furniture but were designed to be used with the plastic furniture from the late 1940s. The set contained rooms for a bedroom, kitchen, dining room, living room, and bathroom. Most of the rooms were 10 1/2" by 7 3/4" by 7" high. The living room was larger, measuring 15 1/4" by 7 3/4" by 7" high. In this cardboard house, the box also formed the foundation and the three-sided rooms were printed to picture the three walls of the houses.

Cardboard houses also lend themselves well to being marketed with cardboard characters. During the 1930s, houses were produced to accompany figures from the Mother Goose Nursery Rhymes. Another cardboard house from the New York firm of A. P. Newberg and Co. which sold at about this same time was full of candy. It also included a cardboard family of figures. An even more interesting house of this type was produced by the Deluxe Game Corp. circa 1950s. It was called "Sparkle Plenty Playhouse." This boxed set included cardboard figures of the main characters from the Dick Tracy comic strip.

Many other cardboard houses were featured in the Christmas catalogs from the 1920s through the 1940s. These houses sold cheaply, and some were priced as low as $1.00 each. Most of these dollhouses were of Colonial design and some models contained an added garage or sunroom. The cardboard or fiberboard used to construct these houses varied from thin to a quite sturdy material. Many of these houses are very similar to the houses marketed by Dowst under the Tootsietoy and Daisy labels. The Sears, Roebuck and Co. Christmas catalogs continued to feature fiberboard houses through 1949. This later house contained six rooms and measured 34 1/2" by 16" by 20" high. The house, of course, came unassembled.

The production of cardboard houses to be used as premiums to promote products is a practice that has continued all through the years. Several recent houses have been marketed in this fashion that are now collectible. Chesebrough Ponds, Inc. sold a house in 1983 as a premium for their products. The three-story house measured 26" by 18" by 9". Bakers' Coconut also recently marketed a premium house. Their two-story, four-room house is made of heavy cardboard and is nicely constructed. The walls are printed with furniture, windows, and doors. The Little Debbie house is another premium house that was offered in the 1980s. This three-story house is made of a lighter cardboard and came with a Snack Shop and Little Debbie figure. It was used to promote Little Debbie Snack Cakes in 1986.

By the end of the 1940s, metal houses and their plastic furniture could be produced as cheaply as a furnished fiberboard house. The metal houses were far more sturdy than the fiberboard models and fewer cardboard dollhouses were produced. It is surprising that so many of these cardboard houses can still be found. They are especially desirable when discovered in their original boxes. The houses make fine additions to dollhouse collections and are in constant demand when found in good condition.

See also the chapters on Built-Rite, Strombecker, and Tootsietoy for more information on cardboard dollhouses.

"Dolly's Play House", a strawboard house produced by McLoughlin Brothers circa 1900. The house contains two rooms. Courtesy of The Toy and Miniature Museum of Kansas City.

Another McLoughlin Brothers two-story house circa 1909. The house measures 16 3/4" by 10" by 16". This "Garden House" has a front that drops down to reveal a lithographed garden. Photograph and house from the collection of Patty Cooper.

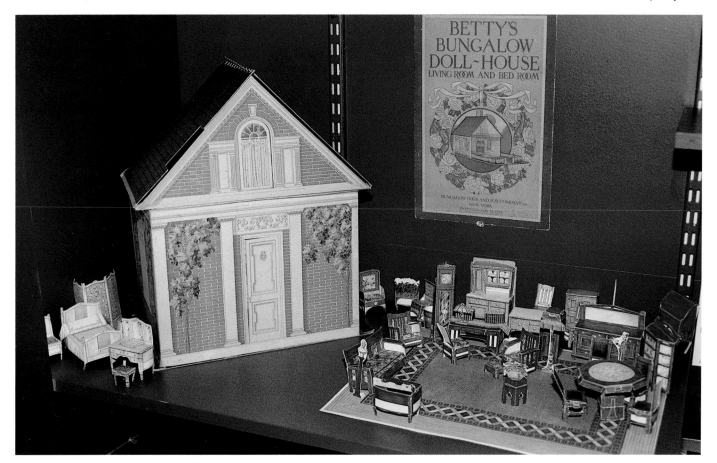

Betty's Bungalow Doll-House dating from 1913. Marketed by the Bungalow Book and Toy Co. of New York. The house came complete with heavy paper furniture. Courtesy of The Toy and Miniature Museum of Kansas City.

Box labeled "Toys That Teach Furniture" made by The Embossing Co. of Albany, New York. The box contained wood furniture and walls for the living room. Pictured are the pieces of wood living room furniture. The assembled living room measures 12" by 24" by 10 1/2" tall.

The boxes for the Embossing Co. furniture were made to be used as rooms for the furniture. Boxes were produced for the living room, bedroom, and dining room. A separate cardboard wall piece was included in each box. This concept was similar to that later used by the plastic dollhouse furniture makers of the 1950s.

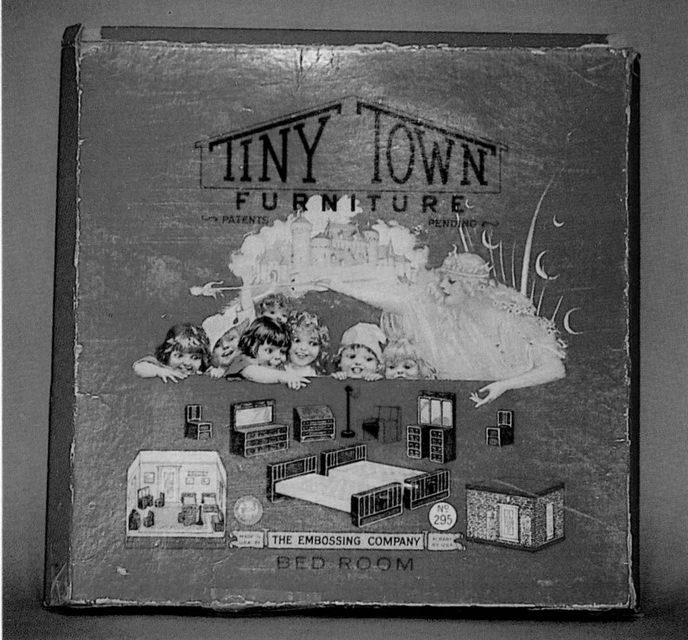

The box that contained the Tiny Town bedroom and its wood furniture. Both box styles were produced by the same company. This room measures 12" by 12" by 10 1/2" tall.

The wood furniture produced by The Embossing Co. could be snapped together or taken apart. Pictured are the wood dining room pieces. The dining room measured 12" by 12" by 10 1/2" tall when assembled.

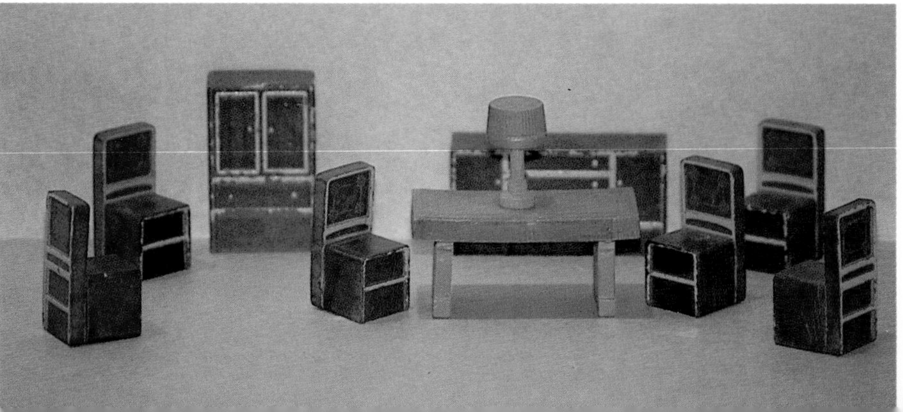

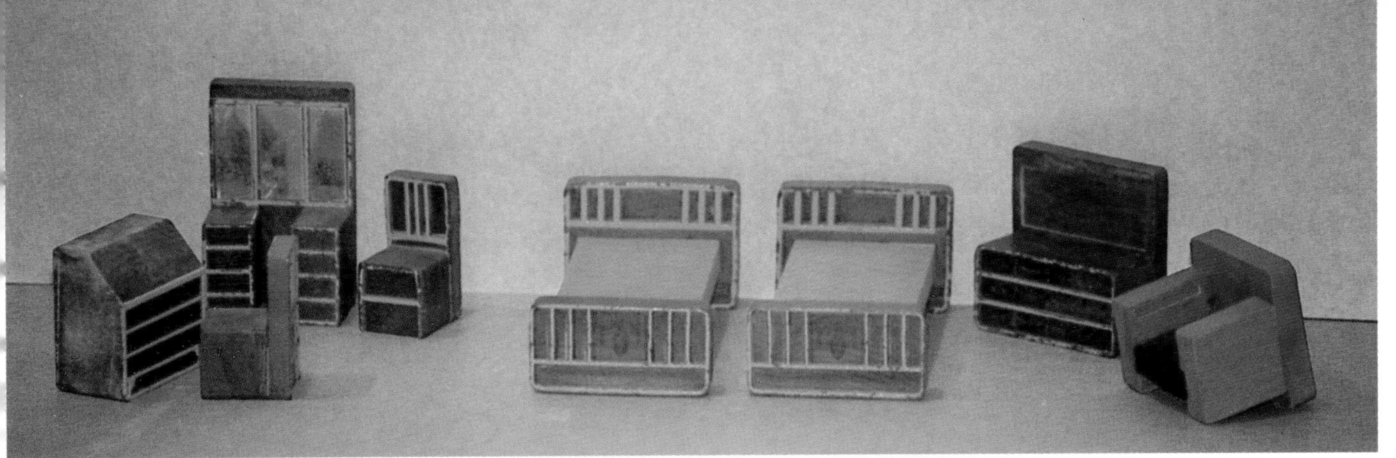

The wood bedroom furniture marketed by The Embossing Co. The chair pieces have been glued together incorrectly.

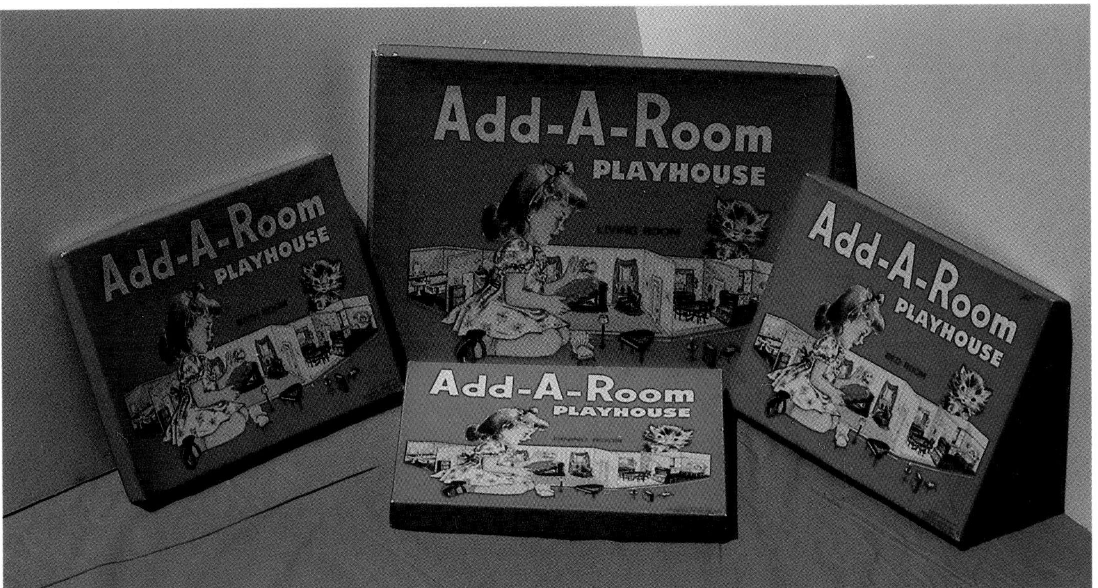

The Add-A-Room Playhouse used a similar concept of producing a box that could also be used as part of a room. Boxes from the collection of Marcie Tubbs. Photograph by Bob Tubbs.

The Add-A-Room dining room is pictured after it has been assembled. Room from the collection of Marcie Tubbs. Photograph by Bob Tubbs.

Cardboard houses could be enhanced with the addition of figures to make the house a little different. Pictured is a house circa 1930s which came complete with figures from the Mother Goose Nursery Rhymes. The house measures 11" by 8" by 7 1/2" high. The house contains one room and the 3 1/2" to 4" high cardboard figures seem too large to inhabit the house. House and photograph from the collection of Marilyn Pittman.

This little house was once used as a candy container. It came complete with a family. The house measures 7 3/8" by 5" by 5" tall. It is marked "Packed by A. P. Newberg and Co., New York." Photograph and house from the collection of Marilyn Pittman.

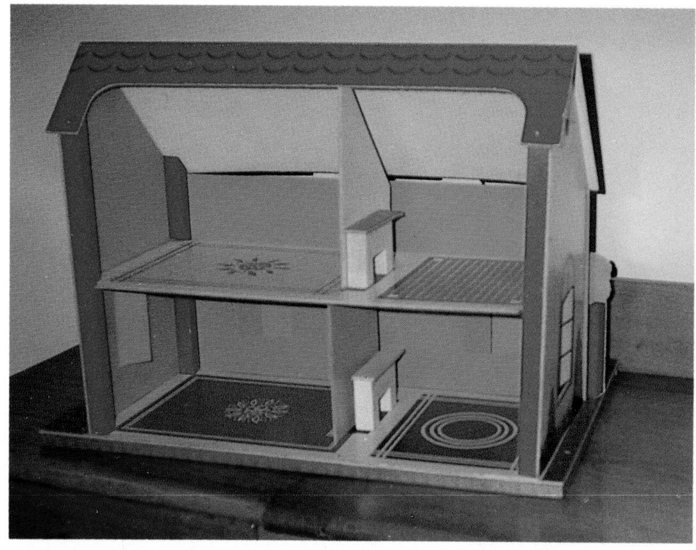

The inside of the Sparkle Plenty house contained four rooms along with a Sparkle Plenty cradle, B.O. Plenty, Dick Tracy, Sparkle Plenty, and Gravel Gertie figures. Photograph and house from the collection of Marilyn Pittman.

One of the most collectible of these cardboard houses, complete with cardboard figures, is the Sparkle Plenty Playhouse. It was produced by the Deluxe Game Corp. circa 1950s. The Louis Marx Co. also marketed a cardboard house which featured Blondie and Dagwood figures circa 1970s. Photograph and house from the collection of Marilyn Pittman.

Trixy Two Room Portable Doll House. Cardboard house with base measuring 8 3/4" by 14". The inside of the Trixy house contains two rooms. A similar house was still being featured in the Sears, Roebuck and Co. catalog as late as 1928. House and photograph from the collection of Marilyn Pittman.

Although this house looks very much like the Tootsietoy house, it is marked "Wayne Paper Products." It measures 18 1/8" by 12 1/4" by 16" tall. Circa 1930. It is probable that this firm also produced houses for Tootsietoy (see Tootsietoy chapter). House and photograph from the collection of Patty Cooper.

The inside of the cardboard Wayne Paper Products house contains a stairway and features six rooms. It has been furnished with Tootsietoy furniture. Photograph and house from the collection of Patty Cooper.

The Concord house included trees and a side porch. It is a two-story house with four rooms. The inside walls are plain with printed floors. The house measures 12 1/4" by 27 1/2" by 14 3/4". House and photograph from the collection of Barbara Staiger.

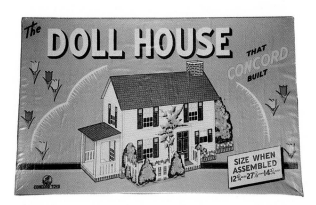

Box that contained the cardboard Concord doll house. It was manufactured in New York City circa 1940. Photograph and box from the collection of Barbara Staiger.

The Nels Doll House was a cardboard house circa 1930s. Courtesy of The Toy and Miniature Museum of Kansas City.

The Nels Doll House measures 29" by 16" by 11 1/2" when assembled. The house contains five rooms in a two-story design. Courtesy of The Toy and Miniature Museum of Kansas City.

This lightweight cardboard house also is circa 1930s. Courtesy of The Toy and Miniature Museum of Kansas City.

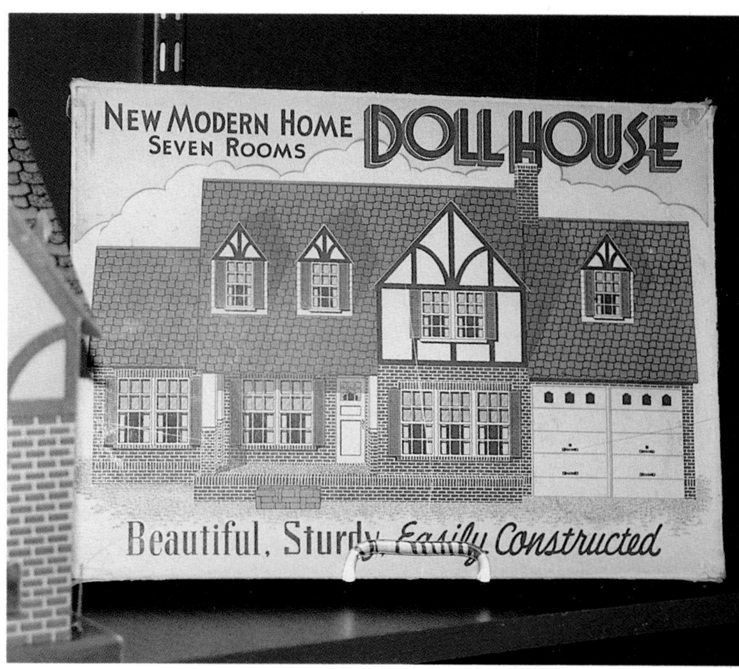

This cardboard house was featured in the Sears, Roebuck and Co. Christmas catalog for 1940. It sold for seventy-nine cents complete with seven rooms of fiberboard furniture. The house measured 31 1/4" by 16 1/2" by 12" and came with sixty-three pieces of furniture. Courtesy of The Toy and Miniature Museum of Kansas City.

This Happitime cardboard dollhouse was sold by Sears, Roebuck, and Co. in 1949. The house contained six rooms and measured 34 1/2" by 16" by 20". The two-story house was made of heavy fiberboard and was sold complete with plastic Ideal furniture at a price of $2.98. House from the collection of Judy Mosholder. Photograph by Carl Whipkey.

The inside of the six-room Sears house was printed with windows, rugs, and furniture. House from the collection of Judy Mosholder. Photograph by Carl Whipkey.

This cardboard dollhouse could be taken apart easily just by removing the brads and folding the roof and walls in a flat position. This concept was used in the Tootsietoy houses in the 1920s. The house measures 19" by 8 1/2" by 13 1/2" high.

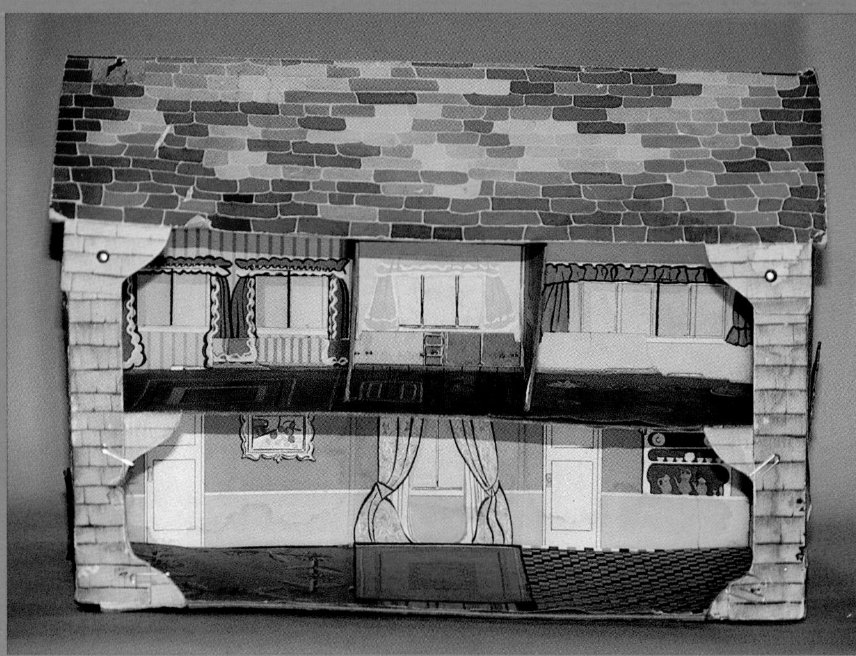

The inside of the house (circa late 1940s) contained six rooms. The partitions are missing from the dining room walls.

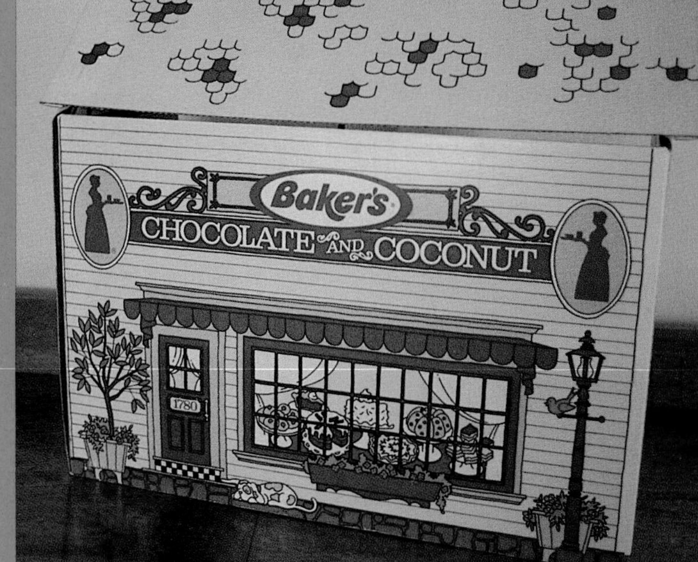

This cardboard Baker's Chocolate and Coconut House is a more recent house. The roof may have been put on backwards with the printing appearing on the wrong side. Photograph and house from the collection of Marilyn Pittman.

This lightweight cardboard house was used as a premium for Little Debbie products. It is three stories tall and came complete with a Snack Shop and Little Debbie figure. The house is approximately 36" wide by 30" tall.

The inside of the Little Debbie house featured printed furniture, windows, and decorations on the walls. It is marked © 1986 McKee Baking Company.

The inside of the Baker's Chocolate house contains four rooms and features walls that are printed with furnishings, doors, and windows. The house is quite large and measures approximately 30" wide. Photograph and house from the collection of Marilyn Pittman.

Plastic Furniture and Metal Houses

Ideal

The plastic dollhouse furniture produced by the Ideal Toy Co. constituted only a small part of the company's large toy operation. Ideal was primarily known for its dolls, toy sets, doll dishes, plastic trucks, cars, and games.

The company began when Morris Michtom and his wife began producing stuffed bears above their candy store in Brooklyn, New York in 1903. The couple called the bears "Teddy Bears" after President Theodore Roosevelt. Michtom went into partnership with Aaron Cone in 1907 and the Ideal Novelty Co. was born. In 1912 this partnership ended and Michtom changed the name of the company to the Ideal Novelty and Toy Co. In 1928 Morris Michtom left the management of Ideal and several of his relatives, including his son Benjamin and his nephew Abraham Katz, took over the company. In 1963 Lionel Weintraub (son-in-law of Katz) became the president of Ideal and he held the position for twenty years. Ideal was sold to Columbia Broadcasting System in 1983 for $58 million. The line merged with CBS's Gabriel toy line and it was called Ideal - Gabriel. Viewmaster then bought the Ideal trademark but after a short time they sold out to Tyco Industries, Inc. in 1989.

The Ideal plastic dollhouse furniture came on the market in 1947. The furniture is considered 3/4" to 1' in scale but some of the pieces are slightly larger. The furniture was featured in the Sears Christmas catalog in 1948. The boxed rooms sold for eighty-nine cents each. The rooms included the dining room, bedroom, living room, bathroom, kitchen, nursery, and patio. The pieces of furniture included in each set ranged from five to eight items. Extra pieces of furniture were also available to the customer and many of these items could be purchased at local "dime stores." The retail prices for each piece of furniture ranged from five cents for a kitchen chair to seventy-five cents for the television set. The early boxes for the room sets of Ideal furniture included inserts that could be used as walls for rooms to house the furniture. There were two sets of kitchen furniture produced, a standard kitchen and a deluxe edition.

The following pieces of plastic furniture were available: Living room: Secretary, floor lamp, piano and bench, fireplace, television, floor radio, tilt top table, sofa, radiator, coffee table, table lamp, wing chair, and club chair. Dining Room: Table, lyre back chairs, china cabinet, and buffet. Bedroom: Twin beds, highboy, boudoir chair, nightstand, vanity with mirror, and bench. Bathroom: Toilet, hamper, bathtub, lavatory, and medicine cabinet. Standard Kitchen: Stove, sink, refrigerator, table and four chairs. Deluxe Kitchen: Sink, stove, regrigerator, table and chairs, and automatic dish washer. Appliance Room: Sewing machine, automatic washing machine, and ironer. Garden: Table and umbrella, bench, dog and dog house, picnic table, chaise lounge, chair, birdbath, pool, and trellis fence. Nursery: Buggy, high chair, crib, potty chair, cradle, playpen (folded), stroller, high chair that folded into a rocker, and baby. Misc. Items: Folding card table and chairs, sofa that made a bed, lawn mower, and vacuum sweepers. A few years later the original bathtub was replaced with a more modern corner model.

The doll produced by Ideal to accompany their furniture pieces was an all-plastic baby doll. There were several different models of the jointed baby doll produced. Some of these babies featured molded diapers while others did not. The dolls are approximately 2 1/2" tall and are not marked. These dolls were included with many of the boxes of furniture. An earlier unjointed plastic doll had been sold with some of the first furniture sets. An unmarked set of family dolls may have also been made as a separate product but the dolls did not accompany the sets of Ideal furniture.

Most of the furniture is marked with "Ideal" in an oval along with an "I" and an identifying number. The pieces that had cardboard backs are marked on the cardboard. Other items carry only an "I" and an identification number.

The Reliable Toy Co. of Toronto, Canada was licensed by Ideal to copy some of its products. Several of the Ideal doll molds were used by Reliable. The same agreement must also have applied to the plastic dollhouse pieces as many of the items identified with the Reliable name have come from the Ideal dollhouse furniture molds. These included the kitchen and bathroom pieces. Reliable also produced other furniture that was of a different design than the Ideal products. Some of these included a sofa, coffee table, rocker, serving cart, washing machine, bunk beds, and desk.

In 1950 the Ideal Toy Co. began producing a much larger set of plastic doll furniture called Young Decorator. This furniture was on a scale of approximately 1 1/2" to 1' and would be compatible with a doll around 6" tall. There were six rooms of furniture in this larger scale. The back of the original box pictures the following pieces: Dining Room: Table, four chairs, buffet, and china cabinet. Living Room: Four-piece sectional, television, coffee table, and floor lamp. Nursery: Bathinet, crib, play pen, high chair, and tricycle. Bedroom: Bed, nightstand, vanity with mirror, bench, and armoire. Bathroom: Toilet, bathtub, lavatory with storage, and hamper. Kitchen: Stove, refrigerator, table and four chairs. A sink was also made for the set but is not pictured on the box. A carpet sweeper and vacuum cleaner were also produced by Ideal that are large enough to be used with the Young Decorator set of furniture.

Many of the drawers and doors opened in the Young Decorator furniture. The pieces are so large that current collectors have a problem finding a house large enough to hold these Ideal items but the Young Decorator series is becoming more in demand by collectors. The furniture is marked with the Ideal name in an oval along with "Made in U.S.A", an identifying number, plus an "I". The furniture apparently was only made for two years.

By the late 1950s, the much-loved plastic furniture of Renwal, Ideal, and Plasco had been pretty much replaced by the cheaper Marx soft plastic pieces which were sold as furnishings in their popular metal dollhouses. While Ideal had stopped production of their furniture in the early 1950s, the other two companies continued marketing furniture through the early 1960s.

After having been out of the dollhouse furniture business for over ten years, Ideal began production of a new line of plastic furniture in 1964. The company took a big gamble when they introduced what was then a very expensive line of dollhouse furniture called Petite Princess. The plastic furniture was mostly in the 3/4" to 1' scale but some pieces were slightly larger. The furniture was made in Japan for the Ideal Toy Corporation. It was called Fantasy Furniture because it was so elaborate it could have been used in a castle. In 1962 a catalog customer could purchase five rooms of Renwal plastic furniture for under $3. In 1964, Ideal marketed its new Petite Princess furniture for from seventy-seven cents (coffee table set) to $2.47 (piano and bench) for each piece. The experiment was doomed for failure. The first year there was no bathroom or kitchen furniture so the dollhouse dolls really had to inhabit a fantasy world with their Fantasy Furniture.

There were lots of pieces of furniture to purchase however. They included the following: Living Room: Curved Sofa (4407-3), Drum Chair (4411-5), Wing Chair (4410-7), Occasional Chair with Ottoman (4412-3), Fireplace (4422-2), Coffee Table (4433-9), Planter (4440-4), Occasional Table (4437-0), Grandfather Clock and Screen (4423-0), Tier Table (4429-7), Palace Table (4431-3), Heirloom Table Set (4428-9), and Pedestal Table Set (4427-1). Several of these items could also be used in a music room. Bedroom: Little Princess Bed (4416-4), Boudoir Chaise Lounge (4408-1), Royal Dressing Table (4417-2), Palace Chest (4420-6), and Fantasy Telephone Set (4432-1). Dining room: Rolling Tea Cart (4424-8), Dining Room Table (4421-4), Host Dining Chairs (4413-1), Hostess Dining Chairs (4415-6), Guest Dining Chairs (4414-9), Royal Buffet (4419-8), Royal Candelabra (4439-6), and Treasure Trove Cabinet (4418-0). Music Room: Royal Grand Piano (4425-5), Fantasia Candelabra (4438-8), Guest Chair (4409-9), and Lyre Table Set (4426-3).

The furniture was marked "© Ideal" (in an oval) "Japan." The moving parts of the furniture and the clever accessories that came boxed with many of the pieces made the furniture especially appealing.

Ideal also produced Fantasy Rooms to house the new furniture. These rooms were made of cardboard and came in pink, blue, or yellow. The rooms sold for only seventy-nine cents each. A three-story six-room cardboard house was manufactured in 1965 to accompany the new line of Princess Patti furniture. Present collectors try to locate the store display cases to use as showcases for their own collections of Petite Princess furniture.

If a parent bought everything listed in the Petite Princess line of furniture in 1964, it would have cost from $50 to $60 depending on the place of purchase. That was a lot of money when a furnished metal Marx dollhouse was still priced at under $10.

In 1965 Ideal tried to correct the mistake of not having produced kitchen and bathroom furniture by adding these items as well as a television in the new Princess Patti line of furniture. The bathroom included an oval tub, sink with mirror, toilet, hamper, waste basket, stool, and linen storage. The kitchen contained a modern sink and dishwasher, refrigerator, stove with hood, round table and clear plastic chairs, and a hutch. Accessories were also included with both sets of furniture.

The expensive kitchen and bathroom sets did not help sales. The furniture was finally sold at reduced prices through the next several years. Many of today's dollhouse collectors began collections by picking up the sale priced Petite Princess furniture in the mid-1960s.

Ideal also marketed a vinyl suitcase type dollhouse, some years later, which came complete with the earlier design of Petite Princess furniture. Although most of the furniture was marked Ideal, there were some less desirable pieces, based on the original designs, that were produced at a later date. These items are labeled "Redbox/Made in Hong Kong." Perhaps this later furniture was made to supplement the left over Ideal stock to furnish the new Ideal vinyl suitcase houses. If a collector wants to collect only the best of the Petite Princess furniture, the name "Ideal" should be on each piece. Sears priced the vinyl house at $10.22 and it included more than twenty plastic pieces of furniture and accessories. The house was designed with a drop front and contained a living room, dining room, bedroom and music room. The house folded up to measure 21" by 8" by 18". Furniture included the bed, dressing table and stool, piano and bench, lyre table, cabinet, two host dining chairs, occasional table set, tier table, curved sofa, occasional chair with ottoman, and lamps. The other furnishings were printed on the walls. The advertising copy says that the house and contents were available only at Sears.

Ideal also provided dolls to be used with their Petite Princess furniture. The dolls were named the Fantasy Family. The family included a father, mother, daughter and son. The dolls were pliable plastic so they could sit or stand. The father was 5 1/2" tall. The set of dolls sold for around $3.

Although the Petite Princess and Princess Patti lines of furniture were not successful at the time they were produced, currently this furniture is at the top of the price list for collectible plastic furniture. The kitchen and bathroom pieces bring especially high prices.

Ideal also produced furniture and a house for larger dolls during this same time period. Because of the success Mattel, Inc. was having with the Barbie dolls and their accessories, Ideal began producing their own products to capture some of this market. Ideal introduced the Tammy doll in 1963. The vinyl Tammy family included Tammy (12"), brother Ted (13"), sister Pepper (9"), Dad (13"), Mother (12"), and Pete (8"). Tammy's family also needed a place to live and soon they moved into Tammy's Ideal House. The house was produced using the same principle as the Barbie Dream House. It was made of chipboard and folded up into a suitcase. The furniture also came unassembled. The house sold for $4.99 in the Sears Christmas catalog in 1964. Tammy also had a separate bedroom set designed for her. It consisted of a bed, vanity, and chair. The Tammy products were successful, at the time, but they did not continue to be popular after the first few years of production. Ideal soon left this market in the capable hands of Mattel, Inc.

The Ideal Toy Corporation brought pleasure to both boys and girls during their nearly eighty years of toy production. Now these toys, dolls, and furniture products are bringing just as much pleasure to collectors. Perhaps these toys are a reminder of a long ago Christmas morning when a child was surprised with an Ideal Shirley Temple doll or a room full of plastic Ideal dollhouse furniture. These memories, as well as the toys, will remain for many years to come.

Box front of the early Ideal dining room furniture. The illustrations on the sides of the box picture the furniture included in the living room, bedroom, bathroom, kitchen (both regular and deluxe), garden, and bathroom sets. The nursery furniture was not produced until a little later. Photograph and box from the collection of Betty Nichols.

Inserts to make walls for the rooms were contained in these early boxes of Ideal furniture. This dining room set also included the radiator. Photograph and furniture from the collection of Betty Nichols.

Ideal dining room pieces were made with different colored chair seats. Pictured are blue, yellow, and red-rose. Photograph and furniture from the collection of Roy Specht.

The outside of the box was made in the shape of a doll house and the box lid could be bent to make a roof. Photograph and box from the collection of Betty Nichols.

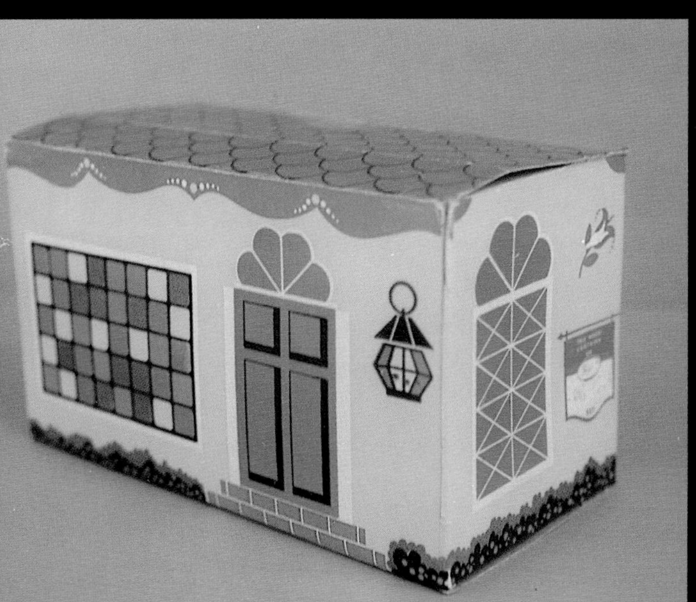

Boxed Ideal bathroom furniture showing another interesting box designed to market the Ideal furniture. Photograph and furniture from the collection of Betty Nichols.

The inside of the Ideal box formed the bathroom itself. The doll that came with the set of bathroom furniture is not the one usually associated with the Ideal furniture. Since the nursery furniture was not produced until a little later, the jointed doll probably was not marketed until that time. Photograph and furniture from the collection of Betty Nichols.

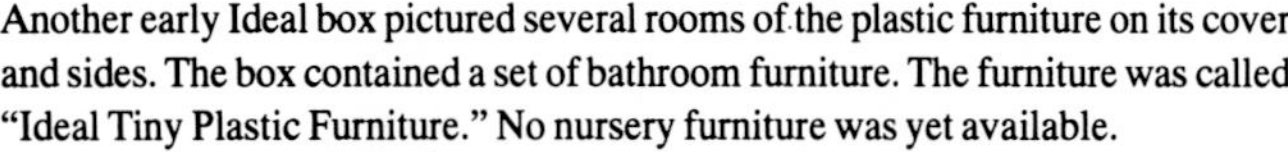
Another early Ideal box pictured several rooms of the plastic furniture on its cover and sides. The box contained a set of bathroom furniture. The furniture was called "Ideal Tiny Plastic Furniture." No nursery furniture was yet available.

This Ideal cellophane front box probably dates from around 1949. Shown is the nursery set of furniture.

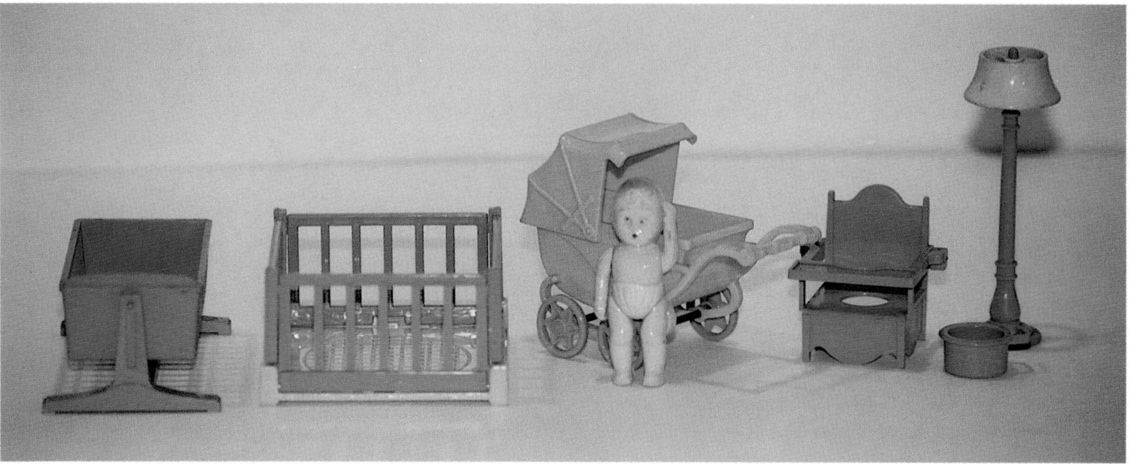

The nursery furniture also included a jointed Ideal baby doll. It is approximately 2 1/2" tall. The playpen from the nursery folded flat. A high chair was also made for the nursery.

Other nursery pieces were also produced by Ideal. These included a folding high chair that made a rocker, a crib, changing table, and a stroller. Photograph and furniture from the collection of Roy Specht.

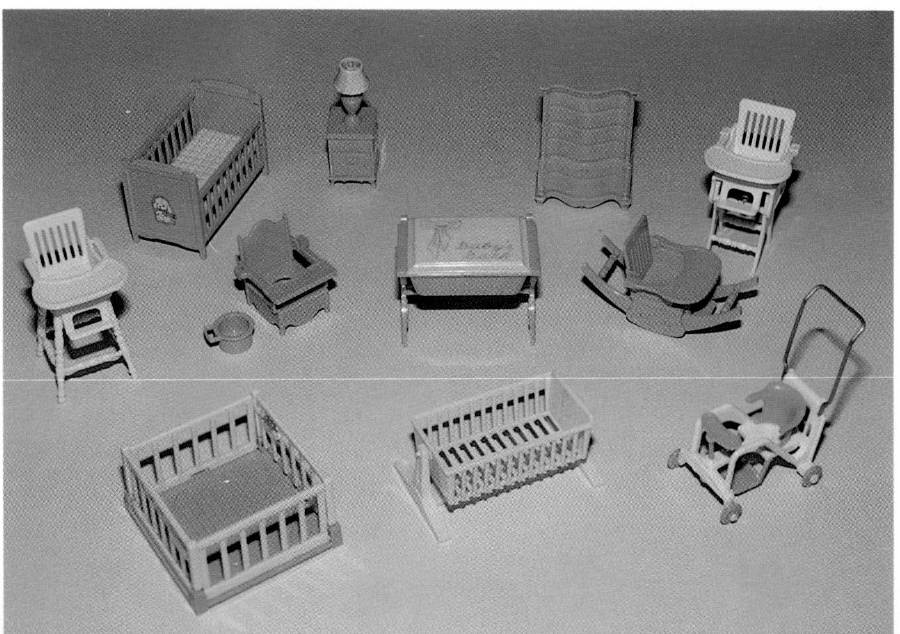

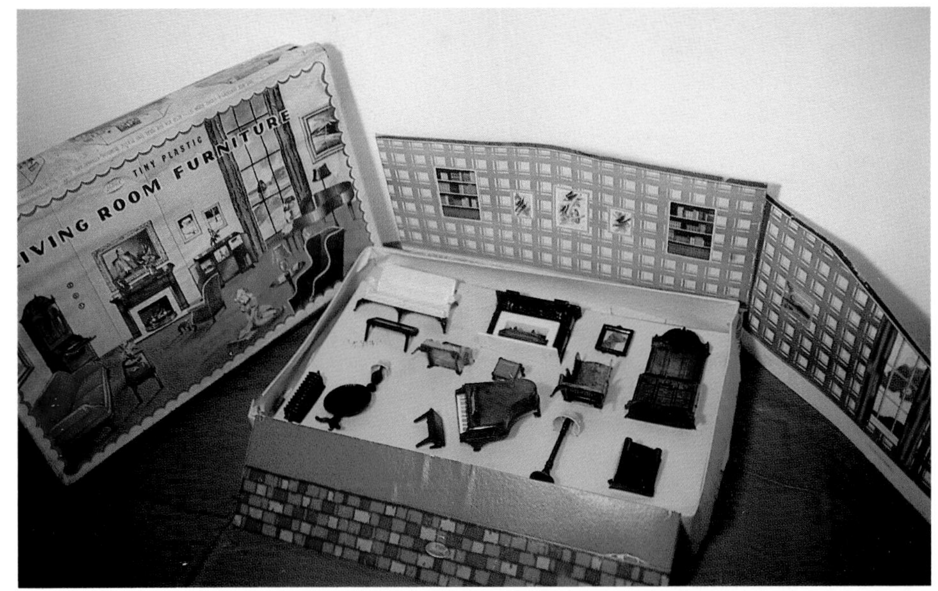

Boxed living room furniture which included the piano, tilt top table, secretary, radiator, and fireplace as well as the regular Ideal pieces. The room background is also pictured. Furniture from the collection of Marcie Tubbs. Photograph by Bob Tubbs.

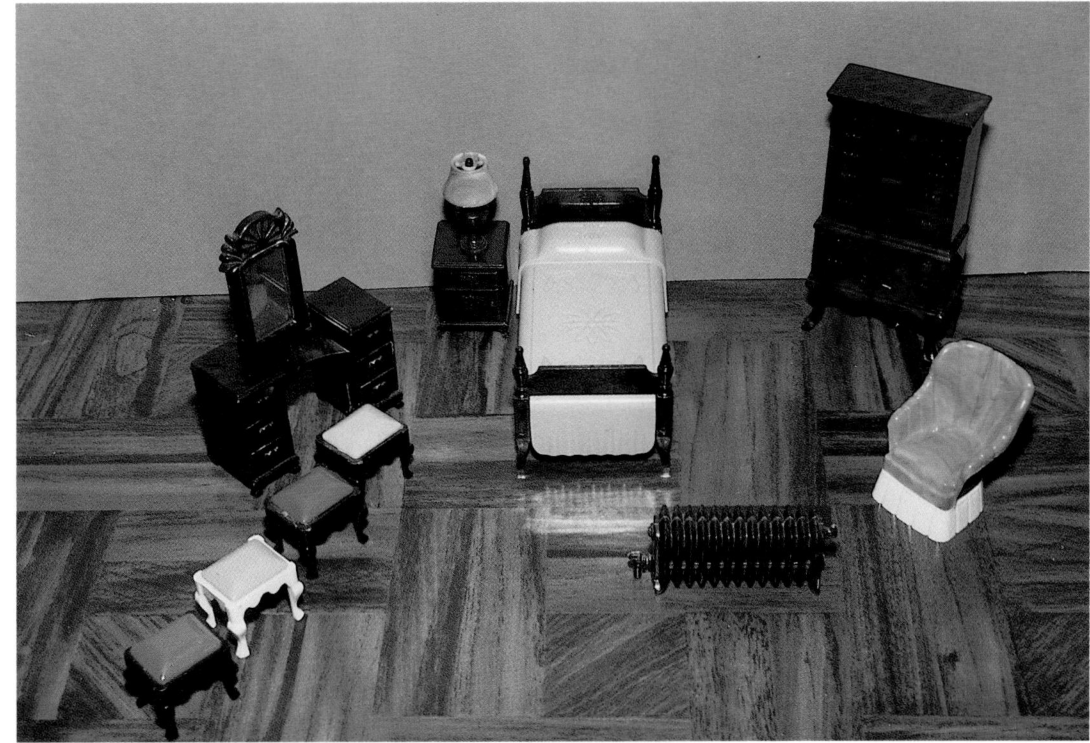

The Ideal bedroom set came with several color combinations. Pictured are different colors of the benches. One of the most popular of the bedroom sets is the one which was made in blue and white. Furniture and photograph from the collection of Roy Specht.

Many of the Ideal living room pieces are pictured in a room setting. This set includes a radio. Photograph and furniture from the collection of Roy Specht.

Early boxed set of Ideal kitchen furniture. The "baby" doll was also included with this furniture.

The kitchen furniture also came with different colors on the seats of the chairs. Pictured are green and two shades of blue. Photograph and furniture from the collection of Roy Specht.

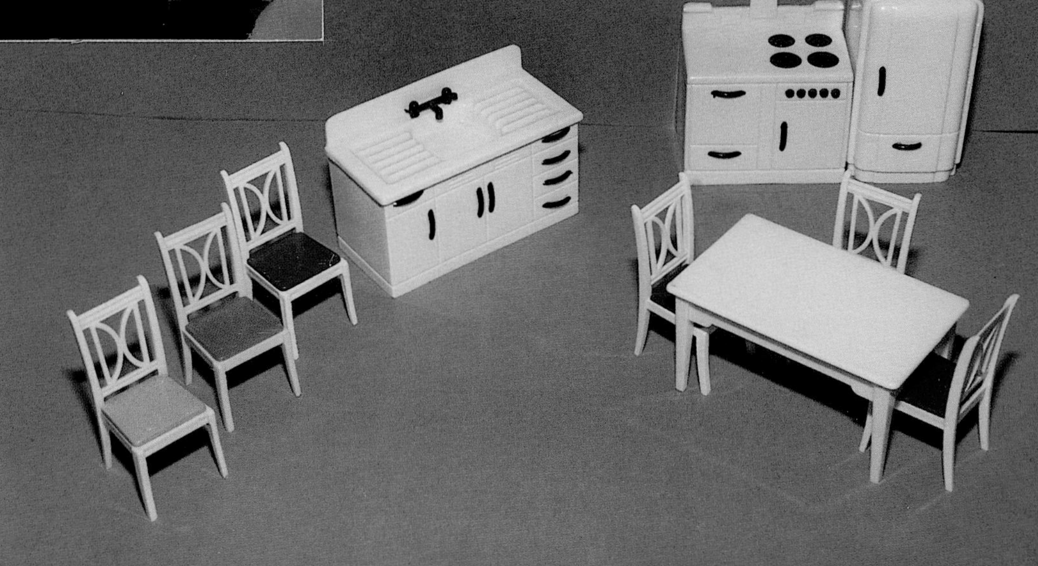

The Ideal garden pieces are especially appealing. The set also included a pool, birdbath, and trellis fence. Photograph and furniture from the collection of Roy Specht.

Ideal also produced an attractive folding card table and chairs. They are shown along with another view of the sofa bed. Photograph and furniture from the collection of Roy Specht.

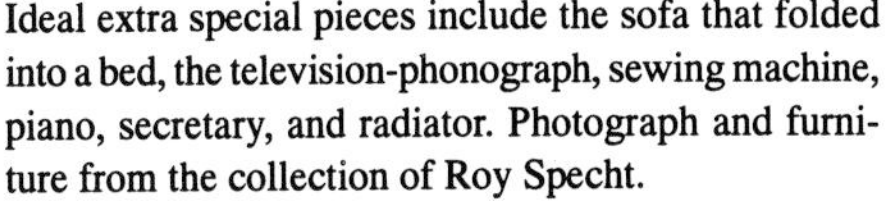

Ideal extra special pieces include the sofa that folded into a bed, the television-phonograph, sewing machine, piano, secretary, and radiator. Photograph and furniture from the collection of Roy Specht.

The older bathroom was modernized with a new corner tub design as well as a more modern toilet. Photograph and furniture from the collection of Roy Specht.

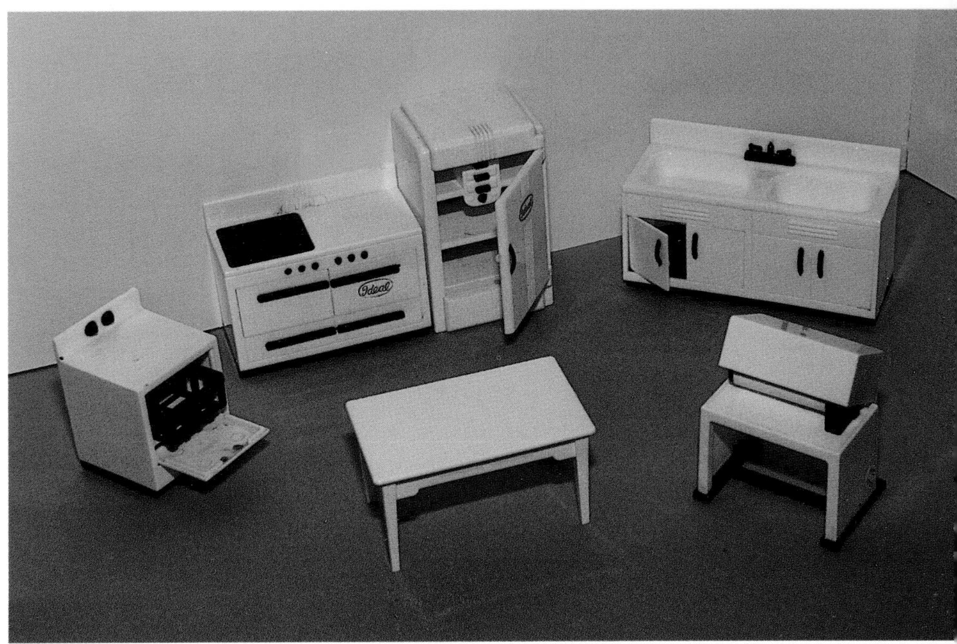

The "Deluxe" Ideal kitchen pieces were made in a little larger scale, approximately 1" to 1'. Photograph and furniture from the collection of Roy Specht.

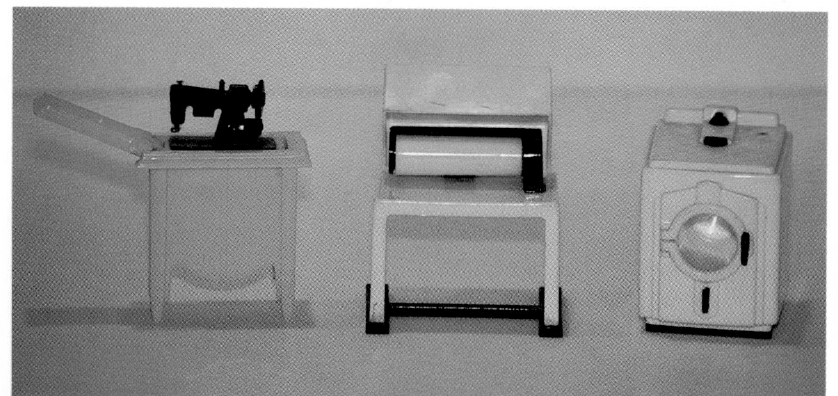

The sewing machine, ironer, and washing machine were originally packaged in an Appliance Room set. The sewing machine can be lowered into the cabinet for storage.

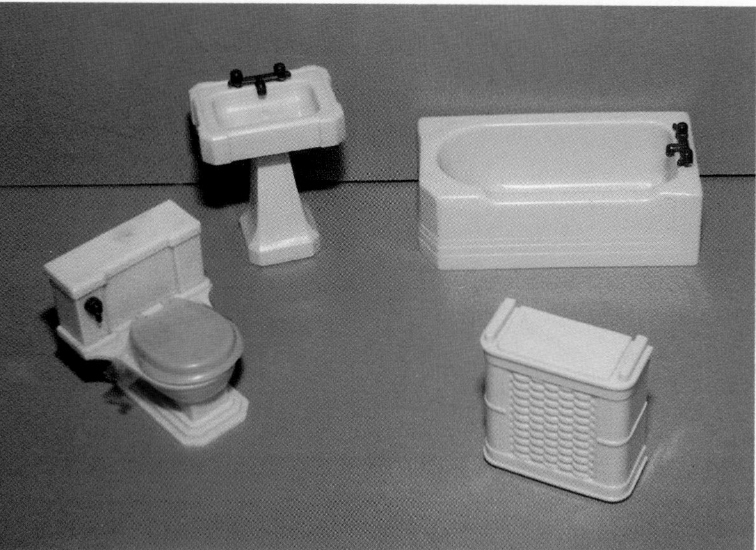

This bathroom furniture made by the Reliable Toy Co. in Canada was based on Ideal furniture molds. Photograph and furniture from the collection of Roy Specht.

The outside of the box showing the different pieces of plastic Young Decorator furniture made by Ideal. Photograph and furniture from the collection of Kathleen Neff-Drexler.

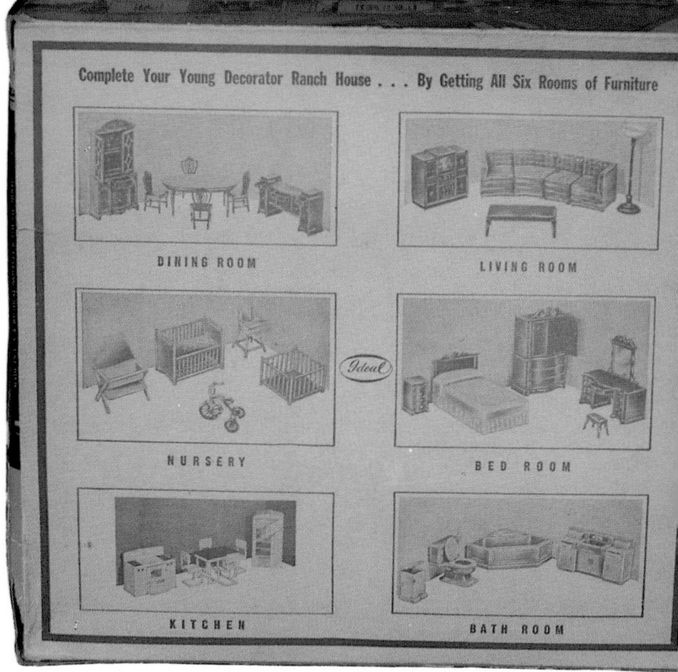

Boxed Young Decorator dining room furniture made by Ideal in 1950. Photograph and furniture from the collection of Kathleen Neff-Drexler.

The Young Decorator bedroom furniture. Photograph and furniture from the collection of Roy Specht.

The living room furniture from the Young Decorator series. The sectional also was produced in green and blue.

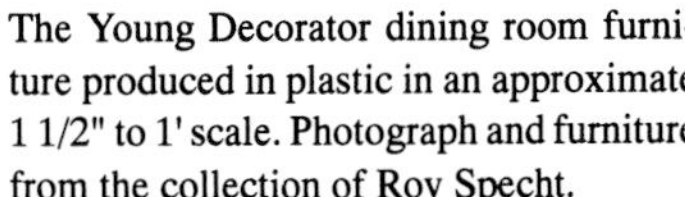

The Young Decorator dining room furniture produced in plastic in an approximate 1 1/2" to 1' scale. Photograph and furniture from the collection of Roy Specht.

The Young Decorator bathroom set is pictured. Photograph and furniture from the collection of Roy Specht.

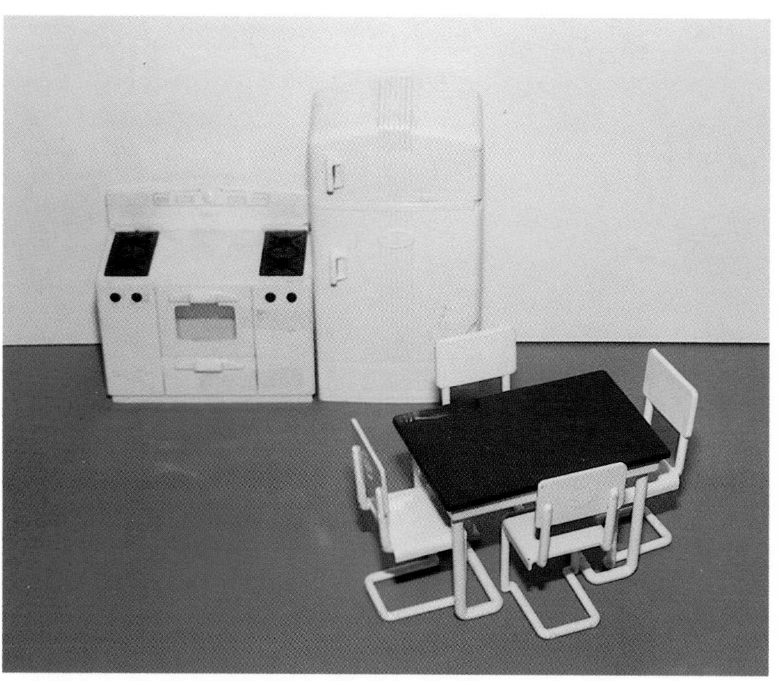

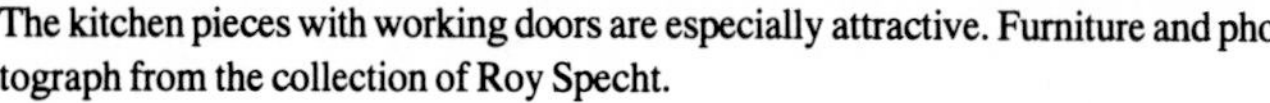

The kitchen pieces with working doors are especially attractive. Furniture and photograph from the collection of Roy Specht.

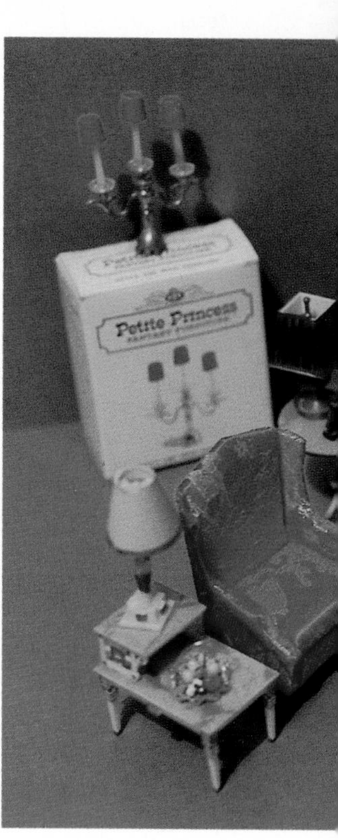

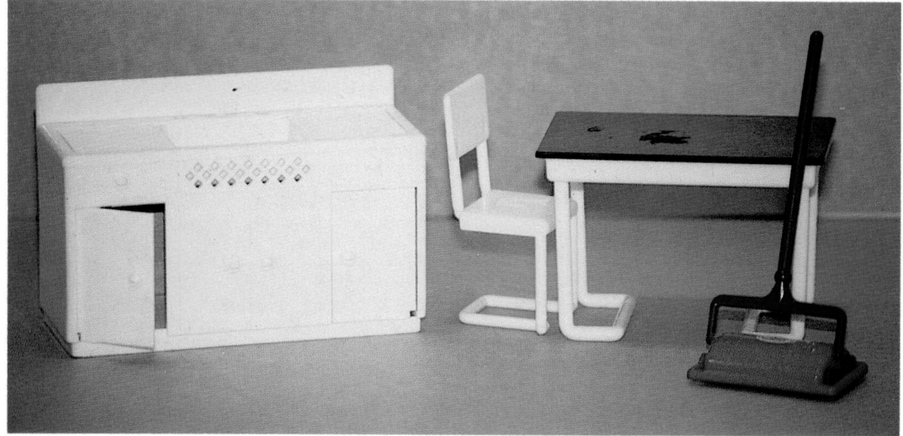

This sink was not pictured on the box of the Young Decorator furniture so it may have been added to the line later. The faucets are missing. Ideal also produced carpet and vacuum sweepers that can be used as accessories for the Young Decorator furniture.

Petite Princess Ideal furniture which could also be used in a living room. The larger table is the Pedestal model. Also included in this room setting is the portable television on the stand from the 1965 Princess Patti furniture. Photograph and furniture from the collection of Roy Specht.

This Young Decorator nursery set of furniture also included a bathinet changing table. These pieces are probably the most difficult to find. Photograph and furniture from the collection of Roy Specht.

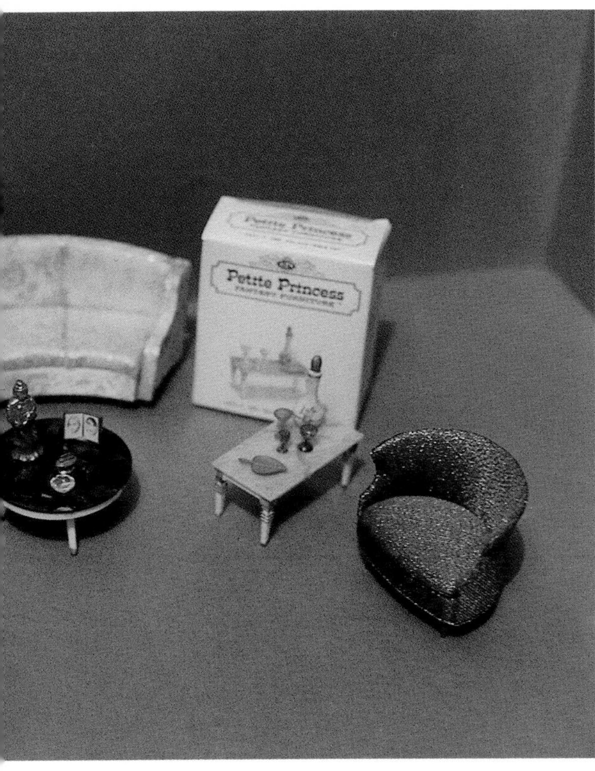

Petite Princess living room furniture from 1964. The various tables came with accessories. Reading from left to right they include: Tier table, Heirloom table, Occasional table, and Palace table. Photograph and furniture from the collection of Roy Specht.

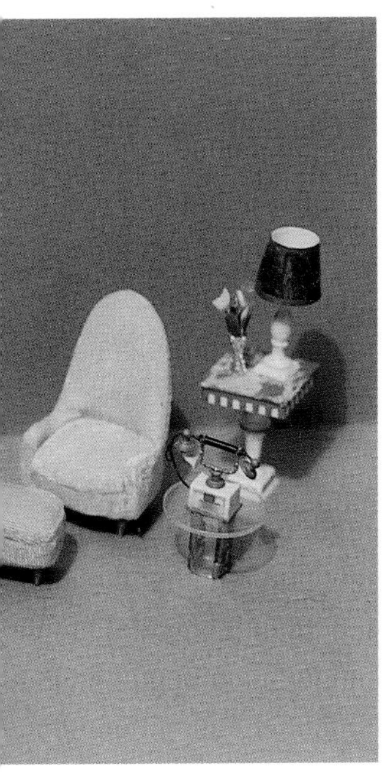

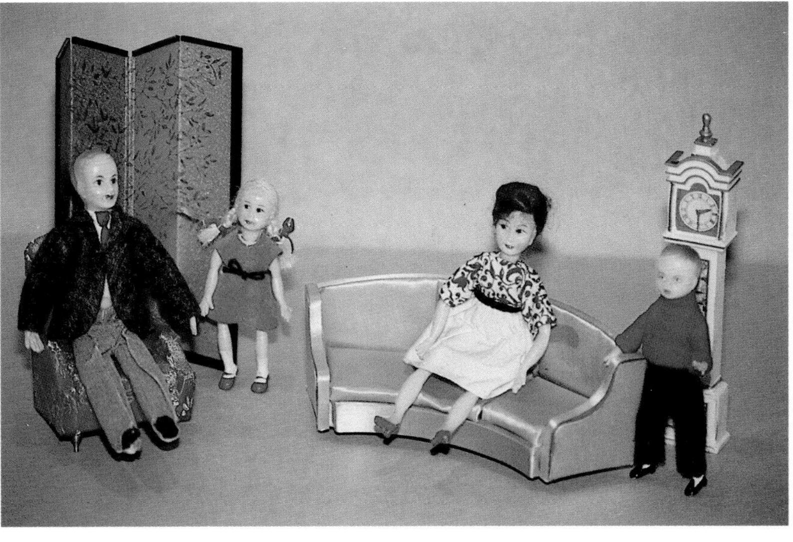

Petite Princess family of dolls which Ideal made to accompany their furniture. The father doll is 5 1/2" tall. The dolls could be bent so they could sit easily.

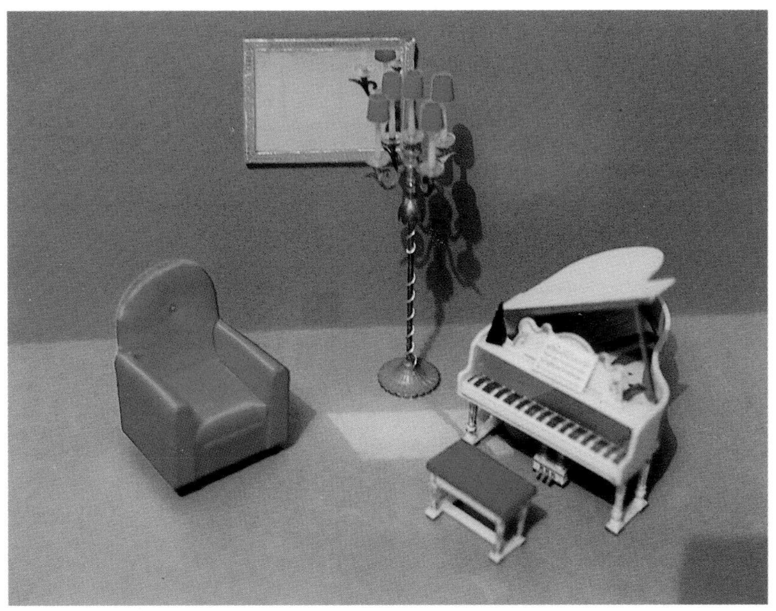

One of the nicest of the Petite Princess pieces is the grand piano. It came with a metronome. The mirror was included as part of the fireplace set. Photograph and furniture from the collection of Roy Specht.

The dining room furniture included a rolling tea cart as well as a high cabinet and a buffet. Photograph and furniture from the collection of Roy Specht.

The bedroom furniture was really fit for a princess. The plastic small table was called a coffee table. The other table is a lyre table and the picture came with it in the original set. Photograph and furniture from the collection of Roy Specht.

Pictures of all the furniture and accessories included in the original Petite Princess Ideal furniture line. Photograph and advertisement from the collection of Roy Specht.

The Princess Patti bathroom furniture marketed by Ideal in 1965. These pieces were produced for such a short time that they are very hard to locate and are expensive when found. Furniture and photograph from the collection of Roy Specht.

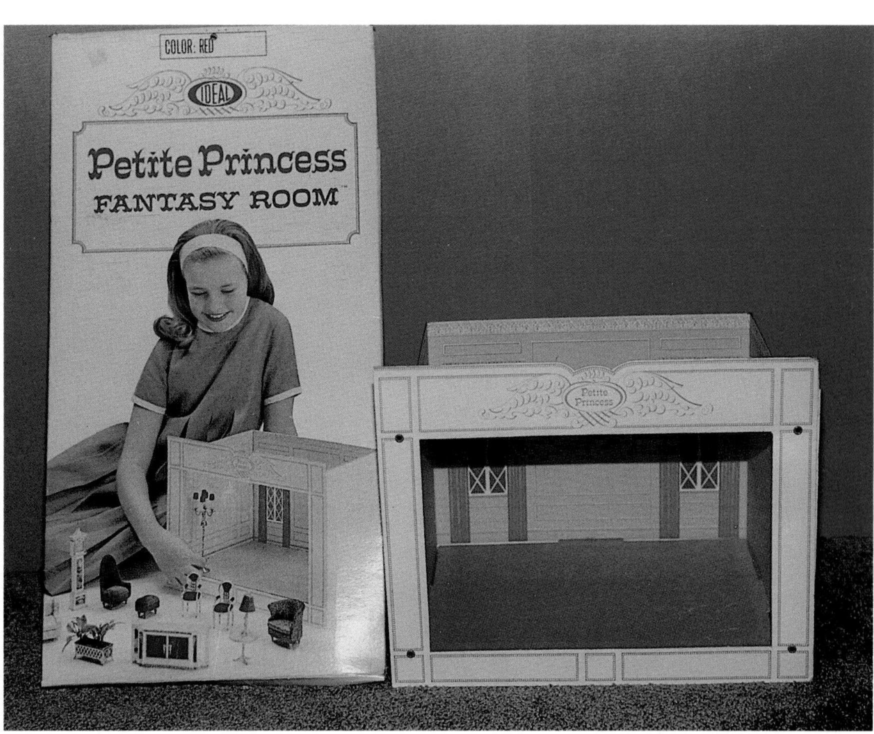

One of the Petite Princess Fantasy Rooms. The rooms were made of cardboard and were to be used to house the Ideal furniture. The rooms also were made in yellow and blue. Photograph and room from the collection of Roy Specht.

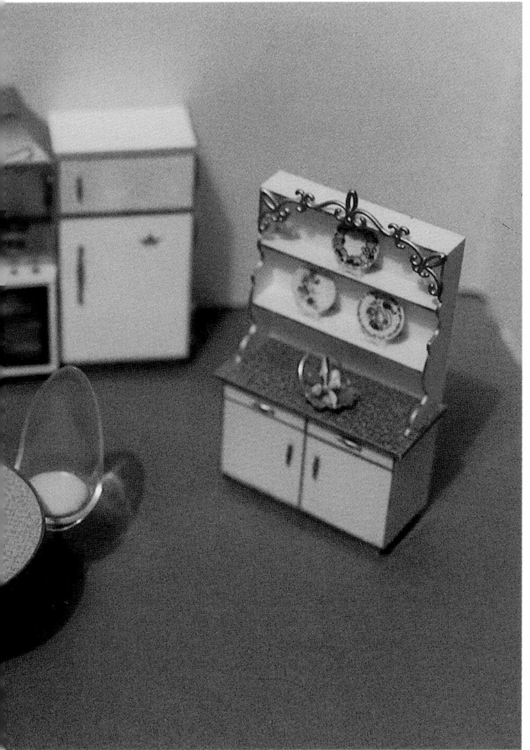

The kitchen furniture was also a part of the Princess Patti line and is also quite scarce and expensive. Photograph and furniture from the collection of Roy Specht.

A Petite Princess store display which featured space for all of the Petite Princess furniture. Photograph and display from the collection of Marilyn Pittman.

Ideal's vinyl suitcase house sold by Sears, complete with over twenty pieces of furniture. Some of the items included in the set were the cheaper furniture pieces made in Hong Kong and labeled as Redbox. Photograph and house from the collection of Roy Specht.

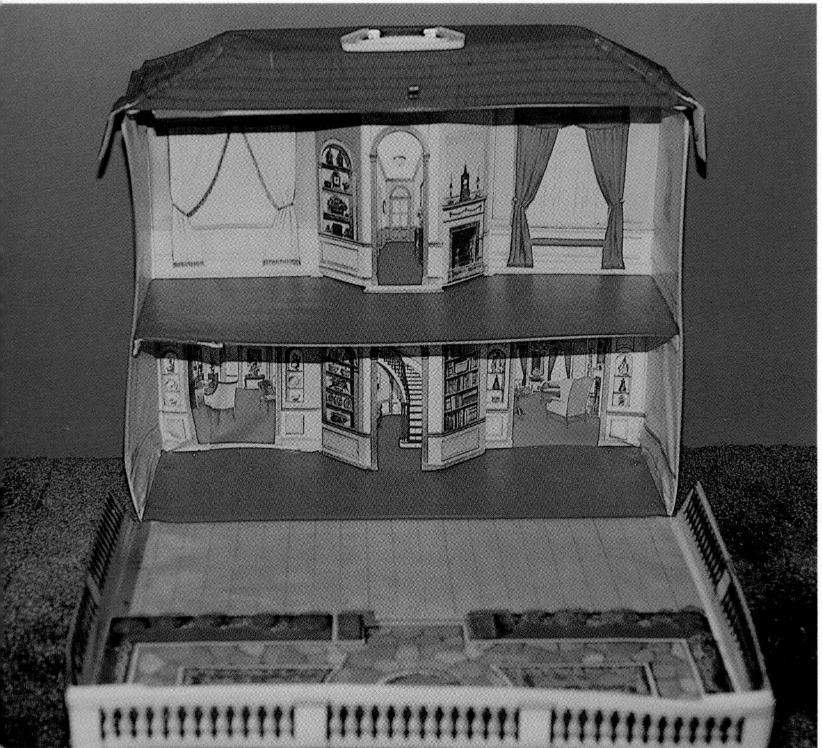

The inside of the vinyl Ideal house has many furnishings and decorations printed on the walls and floor. Photograph and house from the collection of Roy Specht.

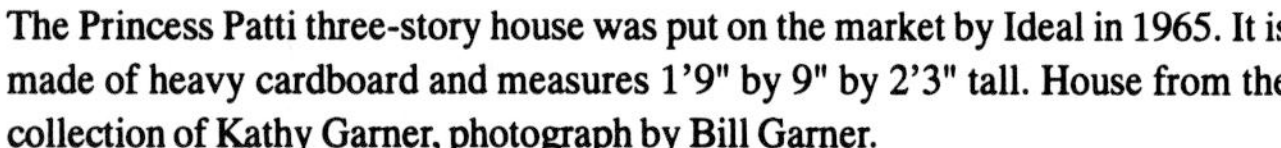

The Princess Patti three-story house was put on the market by Ideal in 1965. It is made of heavy cardboard and measures 1'9" by 9" by 2'3" tall. House from the collection of Kathy Garner, photograph by Bill Garner.

The inside of the Princess Patti house contains six rooms and the walls and floors are printed with very fancy decorations. From the collection of Kathy Garner, photograph by Bill Garner.

Pictured are several pieces of the more cheaply made furniture marked "Redbox/ Made in Hong Kong." This furniture used the same designs as the earlier Ideal Petite Princess pieces but the furniture is inferior to the original Ideal items.

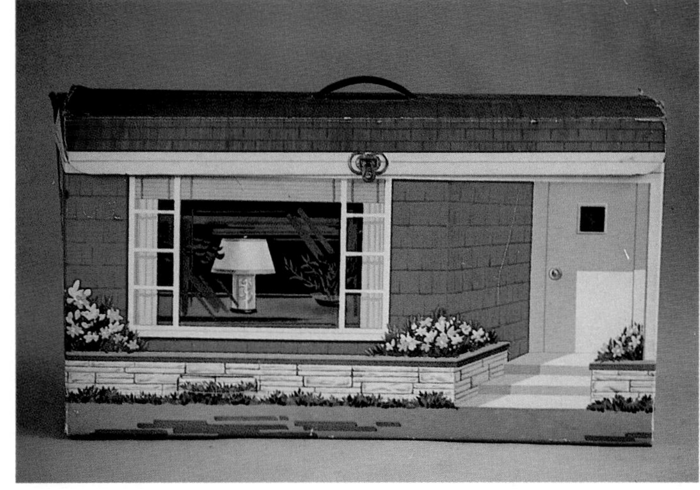

This chipboard Tammy house, produced by Ideal, was marketed in 1964. It was made in conjunction with the Ideal Tammy doll family which also included a mother, father, brothers, and sister. The dolls ranged in size from 9" to 13" tall.

The Ideal Tammy house included a soda fountain, juke box, television set, desk, closet, and chaise lounge. A Ping-Pong table was also part of the house's furnishings. The house sold for $4.99 in 1964. Pictured with the house are the Ideal vinyl Tammy and Pepper dolls.

Marx

The company that became known as Louis Marx and Co., Inc. was begun shortly after World War I when its founder, Louis Marx, purchased toy molds from the well known Strauss Manufacturing Co. As a former employee of Ferdinand Strauss, Marx was familiar with their production of lithographed tinplate mechanical toys. The new company made some minor changes in the toys before they were marketed under the Marx name. The Louis Marx company proved successful and remained in business for over fifty years. Most of the early toys were wind-up mechanical models or metal trucks and cars.

In 1948 the company expanded to include a plastic division. Soon the Marx company became the largest toy manufacturer in the world. Collectors are especially interested in the play sets, service stations, and dollhouses made during these years. At the height of their popularity, over 150,000 dollhouses were made by Marx in only one year.

In the early 1970s, the Marx company was sold to the Quaker Oats Company. Toy sales decreased under the new ownership and the company was sold again in 1976. The buyer was a large English toy company, Dunbee-Combex. The English firm was no more successful in reviving Marx's former glory than Quaker Oats had been. In 1980 Marx filed for bankruptcy. In 1982 the assets were purchased by American Plastics Equipment Inc.

Louis Marx outlived the company he founded. He died in 1982, leaving behind a legacy of collectible toys that may never be equaled by any other firm.

One of the most valued Marx toys for the dollhouse collector is the Newlywed house which the firm began producing during the 1920s. The house could be purchased complete with four furnished rooms or each room could be bought separately. Six rooms were marketed in the set. These included the kitchen, bathroom, bedroom, dining room, parlor, and library. The rooms represented in the house were the parlor, dining room, kitchen, and bedroom. The house was made of cardboard printed on the outside to imitate brick. It measured 8 5/8" by 2 9/16" by 10 1/4". The front of the house opened with two large doors so a child could play with the furniture inside. The metal rooms were housed in the cardboard house container.

Each of the rooms was made of lithographed sheet steel. The walls were decorated to represent the different rooms. The tiny furniture was also lithographed metal. The rooms consisted of three sides and a floor and measured 5 1/2" wide by 2 1/2" deep by 3" high. Sears, Roebuck and Co. was still featuring the Newlywed house in its catalog as late as 1930. At that time, the entire house and its furnishings sold for forty-four cents. The rooms in the house contained the following pieces of furniture: Parlor: sofa, chairs, buffet, and table. Bedroom: bed, dresser with mirror, wardrobe, and two chairs. Kitchen: cabinet, sink, stove, table, and two chairs. Dining room: table, two chairs, china cabinet, buffet, and grandfather clock. The other two rooms contained the following furniture items: Library: sofa, table radio, two chairs, table, and bookcase. Bathroom: bathtub, toilet, stool, sink, and medicine cabinet.

Although the Louis Marx Co. continued to make wonderful metal toys throughout the 1930s and into the 1940s, it wasn't until 1949 that the firm again made products for the dollhouse collector. In the late 1940s, several firms began to manufacture metal dollhouses and service stations. Marx soon followed the lead by producing its first metal dollhouses in 1949. The Sears Christmas catalog for 1949 featured the Marx "Disney" house complete with furniture and electric lights for $4.98. The house contained five rooms plus a garage and patio. The success of the new product was immediate. The Sears and Montgomery Ward Christmas catalogs from the 1950s no longer featured the Rich and Keystone Masonite houses; instead they carried the new Marx metal dollhouses. These houses shared a part of the toy market for over twenty years.

Marx began producing dollhouses as they did their other toys - using the same model for many years. The dollhouses were made of different pieces that could be put together and the same basic dollhouse could be changed simply by using a different outside design. Collectors can find the same basic house made in a five-room model, or with an added garage, playroom, or utility room. Later the larger houses were given dormer windows, staircases or an extra "Florida" room complete with jalousie windows. In this way the company could distribute various models of the same house to meet different price needs. The furnished houses featured in the 1950 Sears and Montgomery Ward catalogs ranged in price from $2.95 to $6.95. Houses were produced in both the 1/2" to 1' scale and the larger 3/4" to 1'scale. Two different sets of furniture were manufactured to fit these scales. Most of the furniture is marked with the Marx trademark in a circle.

The cheapest furnished house pictured in 1950 was priced at $2.95. The house measured 19 1/2" by 9" by 15 1/2" and came with thirty-six pieces of furniture. The house was a two-story Colonial which featured five basic rooms. The baby's room was decorated with a toy soldier motif. The 1/2" scale living room included a sofa, lounge chair, barrel chair, coffee table, end table, television, and floor lamp. The dining room contained a table, four chairs, hutch, and a buffet server. The kitchen was equipped with a sink, refrigerator, stove, table, and four chairs. The nursery contained a crib, playpen, chest, and potty chair. Several of these pieces of furniture were embossed with Disney characters when sold with the house which included the Disney decorated nursery. The bedroom furniture included a bed, highboy, vanity, bench, nightstand, and boudoir chair. The bathroom contained a bathtub, lavatory, toilet, and hamper.

The two-story "Disney" house sold for $3.95 in the Montgomery Ward catalog and for $5.29 in the Sears catalog. There were several differences in the houses that explained the higher cost for the Sears model. It had an added breezeway, recreation room, and two electric lights. The recreation furniture included a jukebox, piano, bench, ping pong table, sofa, round table, two chairs, a counter, two stools, and coffee table. The electric lights operated with a flashlight battery. The house measured 37 1/2" by 9 1/2" by 15 1/2". Both houses were the basic two-story five-room houses which featured the Disney characters on the walls of the nursery. A garage and sun deck were added to make a more elaborate house. Besides the furniture sold with the cheaper house, five pieces of sundeck furniture were included (table, umbrella, chaise lounge, two lounge chairs). Also added to the more expensive model was a fifteen piece play yard which included four children, four sections of picket fence, sand box, sand pail, shovel, see-saw, wading pool, two-piece sailboat, slide, and car. The smaller house measured 25 1/2" by 9" by 15 1/2".

The most expensive Marx house pictured in both catalogs was the $6.95 model which included the 3/4" to 1' scaled furniture. The house was a larger model of the basic five-room two-story Colonial house with a different design for both the inside and outside of the house. A garage and patio were also included in this house. The outside was lithographed in red and white with a grey roof. The baby's room was decorated in a clown motif. The house came with forty-eight pieces of furniture, a car, five detachable metal awnings, and two detachable plastic flower boxes. The house measured 33 1/2" by 12" by 18 3/4". The furniture included the following pieces: Living room: sofa, wing chair, barrel chair, coffee table, step end table, television console, floor lamp, and table lamp (would light with batteries). Dining room: table, two side chairs, two host chairs, buffet, and breakfront. Kitchen: sink, refrigerator, stove, base cabinet, table, and four chairs. Nursery: crib, playpen, chest, potty chair, and high chair. Bedroom: bed, highboy, vanity, stool, boudoir chair, hassock, night table, floor lamp, and table lamp. Bathroom: corner tub, washstand, hamper, and toilet. Sundeck: umbrella table, chaise lounge, and two lounge chairs.

All of these basic house designs were used over and over by the Marx company for nearly twenty years. This same furniture was also produced for many years although extra pieces such as utility room pieces were added as needed. Innovations to these basic house designs included adding a utility room in place of the garage, adding dormer windows to the roof, adding a "Florida" room to the largest Colonial two-story house, using the garage for a fallout shelter, adding plastic opening windows to some models, and restyling both the insides and outsides of the houses with occasional new designs. The Colonial styled house was the model most often produced by the Louis Marx Co.

Besides dollhouses, Marx also produced other items during the 1950s that are desirable to dollhouse collectors. Collectors are familar with Renwal's school house and hospital nursery sets. Marx also manufactured similar items. The Marx nursery has an advantage because it is made of metal and is both durable and attractive. The school furnishings, although not as attractive as those made by Renwal, have more pieces and are very collectible.

In 1953, the Marx company produced a metal "L"-shaped ranch house as well as the more usual Colonial dollhouses. The house contained a combination living room/dining room, kitchen, bedroom, bathroom, child's room, and a patio. The house, in keeping with its original counterpart, included a television antenna and a picture window. The house measured 32" by 16 1/4" by 13 1/8". Sears' Christmas catalog for 1953 priced the house for $7.29 furnished. Most of the furniture was of a different design than the earlier Marx pieces. The living room came with a three-piece sectional sofa, corner table, occasional chair, television console, table lamp, floor lamp, and coffee table. The dining room furnishings included a table, two side chairs, two arm chairs, and a buffet server. In the kitchen were a sink, refrigerator, stove, breakfast bar, and two stools. The bedroom had a new modern bed with headboard, lamps, dresser, vanity, and bench. The bathroom came equipped with a corner tub, commode, hamper, vanity type wash stand, and vanity bench. The child's room was furnished with a crib, two-piece high chair, and hobby horse. For the patio, a chaise lounge, coffee table, chair, settee, table, and umbrella were provided. Also included in the set were a car and fourteen plastic people figures. The furniture for this house is in the scale of 1/2" to 1'. This furniture seems to be made from the same molds as those used for the 1/2" to 1' Superior furniture marketed by the T. Cohn company.

In 1958, the Sears Christmas catalog featured another different Marx dollhouse model when a split level design was sold. The house was one-story on one wing and two-story on the other. Most of the furniture was the same as that featured in the "L"-shaped ranch house. New utility furniture was provided which included a washing machine, ironer, and clothes basket. The house also came with a pool complete with diving board, ladder, and water toys. The house was equipped with a door bell and lights (operated with flashlight batteries). The size of the house was 29 1/2" by 16" by 14". This house was produced in several different models including one that featured printed steps and barbecue instead of the removable plastic ones.

In the same 1958 Sears catalog, the basic 1/2" scale five-room Colonial house had been enhanced with a staircase, dormer windows, a breezeway, and recreation room. Furnished, the house sold for $5.97.

In their 1962 Christmas catalog, Sears featured one of the most expensive Marx doll houses ever made. This white Marxie Mansion was based on the large 3/4" scale Colonial red and white house which first made its appearance in 1950. In order to dress up the new house, dormer windows were added to the roof, an inside staircase was placed in the downstairs, and a "Florida" room was featured with jalousie windows. Other "extras" included a ringing doorbell, living room light, vinyl framed walk with shrubs, swimming pool, play equipment, fence, cloth draperies, awnings, plastic shutters, porch, and window boxes with flowers. The cost of the furnished house was $15.88. The furniture was basically the same as that produced for the red and white 1950 house except that it was made of softer plastic instead of the brittle hard plastic. Utility furniture was added which included a washer, sink, dryer, ironing board, iron, and basket. Also added was the furniture for the Florida room. These pieces included a sofa, two chairs, poker table, television, and an ottoman. Also featured the same year was the 1950 red and white large Colonial house with an added Florida room. It sold for $9.99.

The 1962 Line showed the basic small Marx Colonial with a fallout shelter in place of the garage and with a breezeway and family room added. The cost was $6.33. Also shown in the 1962 line of Marx products was a ranch house of a different design. Instead of being "L"-shaped, the house was a straight line model which featured a bedroom, bathroom, kitchen, living room/dining room combination, and a patio. This model contained two dormer windows on the roof and used the small scale furniture. It sold for $3.99. This house was also made without the dormer windows.

Most of the Marx dollhouses were sold with plastic figures of both adults and children to be used as the inhabitants of the dollhouses. The company did market one set of more realistic dolls that could be purchased to accompany dollhouses. The set of family dolls included a Father (5 1/2"), Mother (5"), Boy (3 1/2"), Girl (3"), and a Baby (2"). The dolls have hard plastic heads, hands, and feet, and cloth over armature bodies. All the dolls have molded hair and painted features. The dolls are not marked but their original box is labeled "Marx." The dolls were sold during the 1960s and the 1970s.

Another interesting set of dollhouse dolls produced by the Marx firm was called "Doll House Family." The complete family included a mother, father, son, and daughter that were made to be assembled. The one-color plastic pieces could be combined to produce the doll figures with jointed bodies.

In 1964 Marx created new plastic dollhouse furniture that would delight collectors for years to come. The new Marx furniture was called "Little Hostess" and it came equipped with working parts and realistic colors. The pieces were marked "Louis Marx & Co., Inc. MCMLXIV" and, in a circle with an X, "Made in Hong Kong Marx Toys." This new doll house furniture was not made at the Marx plant in Glen Dale, West Virginia, but was manufactured in Hong Kong. The furniture came packaged in individual boxes or in bubble packs. This new furniture was much more expensive than the ordinary Marx dollhouse furniture. Items made in the Little Hostess design included the following pieces: Living room: secretary, piano, bench, rocker, standing candelabra, occasional chair, wing back chair, fireplace, floor lamp, end table, grandfather clock, coffee table, tilt top table, screen, chest, mirror, and television. Dining room: oval dining table, chairs (two styles), buffet, breakfront, and server. Bedroom: canopy bed, double dresser, highboy, vanity, bench, blanket chest, nightstand, chaise lounge, mirror, and chair. Bathroom: vanity lavatory with mirror, tub/shower, hamper, toilet, medicine cabinet, scale, and tuffed bench. Kitchen: table and chairs, refrigerator, sink/kitchen cabinet, stove, and washer.

In 1967 the Montgomery Ward Christmas catalog featured a vinyl dollhouse which folded like a suit case with twenty pieces of Marx furniture. All of it was Little Hostess except the patio furniture which was the earlier plastic design for 1950. The Marx dollhouse family (consisting of a mother, father, boy, girl, and baby) was also included in the package. The total cost was $8.99. Again in 1969 Montgomery Ward offered a bargain on Little Hostess furniture when it sold seventeen pieces for $3.99. Apparently parents had not wanted to pay the extra money for quality furniture when they could still buy a lithographed metal house completely furnished for as little as $3.99.

The boom in adult interest in dollhouses and their furnishings was still a decade away when Marx's Little Hostess furniture was manufactured and the collectible pieces did not find a ready market at the time they were produced. Today's collectors, however, are now finding these plastic items very desirable to enhance a plastic dollhouse furniture collection.

During the rest of the 1960s, Marx relied heavily on their earlier Colonial doll houses to stay in the dollhouse market. The Colonial house made in 1969 had one new innovation — plastic windows that would open and shut. The house also included a plastic bay window and a plastic front door that would open and shut. The lithography on the inside walls was also new. The furniture was much the same as that provided for earlier models but the house came with fewer pieces.

By the late 1960s, the company did produce a doll house that was unique to them — a futuristic house called "The Imagination Doll House." The house had seven living areas with three interchangeable house models. The house came with a life-like family of seven and one hundred pieces of furniture. The plastic furniture was modernized to accompany the new house.

Dollhouses continued to be produced by the new owners of the Marx company during the 1970s. Some of these later houses featured plastic roofs as well as doors and windows. The roofs were used on the older Marx Colonial designs as well as on the Colonnade house. The Colonnade house was featured in the Sears Christmas catalog in 1973. The house was priced at $9.88 furnished with five rooms of furniture. This house featured six white columns and three dormer windows as well as a working porch light. The house was mounted on a metal base and measured 26" by 14 1/8" by 16 1/2" high. The furniture that was sold with these houses had also been updated.

In 1978, with new owners at the helm, Marx experimented with a new concept in dollhouses and furniture when they became involved with the Sindy doll and her furniture. The vinyl doll was 11" tall and her furniture was quite expensive if a child was to acquire all the pieces. The bed sold for $4.94, the dressing table was $4.94, and the bedside table and lamp were also $4.94. That would total nearly $15 just to furnish the bedroom. The house for Sindy was not really a house but a "Scenesetter." The room backgrounds were on plastic backdrops and the furniture was to be placed in front of the correct scene. The furniture designs were some of the most detailed ever produced and included a working television, lights, and radio. The new concept apparently did not find a successful market as the furniture was only sold for two or three years.

The Marx company was at the end of its run and the Sindy experiment did not enhance the prospects for the once giant toy firm. The furniture should prove to be very collectible because it contains so many more working parts than any other furniture that has been made in its size range.

Although the decade of the 1970s saw the decline and end of perhaps the greatest toy company ever established, it was only the final chapter in what was truly a remarkable run of successes. In the dollhouse and dollhouse furniture market alone, Marx outlasted competitors Ideal, Plasco, and Renwal by many years. Although the Marx small plastic furniture was not as attractive as the furniture produced by its three rivals, Marx had an advantage in the market place. With the popularity of the lithographed metal houses, unlike most of its competitors, Marx was able to manufacture both houses and furniture at a reasonable cost. The mail order catalogs which featured the Marx products also supplied the company with a very wide market of consumers who purchased these doll products. Because so many Marx houses were manufactured for so many years, different models can still be purchased by collectors at very reasonable prices. There seems to be no shortage of these Marx houses and that trend should continue for some time to come.

The cardboard Newlywed house opened from the front to provide access to the inside rooms. Photograph and house from the collection of Patty Cooper.

The six pieces of Newlywed dining room furniture included a grandfather clock, table and two chairs, sideboard, and china cabinet. Room and photograph from the collection of Marilyn Pittman.

Six Newlywed rooms were made. These included the kitchen, bathroom, parlor, library, bedroom, and dining room. Pictured is the living room with its six pieces of furniture and original box. Photograph and room from the collection of Marilyn Pittman.

The Newlywed kitchen also contained six furniture pieces including sink, stove, cabinet, table, and two chairs. Photograph and room from the collection of Patty Cooper.

The Newlywed bedroom was furnished with a bed, dresser, wardrobe, and two chairs. Photograph and room from the collection of Marilyn Pittman.

Marx cardboard dollhouse produced to hold four of the small tin Newlywed rooms. The house measures 8 5/8" by 2 9/10" by 10 1/4". The bathroom is also pictured. It contained a tin tub, toilet, lavatory, stool, and medicine cabinet. From the collection of The Toy and Miniature Museum of Kansas City.

This Marx two-story, five-room metal Colonial style house was the basic model that the Marx company used for many years. It first appeared in 1949 and was made to hold the 1/2" to 1' scale of furniture.

The inside of the house contained five rooms. In this model, the child's room is decorated with toy soldiers. The house measures 19 1/2" by 9" by 15 1/2".

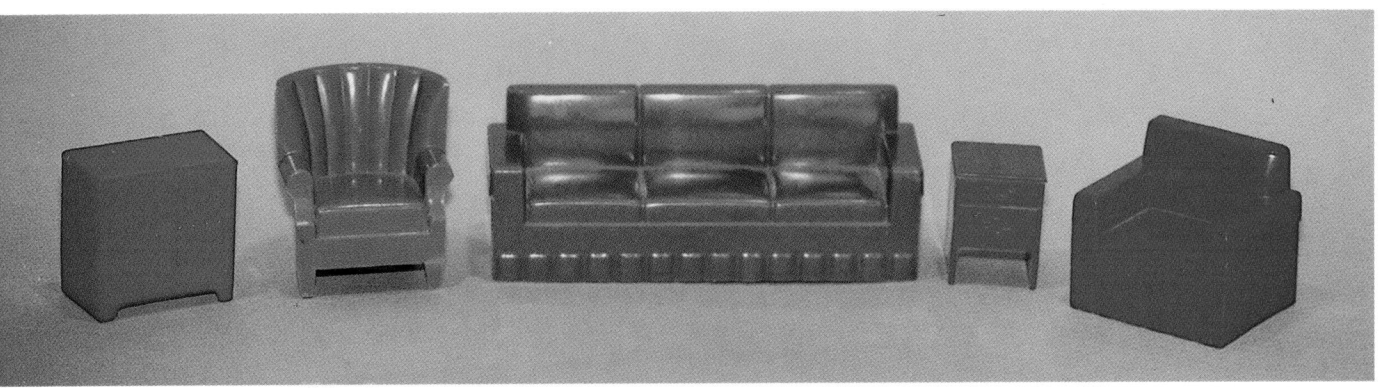

The 1/2" scale hard plastic furniture for the living room included a sofa, lounge chair, barrel chair, coffee table, end table, television, and floor lamp. Missing from the picture is the coffee table and floor lamp.

The dining room furniture included a table, four chairs, a hutch, and a buffet. The furniture was produced in several different colors.

The kitchen pieces consisted of a sink, refrigerator, stove, table, and four chairs. All of the Marx plastic furniture was first made of hard plastic and later produced in a softer plastic. The kitchen stove and sink were eventually combined into one furniture piece.

The early nursery furniture included a crib, playpen, chest, and potty. The crib and playpen were both embossed with Disney characters when sold with the "Disney" houses.

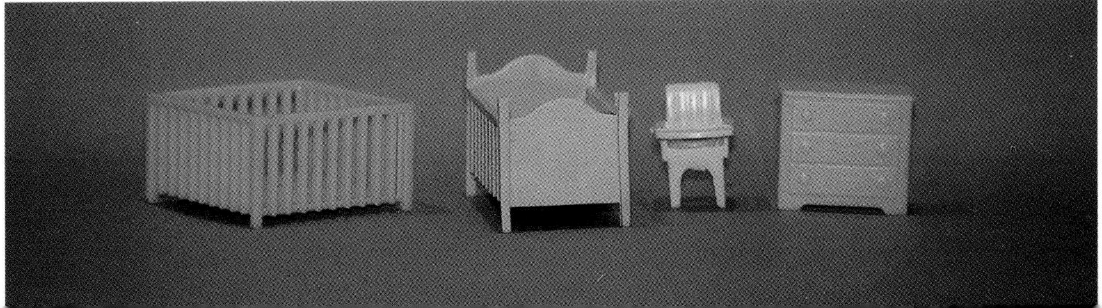

The bedroom pieces included a bed, highboy, vanity, bench, nightstand, and a boudoir chair.

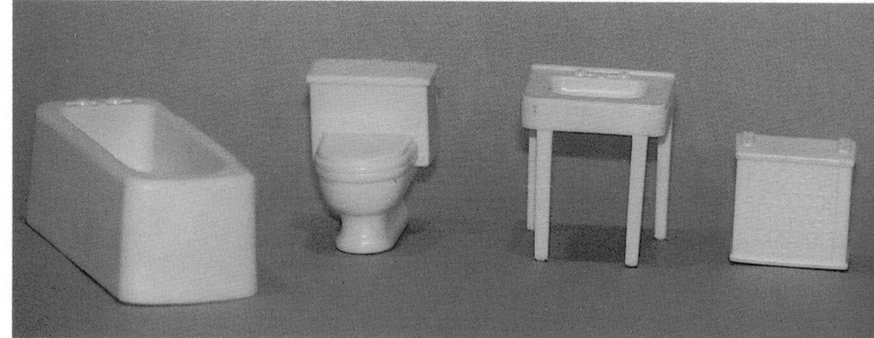

The small bathroom Marx furniture included a bathtub, lavatory, toilet, and hamper.

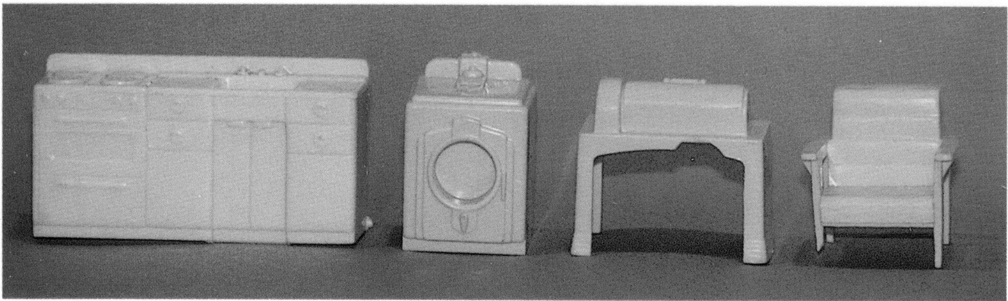

Pictured is the late Marx one-piece kitchen stove and sink as well as the automotic washer, dryer, and chair to be used in the utility room.

The Marx firm added a wing to its basic Colonial house and the wing provided space for either a utility room or a garage. The house measures 25 1/2" by 9" by 15 1/2".

The child's room in this model features an animal decor. The roof of the utility room provided a patio.

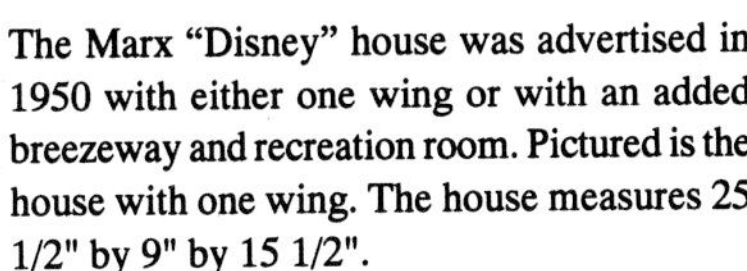

The Marx "Disney" house was advertised in 1950 with either one wing or with an added breezeway and recreation room. Pictured is the house with one wing. The house measures 25 1/2" by 9" by 15 1/2".

The house gets its name from the Disney characters used to decorate the children's room. In this model, one wing is used for a garage.

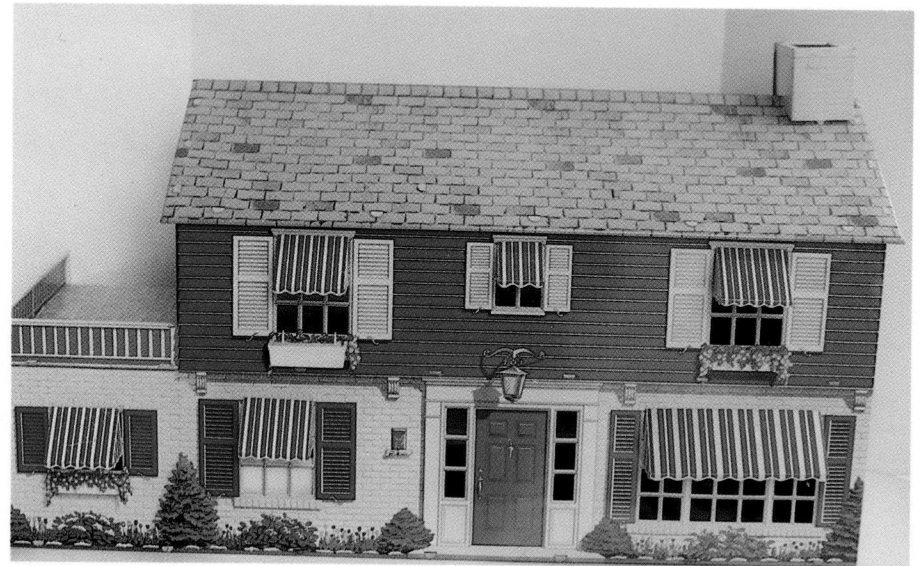

Marx also produced larger houses which were to be furnished with the 3/4" to 1' scale of furniture. In 1950 this Colonial red and white metal house sold for $6.95 complete with furniture. The house measures 33 1/2" by 12" by 18 3/4". The house features awnings and flower boxes on the front. House and photograph from the collection of Patty Cooper.

The inside of the house included the basic five rooms plus a garage with a patio roof. Photograph and house from the collection of Roy Specht.

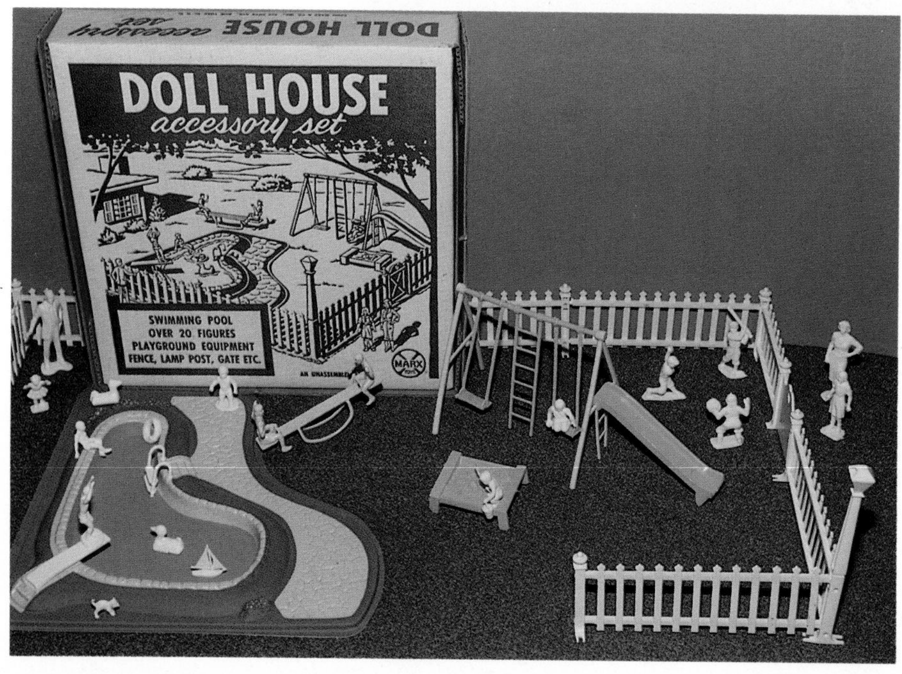

Other "extras" for the dollhouses could be purchased separately. Pictured is a boxed accessory set that included a swimming pool, outdoor play equipment, fence, and plastic figures to add play value to a doll house. Photograph and set from the collection of Roy Specht.

The 3/4" to 1' scale of Marx plastic living room furniture included a sofa, wing chair, barrel chair, coffee table, step end table, television, and floor lamp. The table lamp pictured is from the same era but did not come with the furniture. The proper lamps are missing from this set.

The dining room furniture included a table, two host chairs, two plain chairs, a hutch, and a buffet. Furniture and photograph from the collection of Roy Specht.

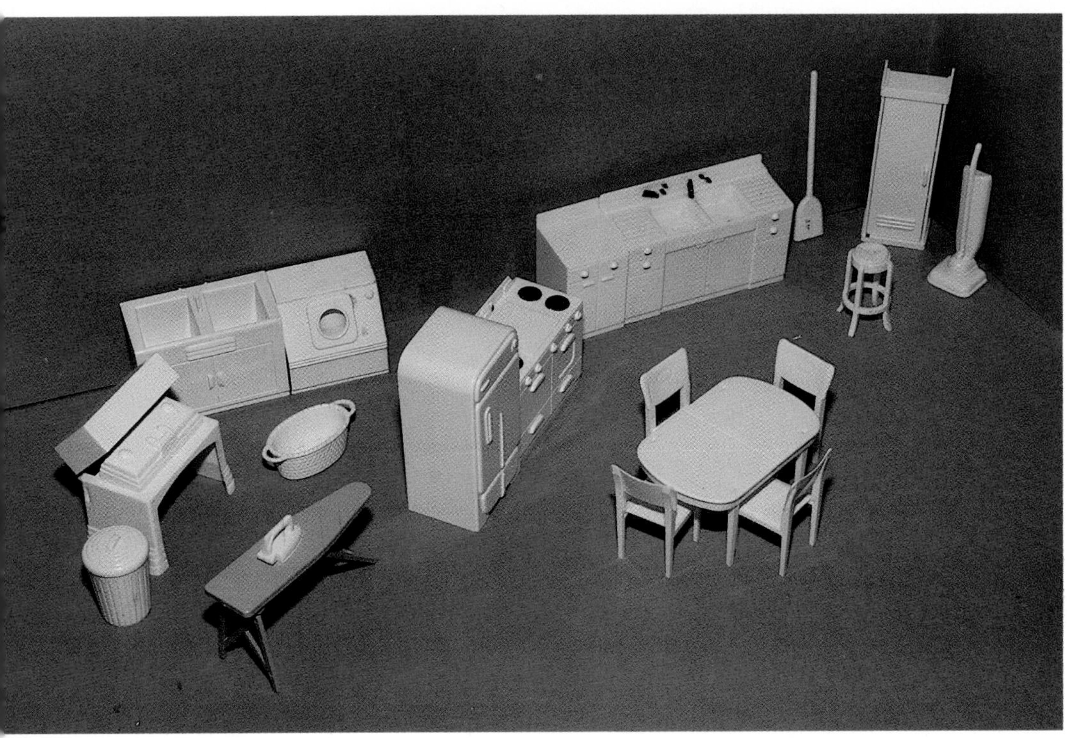

The kitchen was furnished with a sink, refrigerator, stove, base cabinet, table, and four chairs. Also pictured are the Marx laundry and housekeeping pieces which were added to the line at a later date. Furniture and photograph from the collection of Roy Specht.

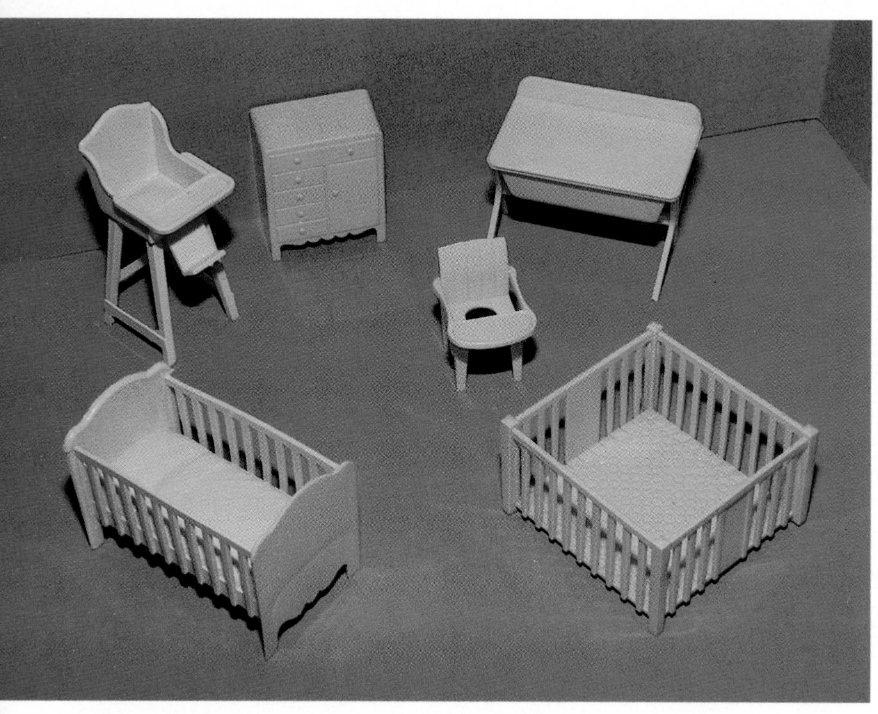

The 3/4" to 1' nursery furniture included a crib, playpen, chest, potty, and high chair. The changing table was added later. Furniture and photograph from the collection of Roy Specht.

Furniture for the sundeck included an umbrella table, chaise lounge, and two lounge chairs. Other items pictured were offered with other houses. Photograph and furniture from the collection of Roy Specht.

The bedroom furniture included a double bed, vanity, bench, highboy, chair, hassock, floor lamp, and night stand. Photograph and furniture from the collection of Roy Specht.

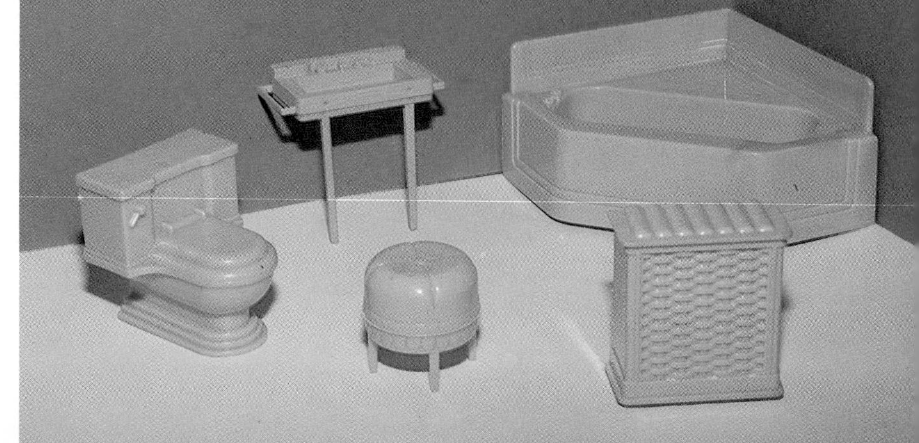

The larger Marx bathroom furniture consisted of a modern corner bathtub, lavatory, hamper, and toilet. Photograph and furniture from the collection of Roy Specht.

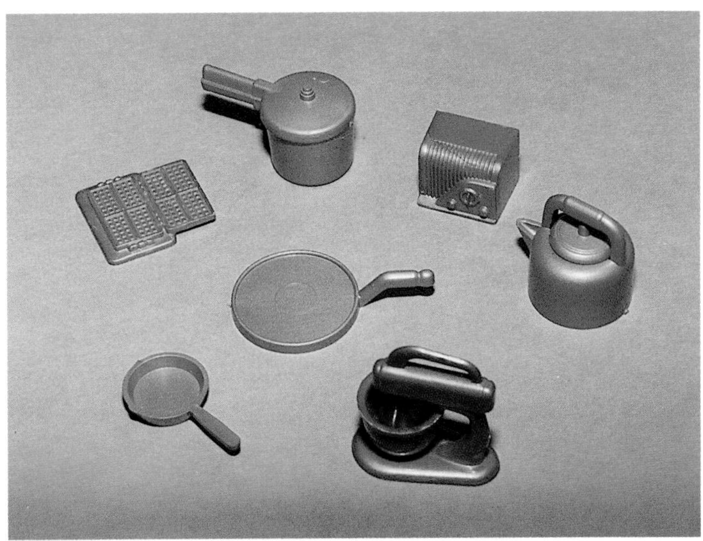

There were also several accessories produced by Marx which could be used in their dollhouses. Pictured are plastic kitchen items. Pieces and photograph from the collection of Roy Specht.

Another version of the 1950 red and white Colonial house was produced a few years later. This one featured dormer windows. House from the collection of Marcie Tubbs. Photograph by Bob Tubbs.

A stairway was added to the inside of the house to make it more realistic. House from the collection of Marcie Tubbs. Photograph by Bob Tubbs.

Marx produced a Babyland Nursery which was housed in a metal building. Nursery from the collection of Marcie Tubbs. Photograph by Bob Tubbs.

Marx also produced plastic furnishings for a school. These items were more varied than the Renwal pieces and included a drinking fountain, movie projector, and globe. Furniture from the collection of Marcie Tubbs. Photograph by Bob Tubbs.

The back of the nursery was open for easy access to the inside furnishings. Nursery from the collection of Marcie Tubbs. Photograph by Bob Tubbs.

Marx "L"-shaped ranch house produced in 1953. The furniture designed for the house was different than that used in other Marx houses. A swimming pool and outdoor play equipment was also included with the house. The house measured 32" by 16 1/4" by 13 1/8". The house came complete with television antenna. House and photograph from the collection of Marilyn Pittman.

The Superior crib and television on the left and the Marx crib and television on the right appear to be made from the same molds. Most of the later 1/2" to 1' furniture from the two companies appear to be identical. This was the furniture used by Marx in the "L"-shaped ranch house.

The ranch house contained a combination living room/dining room, bedroom, bathroom, kitchen, child's room, and a patio.

Box which contained the early Split Level Marx house. The furniture pictured on the box is the same as that used with the "L"-shaped ranch house.

The Split Level house was sold during the late 1950s and early 1960s in several different models. The house measures 14" deep and 14" high on one side and 8 1/2" by 8 1/2" tall on the one-story section. The length is 27". The house also included a chimney.

The inside of the house contained five open type rooms which included a utility room. Missing are the plastic steps and barbecue unit.

The small basic Marx house was enhanced with a staircase, doorbell, and electric lights in the late 1950s.

The house also included a utility room and a plastic front door.

The inside of the large house included a staircase and a utility room. The house measures 44" by 14" by 18" high. Photograph and house from the collection of Roy Specht.

This large 3/4" to 1' Marx metal house was featured in the 1962 Sears catalog priced at $15.88 for the furnished house. The house contained a door bell, "Florida" room, and lights. Photograph and house from the collection of Roy Specht.

The "Florida" room furniture included a sofa, two chairs, ottoman, television, and poker table. These pieces come from that set of furniture.

This metal ranch dollhouse was featured by Marx in the early 1960s. The house included a covered patio and a television antenna. The house measures 8" deep, by 8" high by 27" long. Photograph and house from the collection of Marilyn Pittman.

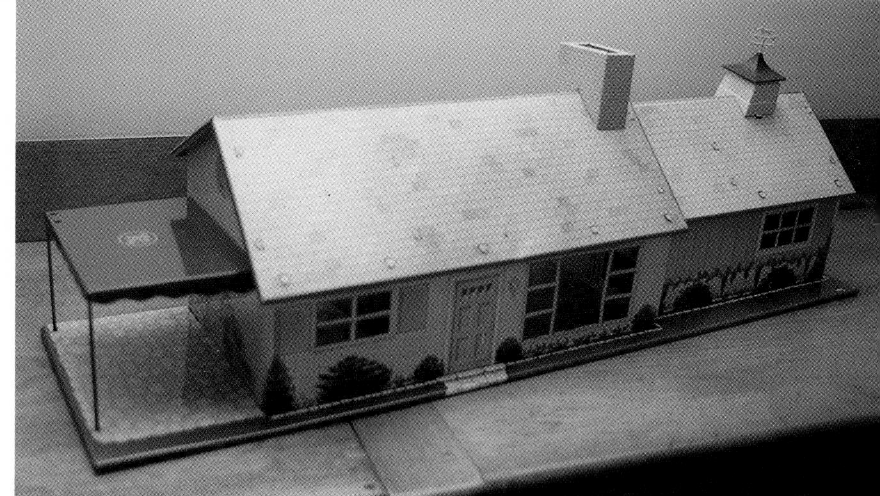

The inside of the metal ranch house included a bedroom, bathroom, kitchen, and living room/dining room combination. Photograph and house from the collection of Marilyn Pittman.

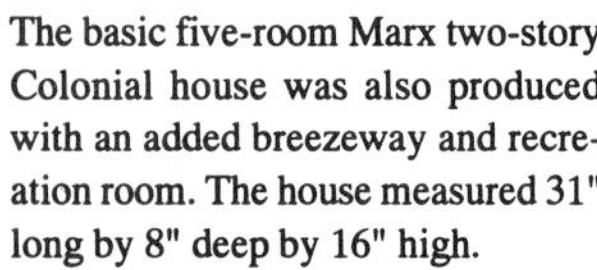

The basic five-room Marx two-story Colonial house was also produced with an added breezeway and recreation room. The house measured 31" long by 8" deep by 16" high.

The Marx recreation room featured proper wall decorations to enhance the furnishings provided for the room. Photograph and room from the collection of Rebecca Kepner.

These dollhouse dolls were marketed by Marx for over a decade during the 1960s and 1970s. The family of dolls ranged in size from 2" (baby) to 5 1/2" (father). The dolls have plastic heads, hands and feet, and cloth over armature bodies.

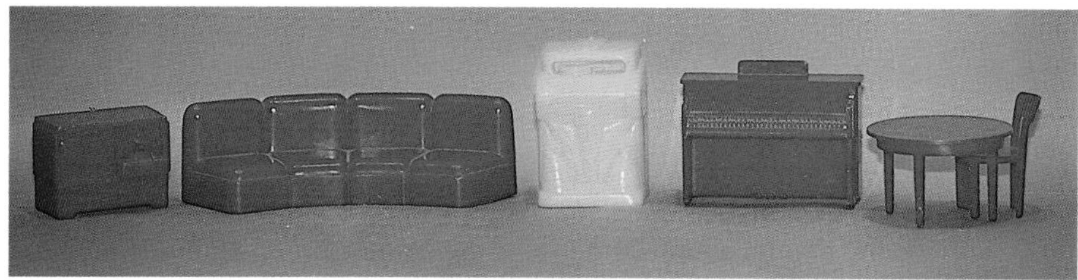

The recreation room in the Marx house included the following pieces of furniture: piano, bench, juke box, rounded sofa, counter, two stools, round table, two chairs, and a table tennis game to be used in the breezeway. The items pictured are an incomplete set of the furniture.

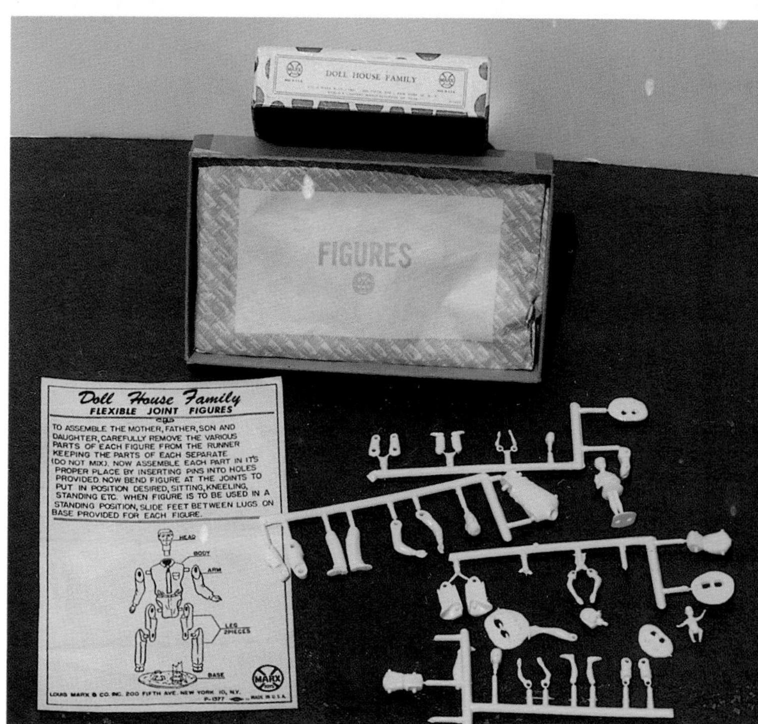

Marx Doll House Family. Flexible jointed figures to be used in the Marx dollhouses. The plastic dolls were made to be assembled by the consumer. Dolls from the collection of Marcie Tubbs. Photograph by Bob Tubbs.

Marx plastic 3/4" to 1' scale Little Hostess furniture first produced in 1964. Pictured are living room pieces from the set. Photograph and furniture from the collection of Roy Specht.

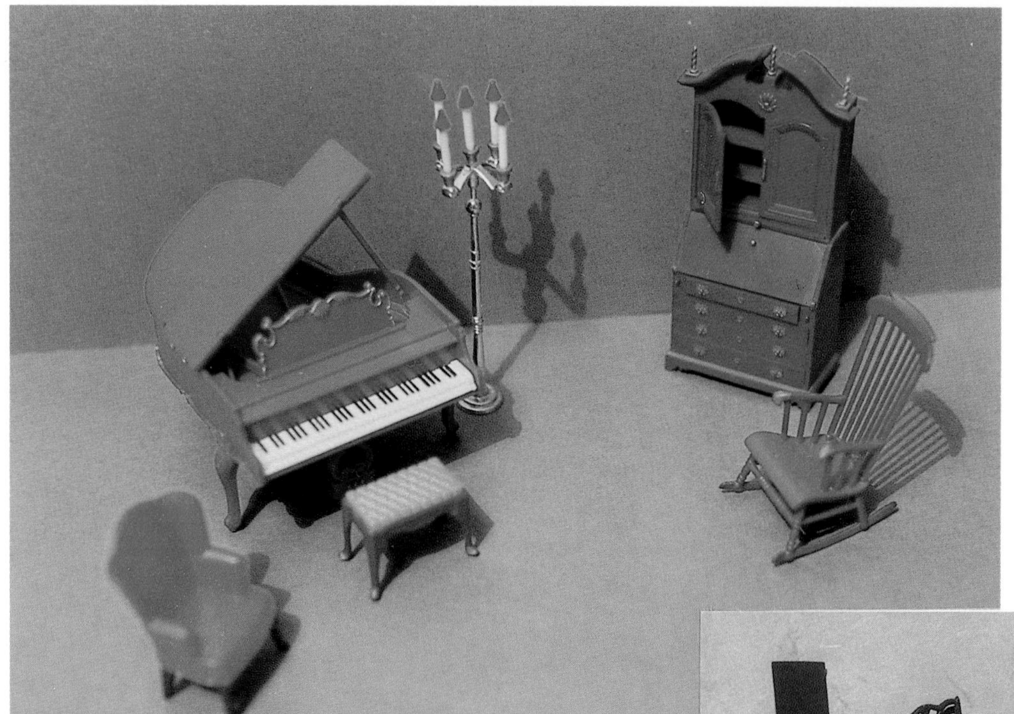

These Little Hostess furniture pieces could be used to furnish a music room. Photograph and furniture from the collection of Roy Specht.

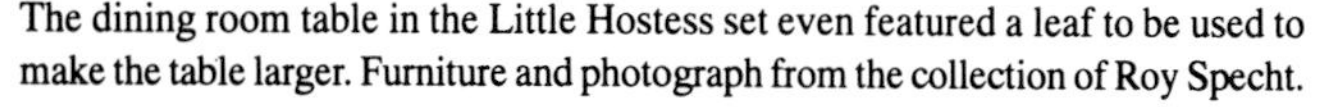

The dining room table in the Little Hostess set even featured a leaf to be used to make the table larger. Furniture and photograph from the collection of Roy Specht.

Many of the drawers and doors were functional in the Little Hostess furniture. Pictured is the bedroom furniture. Photograph and furniture from the collection of Roy Specht.

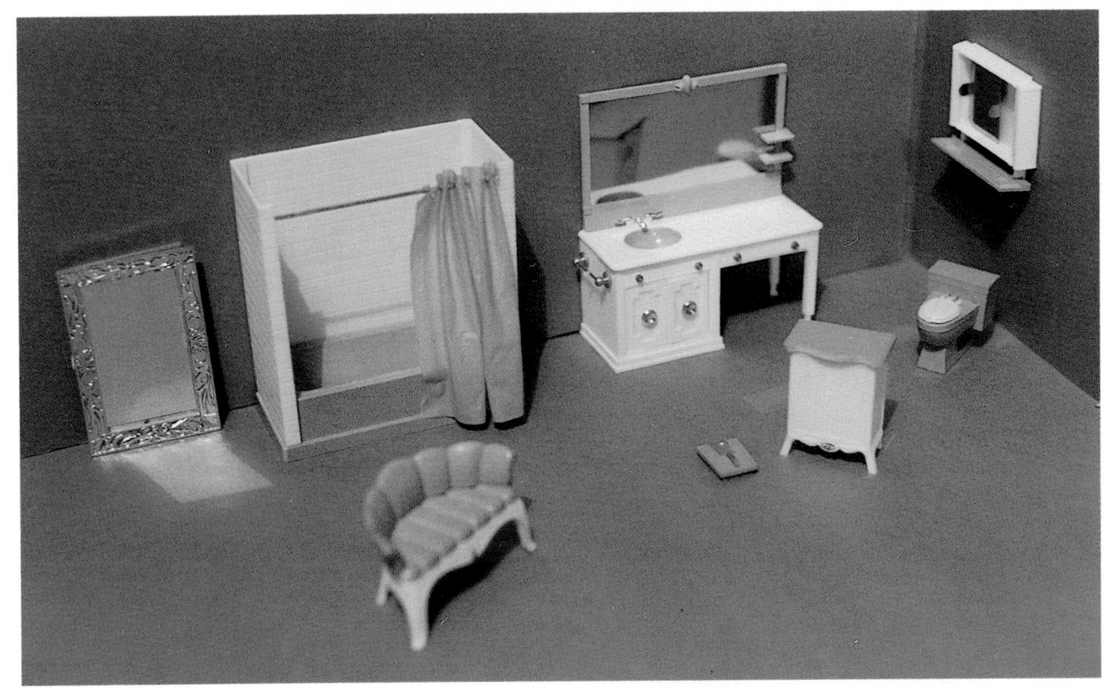

The Little Hostess bathroom pieces include some of the most detailed plastic 3/4" scale furniture ever made. Furniture and photograph from the collection of Roy Specht.

The Little Hostess kitchen furniture still looks very contemporary decades after it was first produced. Furniture and photograph from the collection of Roy Specht.

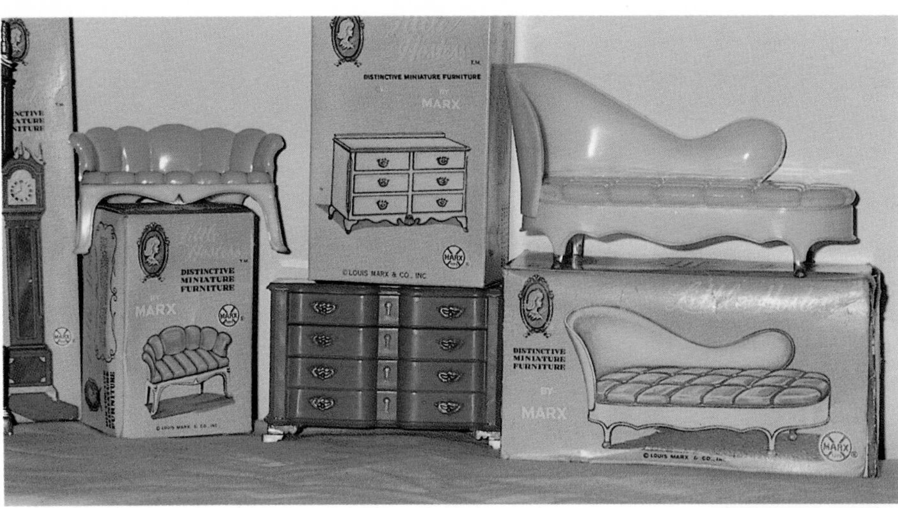

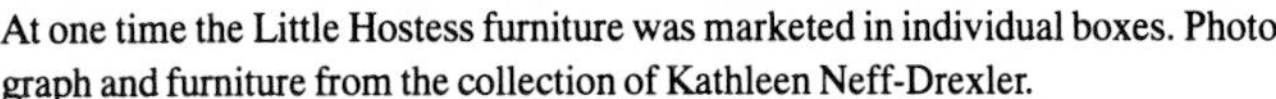
At one time the Little Hostess furniture was marketed in individual boxes. Photograph and furniture from the collection of Kathleen Neff-Drexler.

The later packaging for the Little Hostess furniture was in bubble packs. The television was an especially attractive Little Hostess piece. Photograph and furniture from the collection of Roy Specht.

Box for the basic metal Marx dollhouse still being sold in the late 1960s and early 1970s.

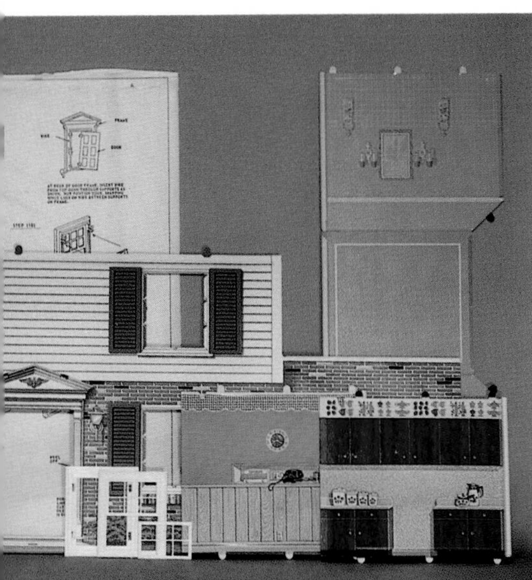

The inside decor had been changed for the house marketed in the late 1960s. Plastic windows which could be raised were a new innovation for the house. The furniture was the same as the earliest 1/2" to 1" Marx furniture but fewer pieces were supplied to furnish the house.

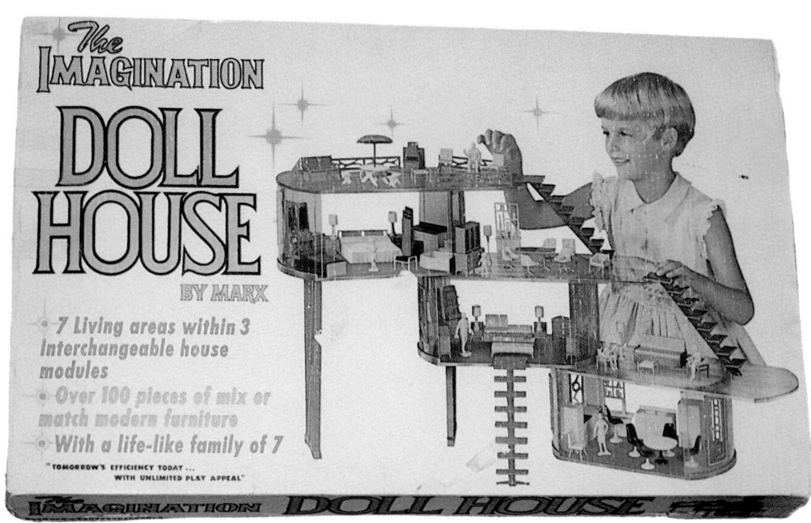

Another unique Marx dollhouse was produced in the late 1960s. It was called the Imagination Doll House and it contained over 100 pieces of mix or match new modern furniture. Although the "family" was described as life-like, the dolls were still just plastic figures. The house contained three interchangable house modules. House and photograph from the collection of Kathleen Neff-Drexler.

The furniture that accompanied the Imagination Doll House was entirely different than the usual Marx furniture. Pictured are pieces to be used in the living room, dining room, and bedroom. The 100 pieces of plastic furniture could be mixed and matched in different designs.

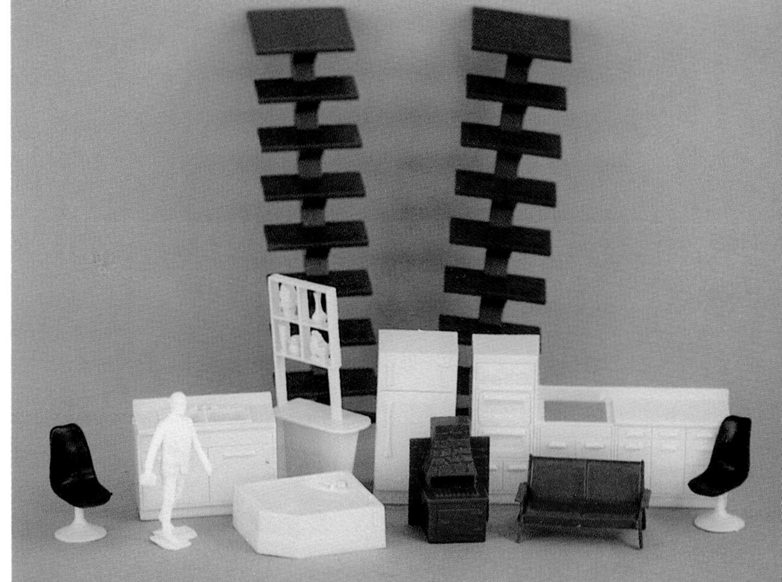

"Imagination" bathroom and kitchen furniture plus ladder type pieces that were to be used as part of the structure of the house.

Another late Marx metal house was produced with dormer windows and plastic porch supports. Photograph and house courtesy of Rebecca Kepner.

The living room has a light fixture on the ceiling. The furniture in the house is from the later period. Photograph and house from the collection of Rebecca Kepner.

Marx also produced a suitcase model dollhouse. The one pictured is made of cardboard and is similar to the ones being made of vinyl in the late 1960s. The house dates from 1968. House from the collection of Marcie Tubbs. Photograph by Bob Tubbs.

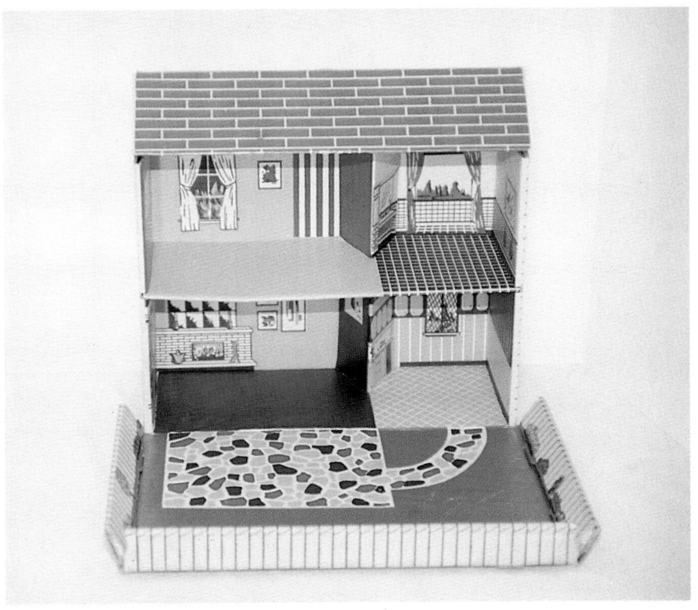

The opened house reveals four rooms, a patio, and yard. House from the collection of Marcie Tubbs. Photograph by Bob Tubbs.

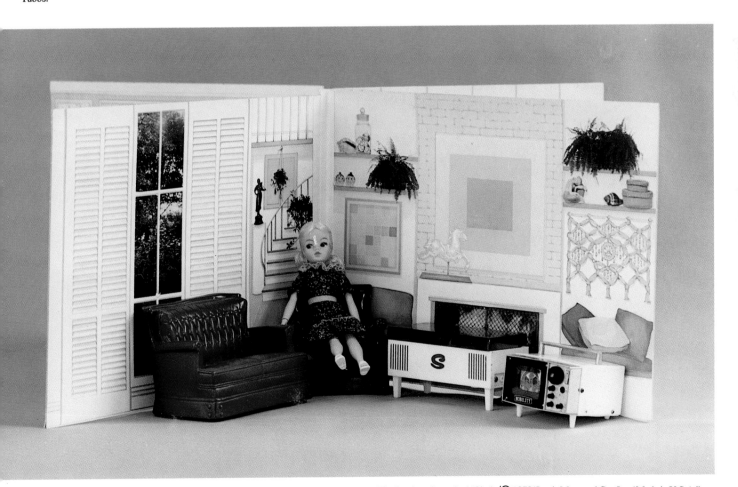

The Sindy doll Scenesetter, produced by Marx in 1978, provided a background for four rooms of Marx Sindy furniture. The Sindy doll is made of vinyl and is 11" tall. She has rooted hair and painted features. She is marked "Sindy" on the back of her head. Her waist, wrists, arms, head, and legs are jointed. Pictured with the Sindy doll are the soft plastic living room sofa, chairs, television and radio-phonograph.

The furniture is marked "Sindy/© 1978/ Louis Marx and Co., Inc./ Made in U.S.A." The Sindy radio-phonograph and the television were made to really work when batteries were installed. The radio-phonograph contains a real radio inside. It is marked with a large "S" on the front. The bottom says it was made in Hong Kong. The bottom of the television is marked "Japan."

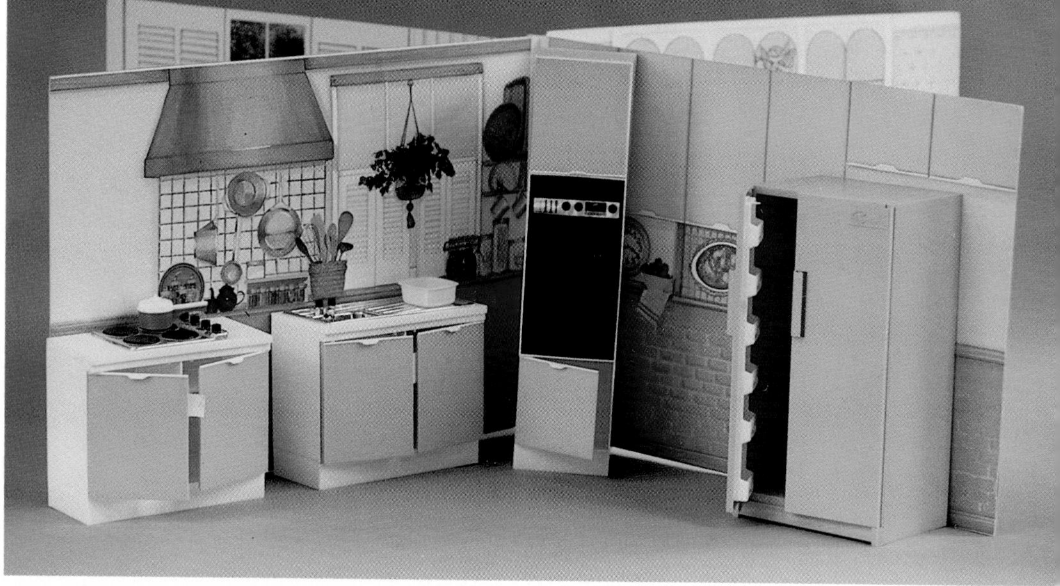

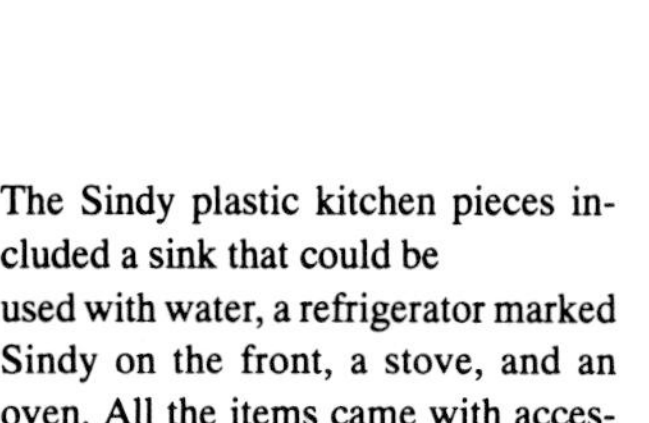

The Sindy plastic kitchen pieces included a sink that could be used with water, a refrigerator marked Sindy on the front, a stove, and an oven. All the items came with accessories.

The Sindy plastic dining room furniture included a table, four chairs, and a china cabinet. All the doors and drawers in the Sindy furniture are functional. Sindy bathroom pieces are also pictured. Both the sink and bathtub contained usable stoppers so the pieces could hold water. The lavatory is labeled "Sindy."

The bedroom featured a bed, vanity with mirror, and a nightstand with a workable lamp. All the Sindy furniture was marketed by Marx in the late 1970s. The bed is not pictured.

Plasco

Plastic Art Toy Corporation of America was begun in the 1940s by Vaughan D. Buckley. The company was located in East Paterson, New Jersey. The trademark for the firm was the figure of a uniformed band member playing a bass drum. On the drum were the words "A Plasco Toy." The Plasco toys included dishes, children's record players, and other plastic toys along with their very collectible dollhouses and furniture.

The dollhouse furniture was first manufactured around 1944 during World War II and was soon marketed under the trade name of "Little Homemaker." In the late 1940s the furniture came in boxed rooms with the insides of the boxes printed to represent the rooms themselves. The boxes could be used in place of a dollhouse to fashion rooms for the furniture.

The furniture was advertised in 1948 for $1.00 per room and included seven different sets of furniture. The furniture pieces were listed as follows: Living room: Davenport, grandfather clock, fireplace, television set, wing and club chairs, and coffee table. Kitchen: Refrigerator, stove, sink, table and four chairs. Bedroom: Highboy with real drawers, twin beds, dresser, and stool. Nursery: four-drawer highboy (moving drawers), crib (sides lower), layette dressing stand (top opens), stool, baby doll, mirrored vanity dresser, and mirrored dresser (moving drawers). Garden furniture: Two tables, one with parasol, chaise lounge, birdbath, four chairs. Dining room: Dining table, two serving tables, buffet and four chairs (two were arm chairs). Bathroom: Bathtub, dressing table, lavatory, toilet, clothes hamper (opens), and stool.

A few extra pieces were also produced to make even larger sets of furniture. These included kitchen cabinets (one piece with open shelves). For the bedroom the set was expanded to also include a nightstand and a vanity with a mirror along with a matching stool. A large china cabinet was also made to accompany the dining room furniture. A nightstand was also added to the nursery sets. Most of the plastic furniture is marked with the bass drummer trademark reading "A Plasco Toy." The first furniture was also marked "Patents Pending." Sometimes the furniture also carries a number and letter indicating the number of the furniture piece as well as the room it comes from as in B-2 (Bathroom-2).

Several different designs of boxes were used to display the Plasco furniture. After the boxes containing the room designs were dropped, containers for the furniture were made with a cellophane front so the items could be easily seen by the customer. Later, the sets of furniture each included a baby doll. The doll is marked "Plasco/U.S.A." The plastic doll is 3" tall and is jointed at the shoulders and hips. A diaper and shirt are lightly molded on the doll.

One of the nicest pieces of Plasco furniture is the television. It has a small screen which shows a picture. This cardboard picture can be changed with a rotation of the cardboard "wheel." The Plasco grandfather clock is also unusual in that the other major plastic dollhouse furniture makers did not include a similar clock in their line.

The Plastic Art Toy Corporation of America also made several different dollhouses. These houses are very hard to find for today's collector. The first model called "Little Homemaker's Open House" was a circular house made of fiberboard. Part of the exterior had transparent walls.

A later ranch house was much more durable. The house, unlike other dollhouses of the period, featured walls made of plastic. The house came unassembled and was to be put together by the consumer. The completed size was 3' wide and 18" deep. The furniture (seven rooms) could also be purchased with the house. The house contained a kitchen, living room/dining room, bedroom, nursery, bathroom, and patio. The doors opened and the roof could be removed for play. The roof and floor were made of a heavy fiberboard material.

In 1951 Plasco advertising featured a smaller all plastic four-room house. This house also had a removable roof for easy play but the house was square instead of the rectangle shape of the ranch house.

The early plastic Plasco dollhouse furniture is of very good quality and certainly rates a place in any collection of dollhouse furniture. If a Plasco house is unavailable, the furniture fits very nicely in a Rich or a Keysone house. It was advertised to be used with a Keystone house in a *House and Garden* magazine in November, 1948.

Apparently the furniture continued to be marketed into the early 1960s. The later sets of furniture used the new bubble packs to display the rooms of furniture. The furniture was marketed as "Little Miss Homemaker Furniture" and the package stated the furniture was in "*House and Garden* colors." This later furniture does not appear to be marked as to origin. Although the designs for this furniture are based on the earlier ones, short cuts were taken to produce the furniture more cheaply. The beds no longer have headboards, mirrors were eliminated, and the plastic work is not done as well as the earlier furniture.

The Plasco plastic dollhouse furniture was marketed for many years but the furniture did not have the exposure given to Ideal and Renwal furniture during the late 1940s and early 1950s. Because of this, fewer pieces of Plasco furniture were sold during these "glory years" and it may be harder to locate an early example of all of the items of furniture produced by Plasco. Because the company produced so many excellent furniture pieces including the television and nursery items, the collector will be justly rewarded when the furniture is found.

Plastic bathroom furniture with original box which could be used for a backdrop room for the plastic pieces. The furniture was marketed under the "Little Homemaker" brand name. Photograph and furniture from the collection of Roy Specht.

Box and dining room furniture which also was packaged with a box printed with background walls. Photograph and furniture from the collection of Roy Specht.

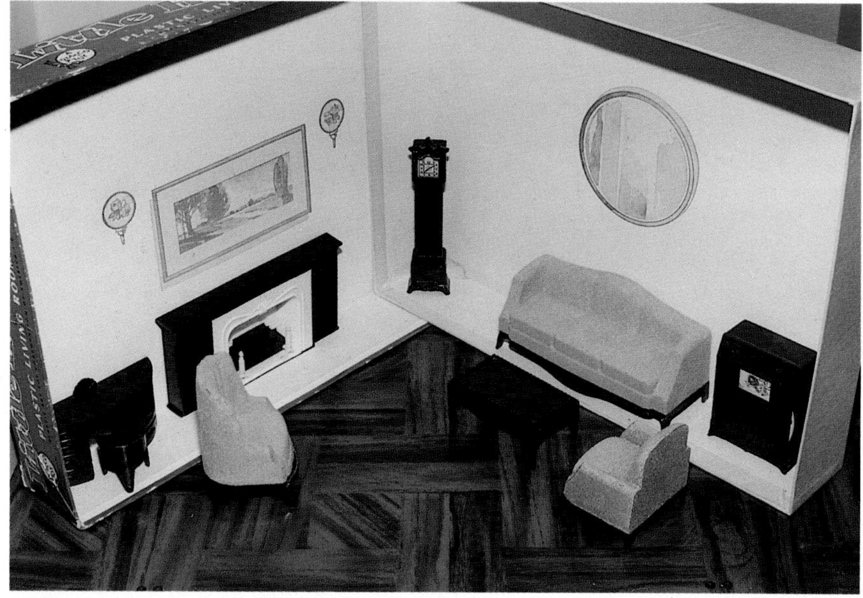

The Plasco living room furniture showing the background made by the inside of the boxes. The living room furniture included the excellent fireplace and television pieces as well as a grandfather clock made by no other company producing plastic dollhouse furniture. Photograph and furniture from the collection of Roy Specht.

Little Homemaker Plasco boxed living room furniture showing the cellophane covered box. The furniture could be displayed nicer in this manner. Photograph and furniture from the collection of Roy Specht.

Large set of boxed Plasco bedroom furniture which also included the printed bedroom background. Photograph and furniture from the collection of Roy Specht.

The bedroom furniture placed in front of the insides of the box to show the bedroom background. Photograph and furniture from the collection of Roy Specht.

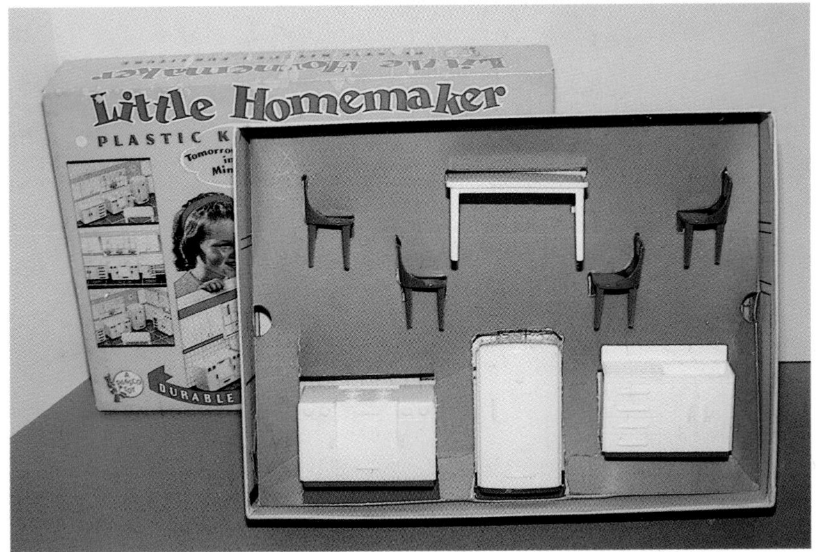

Boxed Plasco kitchen furniture in a set of eight pieces. Photograph and furniture from the collection of Roy Specht.

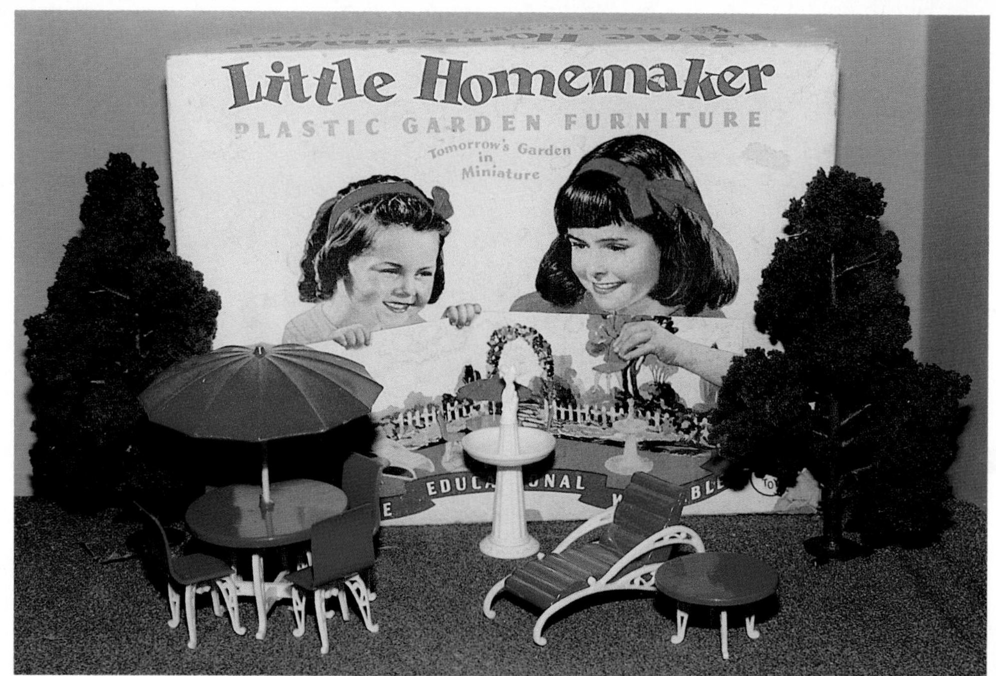

Boxed Plasco garden furniture. These pieces are hard to find in excellent condition. Photograph and furniture from the collection of Roy Specht.

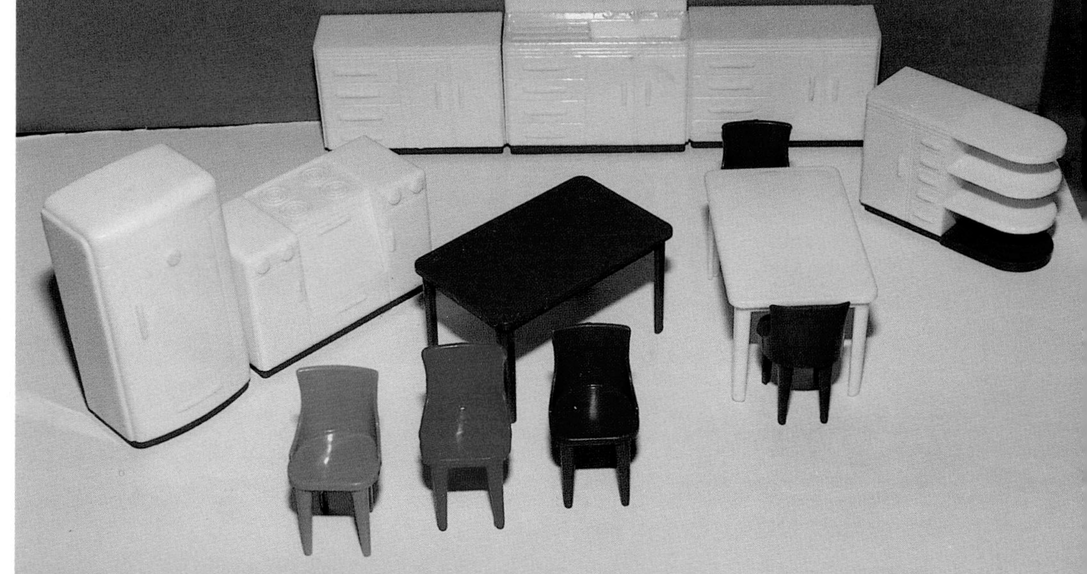

Additional kitchen furniture was produced that was not included in the eight-piece boxed set. These included kitchen cabinets and an open shelf cabinet unit. The table and chairs were also made in additional colors as pictured here. Photograph and furniture from the collection of Roy Specht.

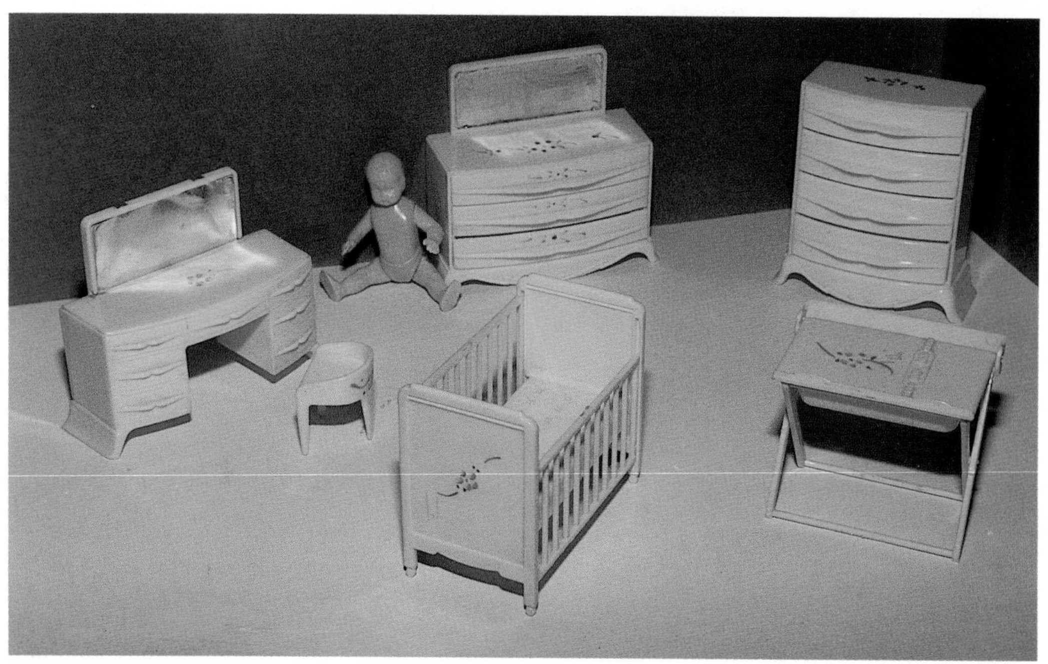

The Plasco nursery furniture is especially hard to find. The moving parts of the chest, dresser, crib, and changing table made this set particularly desirable. A nightstand was also available for the nursery set. Photograph and furniture from the collection of Roy Specht.

This box of Plasco dining room furniture came complete with a 7" plastic record called "Drummer Boy." From the collection of Roy Specht.

Plastic one-story ranch house made by Plastic Art Toy Corporation. The house measures 36" long by 18" deep. It even includes a television antenna. Photograph and house from the collection of Roy Specht.

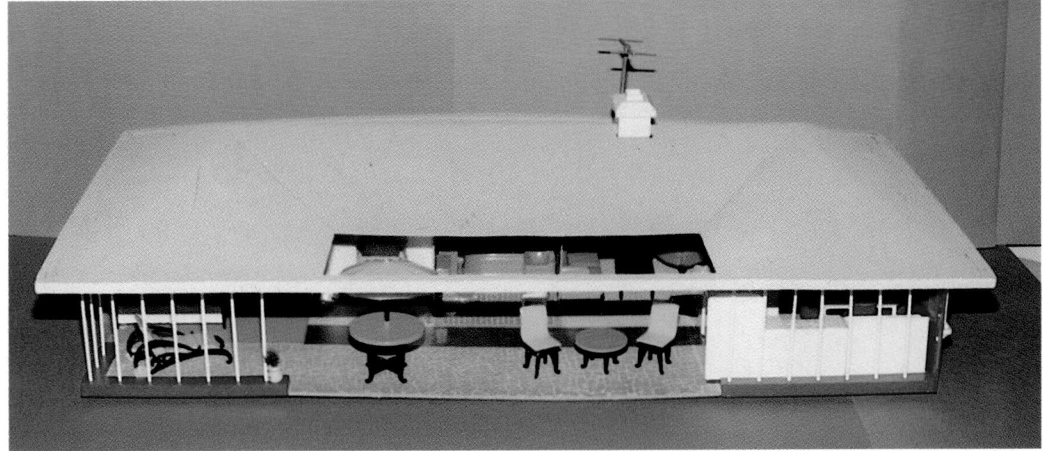

The back of the ranch house includes a patio. Photograph and house from the collection of Roy Specht.

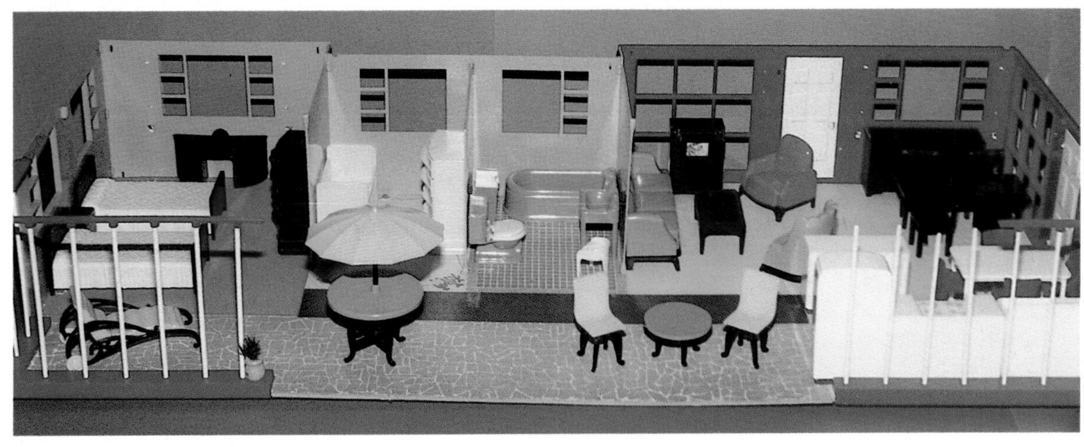

The Plasco house with its roof removed offered an ideal playhouse for children. Easy access was provided to all the rooms. Photograph and house from the collection of Roy Specht.

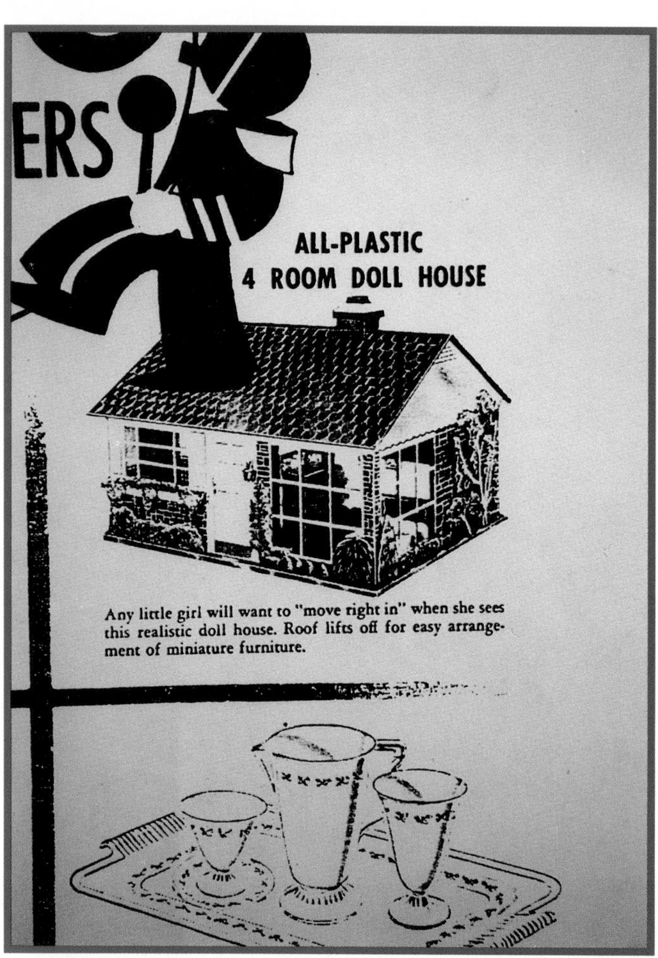

Pictured is a Plasco advertisement from 1951 showing their new all plastic four-room house. The house was made in a square design.

Plasco dining room furniture produced at the same time in the bubble package. The pieces include the table, four chairs, two serving pieces, and a buffet.

Later Plasco "Little Miss Homemaker" furniture probably marketed in the early 1960s. The box is marked "Molded and decorated in *House and Garden* colors." It may have been sold, in part, through that publication. The bedroom furniture included twin beds, a highboy, a vanity (no mirror), stool, and night table. There are no head or footboards on the beds.

Another bubble package of "Little Miss Homemaker" dining room furniture. The company apparently had a name and location change. The maker is listed as Plastic Toy and Novelty Corp. in Brooklyn, New York. The trademark shown on the package is a clown holding a hoop. The package also reads "*House and Garden* Colors." The furniture is not marked.

Renwal

The Renwal Manufacturing Co. of Mineola, Long Island, New York was founded in 1939 by Irving Lawner (backward -- Renwal). The company became famous and is best remembered for the plastic toys and dollhouse furniture that was marketed from the mid-1940s until the early 1960s. The Renwal Manufacturing Co. was sold in the early 1970s.

The Renwal plastic dollhouse furniture was first put on the market in 1946. At that time the furniture was boxed in sets under the name "Jolly Twins." The boxes contained printed inserts that could be used as rooms for the furniture. Individual pieces of the furniture could also be purchased at dime stores. The O. & M. Kleemann Ltd. firm, based in England, also used some of the same molds as Renwal and they began producing their plastic dollhouse furniture in 1947.

The 1947 Sears, Roebuck and Co. Christmas catalog featured the Renwal furniture rooms priced from 74 cents (bathroom) to $1.98 (thirteen piece living room). Other rooms included the bedroom, nursery, kitchen, and dining room. Apparently none of the drawers opened at this time. The same designs of furniture were sold for many years. By 1962 the prices were much cheaper in the Sears Christmas catalog of that year. For only $2.99 a customer could purchase thirty-six pieces of the Renwal furniture. That was enough to furnish a five-room dollhouse according to the advertisement. The pieces were the models which included opening drawers and moving parts.

The Renwal plastic furniture is in the 3/4" to 1' scale. Each item of furniture is marked with the Renwal name as well as the stock number given to that particular piece of furniture. The furniture included the following items:

Living Room: L
18 Radio phonograph
70 Floor lamp
71 Table lamp
72 Coffee table
73 End table (pedestal)
74 Piano
75 Piano bench
76 Club chair
77 Barrel chair
78 Sofa
79 Floor model radio
80 Fireplace

Dining Room: D
51 Table
52 Hutch
53 Chair
54 Small server
55 Buffet

Bedroom: B
81 Bed
82 Vanity with mirror
83 Dresser with mirror
84 Nightstand
85 Chest-on-chest
75 Vanity Bench (L)

Bathroom: T
95 Bathtub
96 Lavatory
97 Toilet
98 Hamper

Kitchen: K
62 Kitchen hutch
63 Chair
66 Refrigerator
67 Table
68 Sink
69 Stove
64 Garbage can and dust pan
12 Kitchen stool

Nursery:
87 Baby push cart
114 Buggy with spread
115 Buggy with doll
118 Playpen
119 Cradle with spread
120 Cradle with baby insert
122 Bathinette
30 High chair
36 Potty chair
85N High boy
84N Nightstand

Misc.:
89 Sewing machine
108 Folding card table
109 Folding chair
116 Carpet sweeper
117 Mop
121 Broom
65 Rocker
7 Tricycle
10 Scale
11 Alarm clock
13 Smoking stand
14 Mantle clock
16 Table radio
19 Swing
20 Slide
21 Teeter totter
27 Kiddie car
28 Telephone
31 Washing machine
32 Ironing board and iron
37 Vacuum sweeper

Besides the regular 3/4" small scale dollhouse furniture, Renwal also made three pieces of larger kitchen items called "DeLuxe." These included a stove, sink, and a refrigerator. The refrigerator was 7" tall while the sink and the stove were approximately 6" wide. The sink could be used with running water and the drawers and doors opened. The other pieces also had doors that opened and each item came with small accessories.

In addition to the basic room sets of furniture and accessories, Renwal also offered several special sets of furniture. These included the Hospital Nursery set (No. 214) with seven cribs, seven dolls, seven blankets, two tables, two chairs, bath sink, night table, nurse, scale, tub, pans, bottle tray, seven bottles, cotton balls, and diaper material. A smaller set containing only five cribs and babies was also produced.

Another interesting set is the "Cook 'n Serve Toy Set" (No. 247). The box contained forty-eight pieces. It included two tables, four chairs, refrigerator, sink, stove, cups and saucers, plates, bowls, silverware, pots and pans, father, mother, and girl dolls.

Another special boxed Renwal package was called "Busy Little Mother Toy Set" (No. 249). It consisted of nineteen pieces, including: Mother, girl, and baby dolls, play pen, table, chairs, ironing board and iron, sewing machine, bench, table lamp, telephone, toilet, bathroom sink, washing machine, garbage can, dust pan, vacuum sweeper, and hamper.

The boxed "Little Red School" set (No. 1200) was another very unusual play set produced by Renwal. The set came with thirteen pieces including four pupils, six student desks, a teacher desk with opening drawers, a swivel chair and a cardboard school. The boxed set retailed for under $3.00 in the late 1940s.

Other boxed sets marketed by Renwal included a music room, several different accessory sets, and an outdoor set. The latter included the swing, slide, teeter totter, tricycle, baby buggy, and babies.

In order to encourage sales for the Renwal furniture, the company made several changes in the furniture through the years. The best change was the making of the furniture with moving parts. Drawers and doors were designed so they could be opened and with the moving parts of the sewing machine and the washing machine, Renwal's furniture was perhaps the best plastic furniture on the market. Other less dramatic changes included marketing the sets of furniture with painted designs to brighten the plain surface of the plastic. Renwal also produced sets of furniture that little girls could decorate themselves. These boxes of furniture came with the paint, a brush, and decal decorations that could be used to change the look of the furniture. Renwal also varied the colors of the furniture so that a collector can find the same piece of furniture made in several different colors.

The black Renwal furniture that is sometimes found by collectors did not come from little girl decorators, however. This furniture was used as a premium by Cross and Blackwell Foods. Customers could send 35 cents and a label from their date rolls or canned soup to receive one piece of furniture. This really doesn't sound like much of a bargain when the furniture in its regular color could be purchased for about the same amount of money. Perhaps the company thought the pieces were more attractive in the black decor with the Pennsylvania Dutch style decoration which was applied by hand. This offer appeared sometime in the mid-1950s. The following pieces are known to have been used as advertising premiums: card table, folding chairs, dining table, dining chairs, hutch, server, rocker, coffee table, lamp table, highboy, and cradle. The decorations included birds, flowers, leaves, and gold highlights.

The Renwal dolls are also important collector items to accompany the sturdy plastic furniture. Two models of Renwal baby dolls were made. The more common baby is the jointed 2 1/4" tall model (No. 8). The doll is all plastic and has a molded shirt and pants. The more unusual baby doll is the 5" (No. 9) model. Since that doll would have been too large to be used with the furniture, it was really not made in connection with the dollhouse furniture. The family dolls were marketed, along with the 2 1/4" baby, to accompany the furniture. The four family dolls are made of all plastic and have jointed arms, legs, and knees so that they are able to sit in the chairs. All of the dolls have their clothes and features molded on their bodies. The sister doll (No. 41) is 3 5/8" tall, the brother (No. 42) is 3 3/4" tall, the mother (No. 43) is 4 1/8" tall and the father (No. 44) is 4 1/4" tall. The dolls were also marketed at a later time in a "Paint the Dolls" set (No. 233). The set came with all the family dolls plus a baby doll, along with paints and a brush. Besides these dolls, Renwal also made similar dolls to represent a nurse (No. 431), a doctor (No. 441), a policeman (No. 442), and a mechanic (No. 421). These dolls could be used with Renwal's various other sets which included a service garage.

Although Renwal marketed their plastic dollhouse furniture for over fifteen years, it did not have full exposure from the major Sears and Montgomery Ward Christmas catalogs after the 1940s. As the metal dollhouses became popular, they were sold complete with furniture. Major firms like Marx could sell a dollhouse completely furnished with plastic dollhouse furniture for less money than the cost of five rooms of the Renwal furniture alone. Since these new houses were always sold furnished, the only customers for the Renwal products were the people who were making a dollhouse at home or furnishing a dollhouse from the past. The other customers who were available to buy the Renwal products were the little girls who purchased one or two pieces at a time from their local dime stores.

Most collectors now consider the Renwal pieces with the moving parts as the most collectible and desirable plastic dollhouse furniture from the past. Therefore, prices keep rising accordingly. The hardest pieces to find include the mop, broom, stroller, swing, and kiddie car. These pieces will fetch top prices. Other more expensive items include the fireplace, piano, iron and ironing board, washing machine, and all of the baby items.

Renwal plastic dollhouse furniture was first sold under the "Jolly Twins" name. Printed on this box of bedroom furniture is "A Renwal Product No. 1100 Pat. Pend. U.S.A. and Canada Made in U.S.A." The sides of the box picture the bedroom, kitchen, nursery, and living room furniture. The furniture was made in the 3/4" to 1' scale. Box and photograph from the collection of Roy Specht.

The furniture and room insert that was contained in the Jolly Twins bedroom box. The "room" folded to fit into the box. Photograph and furniture from the collection of Roy Specht.

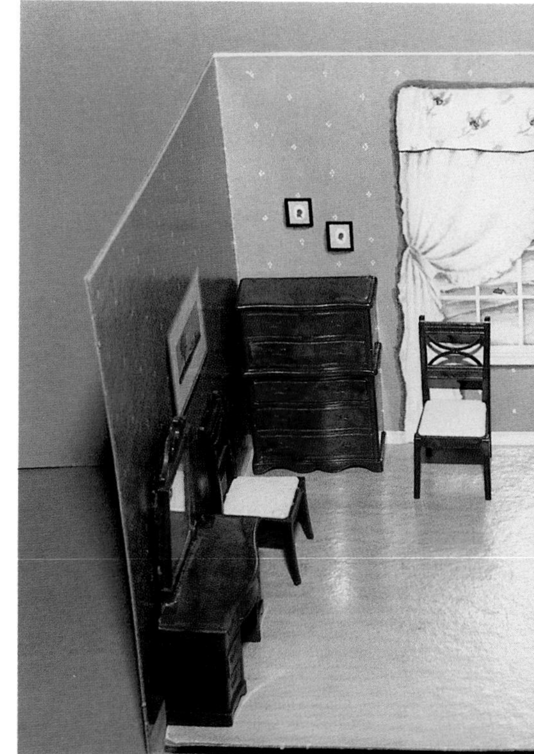

Later the Renwal bedroom was sold with an additional dresser as well as the twin beds, highboy, vanity, and nightstand.

The baby dolls made by Renwal were included in many of the various special sets produced by the company. The hard plastic jointed dolls measure 2 1/4" tall. Another larger 5" tall baby was also produced by Renwal but it is too large to be used with the Renwal dollhouse furniture. Pictured with the doll is the Renwal potty chair that was later included in the set of nursery furniture and the rocking chair.

The boxed nursery furniture included the buggy, bathinet, high chair, playpen, highboy, cradle, nightstand, lamp, and baby doll. Box and furniture circa 1947.

The early Renwal dining room also was packaged with a room insert as were all the sets during this period. Photograph and furniture from the collection of Roy Specht.

A server piece was later added to the Renwal dinning room set of furniture. In addition the set included a buffet, hutch, table, and chairs.

The Renwal kitchen furniture is shown in its original box. From the collection of Judy Mosholder. Photograph by Carl Whipkey.

The basic Renwal kitchen furniture included the sink, refrigerator, stove, table, and four chairs.

The Renwal bathroom included a bathtub, toilet, lavatory, and clothes hamper. This boxed set is from the collection of Judy Mosholder. Photograph by Carl Whipkey.

The living room furniture came in a large set that was priced at $1.98. The pieces included were a fireplace, sofa, two chairs, radio, coffee table, two round lamp tables, two table lamps, floor lamp, piano, and bench. This boxed set is from the collection of Judy Mosholder. Photograph by Carl Whipkey.

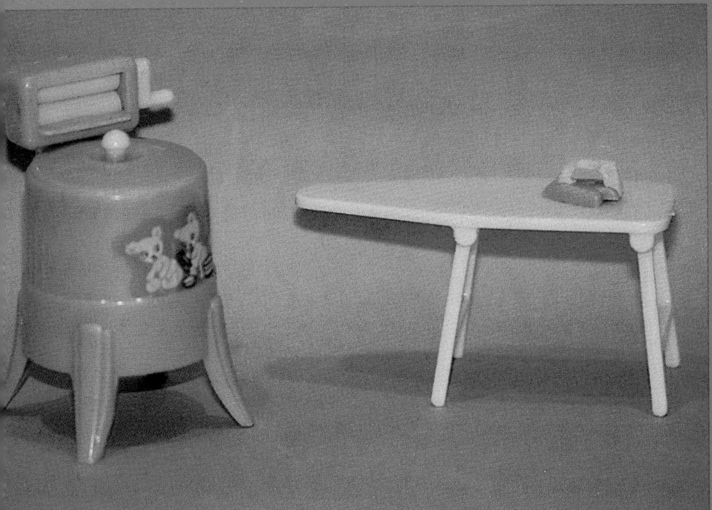

The washing machine with its working parts and the iron and ironing board are very desirable pieces of Renwal furniture and are hard to locate.

A smaller living room set was also available in this later package. One of the Renwal babies was also included in this boxed set. Furniture from the collection of Judy Mosholder. Photograph by Carl Whipkey.

The Renwal furniture came in many different colors. This living room set, featuring red on all the pieces is very unusual. Photograph and furniture from the collection of Roy Specht.

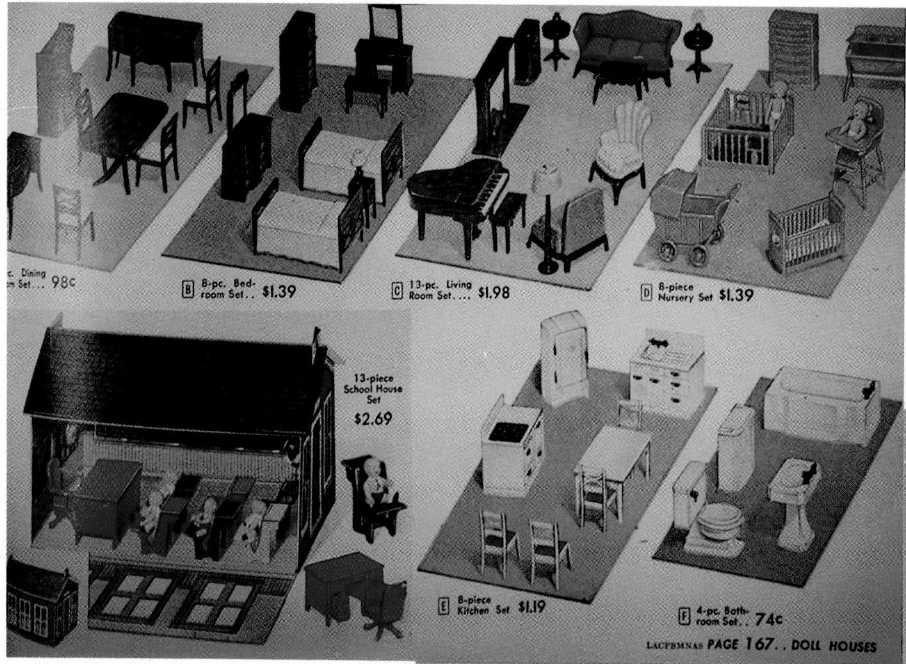

The 1947 Sears Christmas catalog featured the basic sets of Renwal furniture priced from 74 ¢ to $1.98.

In addition to the basic pieces of dollhouse furniture, Renwal also produced many other items that could be used to accompany their rooms of furniture. These toys are some of the hardest Renwal pieces to locate. Included were a slide, teeter totter, kiddie car, swing, and tricycle. This "Play Yard" set, with the substitution of a buggy instead of the kiddie car, was advertised in the Montgomery Ward Christmas catalog for 1948 at a price of $1.15. Photograph and furniture from the collection of Roy Specht.

This Renwal accessory set was also produced as an extra to supplement the furniture. Included in the set were an alarm clock, telephone, dust pan, sweeper, garbage can, smoking stand, stool, scale, and mantle clock.

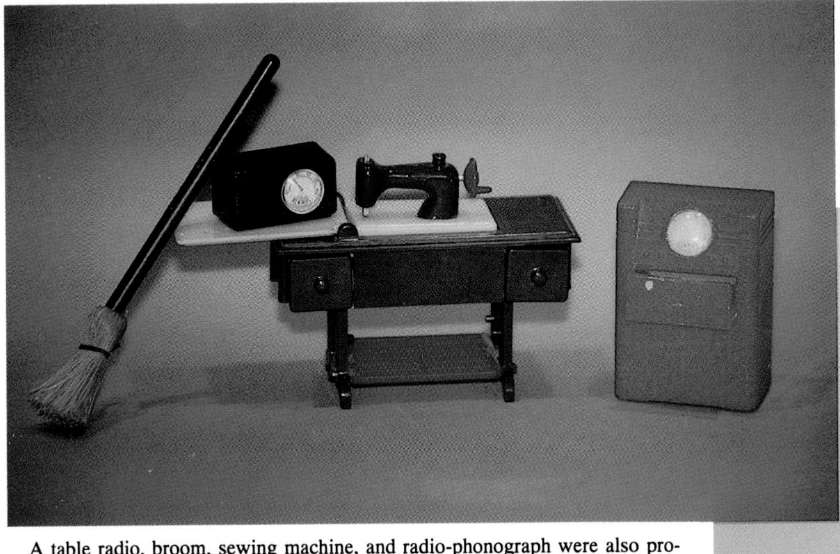

A table radio, broom, sewing machine, and radio-phonograph were also produced to accompany the Renwal furniture. The broom remains an elusive item for many collectors.

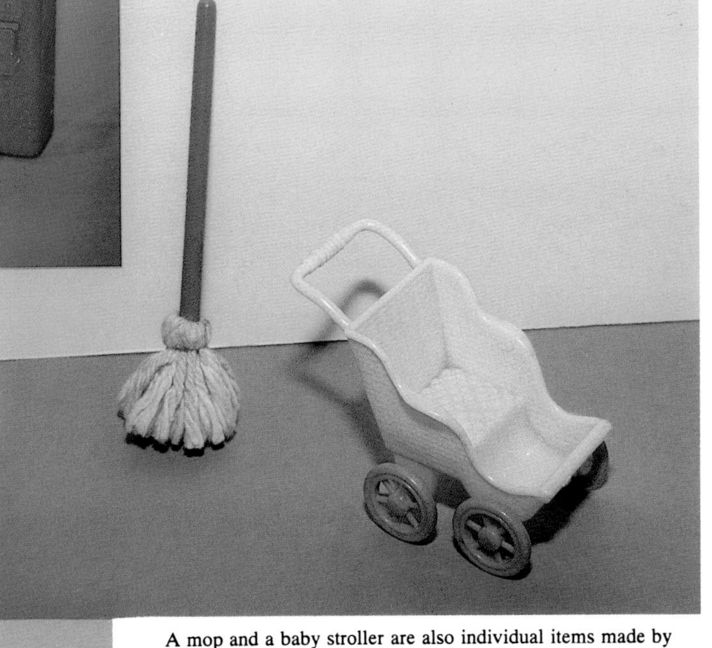

A mop and a baby stroller are also individual items made by Renwal that are hard to find for today's collector. Photograph and furniture from the collection of Roy Specht.

Another of the Renwal products that collectors especially value is the folding card table and chairs. Pictured with the card tables are the piano and bench. Photograph and furniture from the collection of Roy Specht.

Another very desirable set marketed by Renwal is called "Cook 'n Serve Toy Set." The box contained forty-eight pieces of Renwal items. It included: two tables, four chairs, refrigerator, sink, stove, cups, saucers, plates, bowls, silverware, pots, pans, father, mother, and girl dolls. Photograph and boxed set from the collection of Betty Nichols.

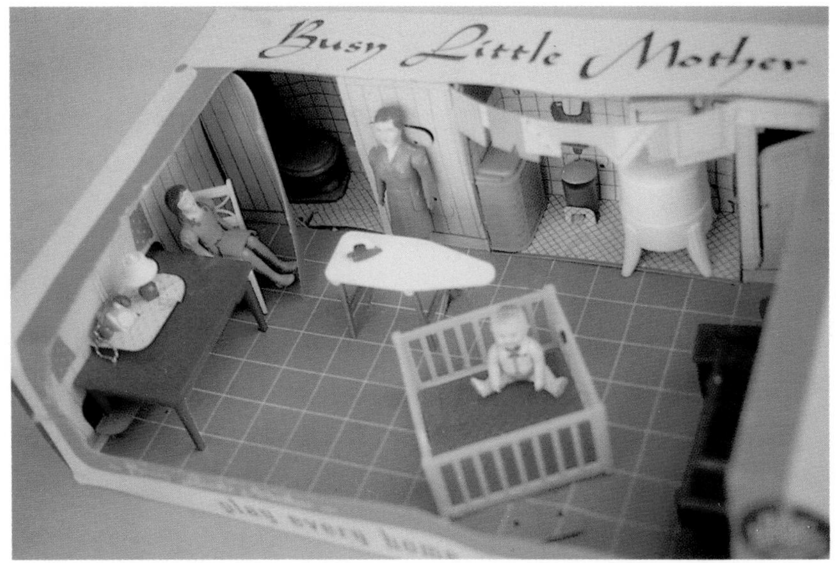

A similar set of Renwal items was boxed with the title "Busy Little Mother." The "Busy Little Mother" set consisted of nineteen pieces including: Mother, girl, and baby dolls, playpen, table, chairs, ironing board and iron, sewing machine, bench, table lamp, telephone, toilet, bathroom sink, washing machine, garbage can, dust pan, vacuum sweeper, and hamper. Set and photograph from the collection of Betty Nichols.

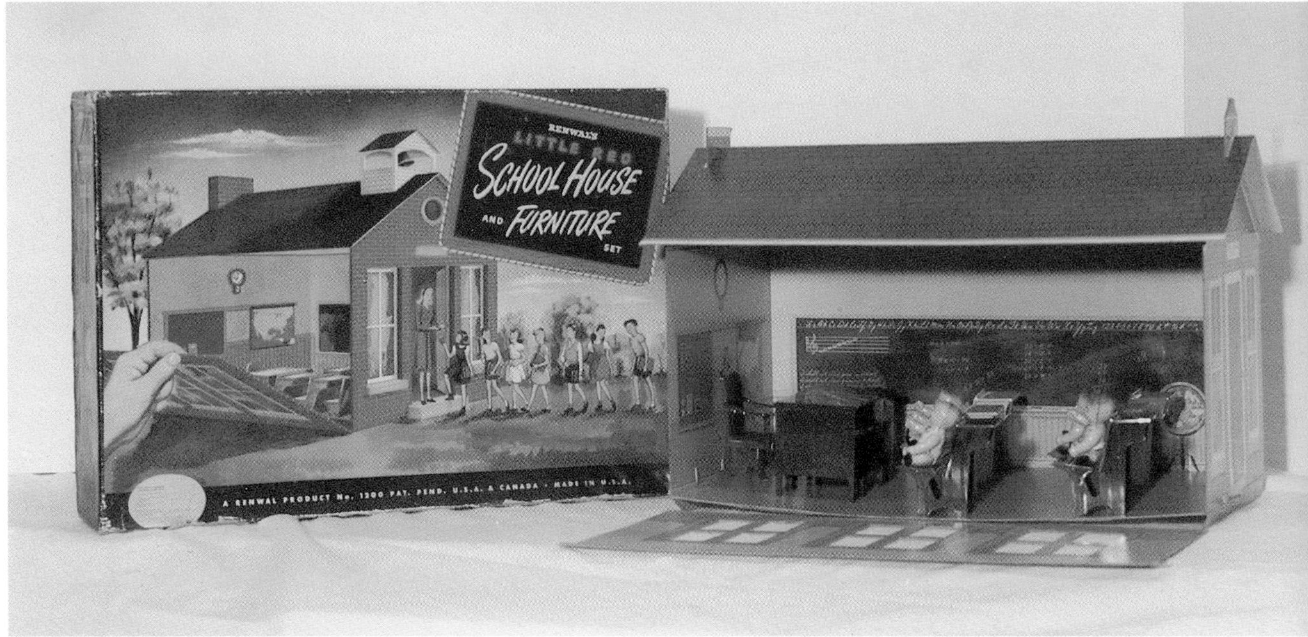

Renwal also produced a school house set that included a cardboard school, a teacher's desk with opening drawers, a swivel chair, six student desks, and four pupils. School house from the collection of Marcie Tubbs. Photograph by Bob Tubbs.

The "Little Red School" was made of cardboard and the set sold for under $3.00 in the late 1940s. School from the collection of Marcie Tubbs. Photograph by Bob Tubbs.

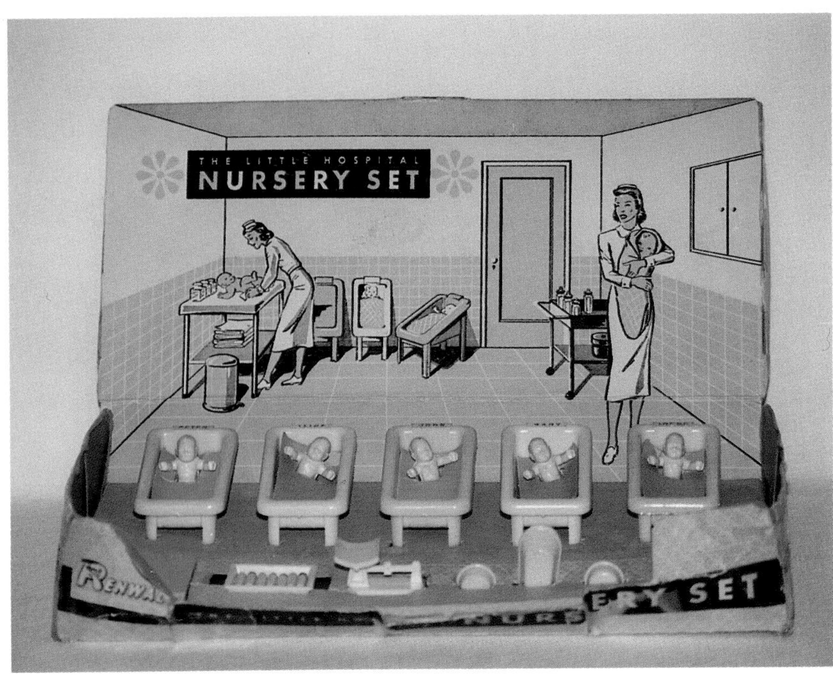

Renwal's Little Hospital Nursery Set is also a very desirable set for today's collector. This boxed set contains five cribs and babies as well as accessories. From the collection of Judy Mosholder. Photograph by Carl Whipkey.

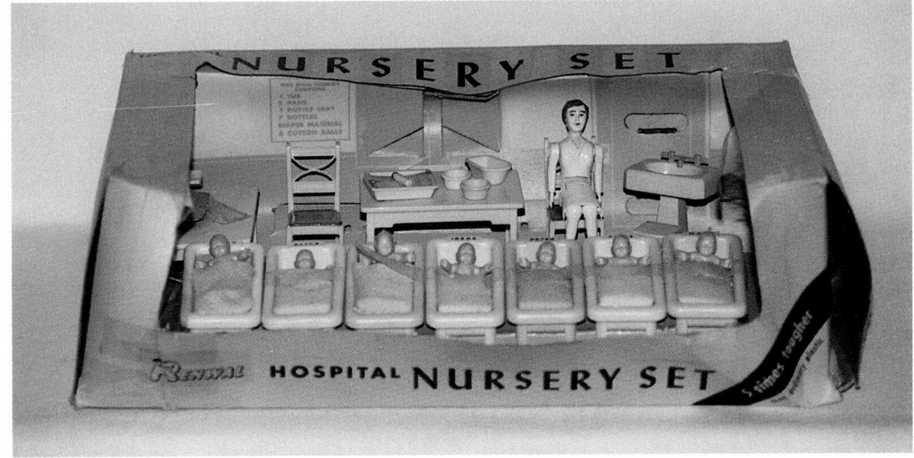

The larger Hospital Nursery Set included seven babies and cribs as well as seven blankets, two tables, two chairs, bath sink, night table, nurse, scale, tub, pans, bottle tray, bottles, cotton balls, and diaper material. From the collection of Judy Mosholder. Photograph by Carl Whipkey.

This black Renwal furniture was used as a premium by Cross and Blackwell Foods circa mid-1950s. The furniture was decorated in the Pennsylvania Dutch style and has become quite collectible. Photograph and furniture from the collection of Roy Specht.

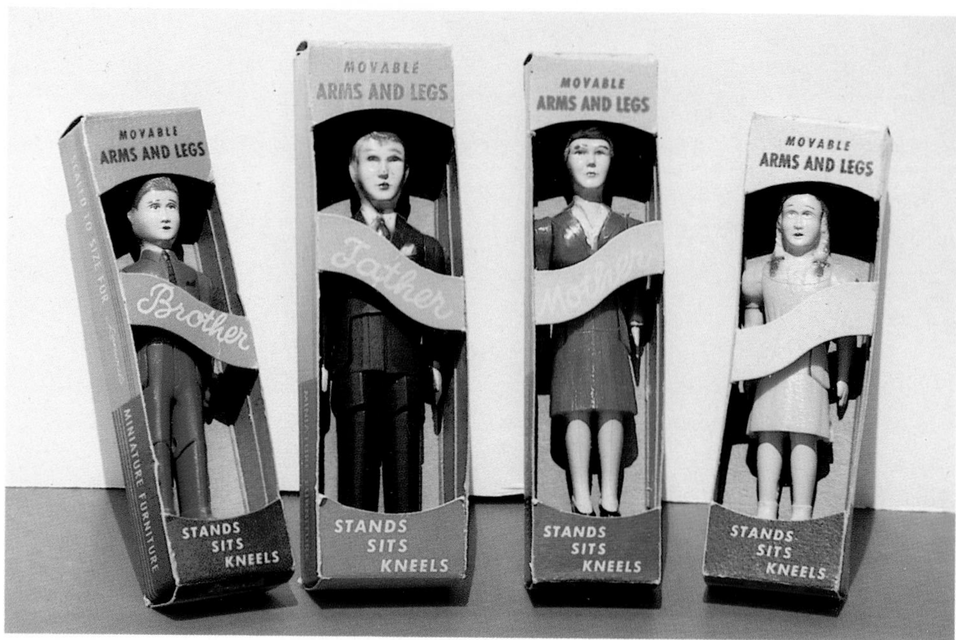

Renwal also produced unique plastic dolls to accompany their furniture. The family of dolls included a mother, father, daughter, and son. The dolls were jointed at the hips, knees, and shoulders so they could easily sit in the Renwal furniture. The dolls ranged in size from 3 5/8" tall to 4 1/4" tall. The paint came off of the dolls very easily so few mint dolls are found. These boxed dolls are from the collection of Marcie Tubbs. Photograph by Bob Tubbs.

The Renwal dolls were also packaged in other ways. Pictured are two different displays which include both the dolls and assorted items of furniture. From the collection of Judy Mosholder. Photograph by Carl Whipkey.

A clear plastic store display is pictured which was used to promote the sale of Renwal plastic dollhouse furniture. The furniture could be purchased by the piece in many stores. From the collection of Judy Mosholder. Photograph by Carl Whipkey.

T. Cohn

T. Cohn, Inc., located in Brooklyn, New York, produced many collectible dollhouses over a period of many years. The company was responsible for what appeared to be one of the first modern metal dollhouses. The Cohn house was featured in the Montgomery Ward Christmas catalog in 1948.

The advertising states that the house was new at that time. The all metal house contained six rooms plus a sun deck. The outside of the house was lithographed with red brick on the lower part and white shingles above. The roof was supposed to represent red tile. The five front casement windows and the door opened. The house was pictured furnished with Renwal plastic furniture in the Montgomery Ward ad but the furniture was priced separately. The house alone sold for $4.75.

Cohn also made another similar model of this house. There were no longer opening casement windows and the design of the front of the house was different. The new model also included an additional wing which was to be used as a garage. The garage included a door that opened. The decor of the inside of the house was also different from that of the first model. T. Cohn, Inc. used the "Superior" trade name for many of its products.

When the company produced the second of their metal dollhouses, they also marketed plastic furniture that could be used to furnish the house. This furniture is in the 3/4" to 1' scale. A flyer came with the house to show where each piece of furniture should be placed. The furniture included the following pieces: Living Room: sofa, coffee table, barrel chair, and club chair. Dining Area: table, four chairs, and china closet. Kitchen: table, two chairs, combination sink and stove, and refrigerator. Bedroom: dresser, stool, chest, and bed. Bathroom: bathtub, toilet, and wash basin. Nursery: chest, crib, and potty chair. Utility Room (behind garage): washing machine. Terrace: umbrella set (two chairs and an umbrella). Swimming Terrace: water slide (pool and slide). Most of the furniture pieces are marked in an oval "Made in U.S.A. TC Superior." Although this furniture does not meet the high quality of the furniture made by Renwal, Ideal, and Plasco, it does match the Marx standards. With no moving parts, many of the furniture designs are similar to the Marx furniture pieces.

Many of the early models of T. Cohn dollhouses can be distinguished from houses manufactured by other companies because of their hipped gable roofs. Most of the Cohn houses were labeled with the company name. The first house was marked above the deck floor, "T. Cohn Inc. Made in the U.S.A."

Several different designs of metal dollhouses were produced by Cohn throughout the 1950s and into the early 1960s. One of the most popular was a house that came in pink, green or blue. This house, too, had a hipped gable roof. These designs were used for several years. The early houses had an opening window and door while the later models had only printed windows and doors. The house contained three rooms upstairs, a large combination living room/dining room and a kitchen downstairs. A deck was also included on the upper level of the house.

The ranch house was also a popular model for the T. Cohn firm. One version featured in the early 1950s was a four-room model. This house was unusual in that it included a large landscaped patio with a swimming pool and slide. There were nineteen pieces of furniture included with the house. The house measured 25" by 16 1/4" by 10 1/2" high. The price at that time was $3.98. The furniture that came with the house was the 3/4" scale Superior line.

A smaller metal ranch house was designed a few years later. The house came in either a three- or four-room model. This house was sold furnished with the small 1/2" to the 1' scale of Superior furniture. The four-room version came equipped with a metal swimming pool and pool accessories. This house was still being sold in 1961 along with several other T. Cohn dollhouses including a metal split level design.

The furniture for these newer houses was made in a 1/2" to 1' scale in a soft plastic. There were many pieces included in the small size of furniture. They were as follows: Living Room: grand piano and bench (made in even smaller scale), four-piece sectional, coffee table, end tables, floor and table lamps, and television. Nursery: crib, high chair, playpen, chest, and rocking horse. Bathroom: toilet, corner bathtub, vanity sink, and hamper. Utility: washer-dryer combination, stool, and ironer. Bedroom: double bed with attached headboard and lamps and nightstands, chair, dresser, vanity dresser, and bench. Kitchen: refrigerator, sink, stove, table, and chairs. Dining Room: table and chairs, buffet, and hutch. Patio: umbrella table, chaise lounges, and small table on wheels. Many of the lamps were only half lamps. The furniture was marked "Superior" inside a pennant design. This furniture appears to have been made from the same molds as those used to produce the Marx 1/2" to 1' plastic furniture used in several of their houses.

Besides the metal dollhouses, the T. Cohn firm also marketed some wood dollhouses. In 1961 they featured a house measuring 2 1/2' by 2 1/2' by 1'. The two-story house included five rooms. The living room and dining room were combined into one room. The house was furnished with the early 3/4" to 1' scale of plastic furniture. It had molded plastic windows and doors and was probably made of gypsum. The house had to be assembled by the consumer, as did nearly all dollhouses of that time.

Although T. Cohn wasn't as large a manufacturer of dollhouses as Marx, the company stayed in the business for many years. No contemporary doll house collection would be complete without at least one doll house produced by T. Cohn, Inc. and furnished with their inexpensive plastic furniture.

Lithographed Steel Doll House

COMPLETELY DECORATED...FURNITURE SOLD SEPARATELY

ew All Metal Doll House—six attractively decorated oms plus a sun deck to help keep the doll family in ood health. Outside is decorated to resemble a typical ontemporary American house. Lithographed with red rick lower part, white shingle upper part, red tile roof, reen and white shutters and window boxes. The 5 ront casement windows and door open.

Rooms: Living Room, Dining Room, Bedroom, Nursery, itchen and Bathroom . . . and the sun deck. Interior is olorfully lithographed to resemble wallpaper, draper-es, rugs, pictures, tile bathroom floor, etc. Strong metal wall partitions. Back of house open. Exposed edges moothly rolled. House shipped flat, assemble with ac-curate tabs and slots. No nuts, bolts or screws needed. Instructions included.

Forty-One-Piece Furniture Set (shown in rooms of house above) is not included—it may be bought separately below (48 T 4521). House: 22x13x18½ in. high to chimney top. Ship. wt. 8 lbs. 8 oz.
48 T 4475 M—Doll House only. *Mailable*......4.75

41-Pc. Plastic Furniture Set. Selected from sets at bottom of this and opposite page; (See (C) Page 104 for Nursery set). Designed to fit 6-room house above and other Doll Houses of approximate size. 41-pc. set includes: 10-pc. Living Room, 7-pc. Dining Room, 5-pc. Bedroom, 5-pc. Nursery, 4-pc. Play set (Car-riage, Tricycle, See-saw and Doll), 6-pc. Kitchen and 4-pc. Bathroom. Ship. wt. 2 lbs. 4 oz.
48 T 4521—41-Pc. Furniture Set..............5.98

Front View of Doll House shown above, showing metal casement windows that open, entrance, and door-yard planting. Shipped flat; easily assembled.
Buy Christmas Gifts now—avoid delay. Wards Monthly Payment Plan—Page 176.

ROOM SETS OF MINIATURE FURNITURE—DESIGNED TO DOLL HOUSE PROPORTIONS

(C) 9-Pc. Kitchen Set. Ivory Refrigerator, Stove and Sink have doors that open; Kitchenette Table, 4 Chairs and Kitchen Clock. Chair seats and Clock are Red. Table 3⅛ in. long.

(D) 8-Pc. Bedroom Set. Walnut colored Twin Size Beds, Night Table, Vanity and Bench, Dresser and Highboy. Top drawers open on Vanity, Dresser and Highboy. Red table lamp. Beds 4 in. long. 1.19

(E) 6-Pc. Bathroom Set. Pink Bath-tub, Wash bowl with tiny faucets, Pink Toilet and Ham-per. Red Scale, Red and Ivory Bath Stool. Hinged seat and cover on toilet. 4-in. tub.
48 T 4517—Boxed. Ship. wt. 1 lb............74c

Pictured is the first T. Cohn, Inc. metal dollhouse which was sold for $4.75 in the Montgomery Ward Christmas catalog for 1948. Photograph and house from the collection of Marilyn Pittman.

The inside of the two-story five-room Cohn metal house produced in 1948. The metal casement windows and the front door opened. The house measures 28" by 13" by 18 1/2" to the top of the chimney. Photograph and house from the collection of Marilyn Pittman.

A later model of the Cohn metal dollhouse, circa 1951. This house featured only one opening window (it has been broken), and an opening door.

The inside of the Cohn house included five rooms plus a garage with a garage door which opened. The house measured 28 1/2" by 9 3/4" by 16" tall including chimney.

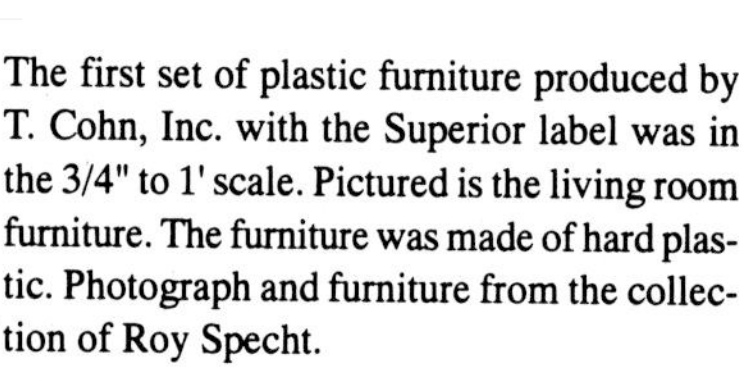

The first set of plastic furniture produced by T. Cohn, Inc. with the Superior label was in the 3/4" to 1' scale. Pictured is the living room furniture. The furniture was made of hard plastic. Photograph and furniture from the collection of Roy Specht.

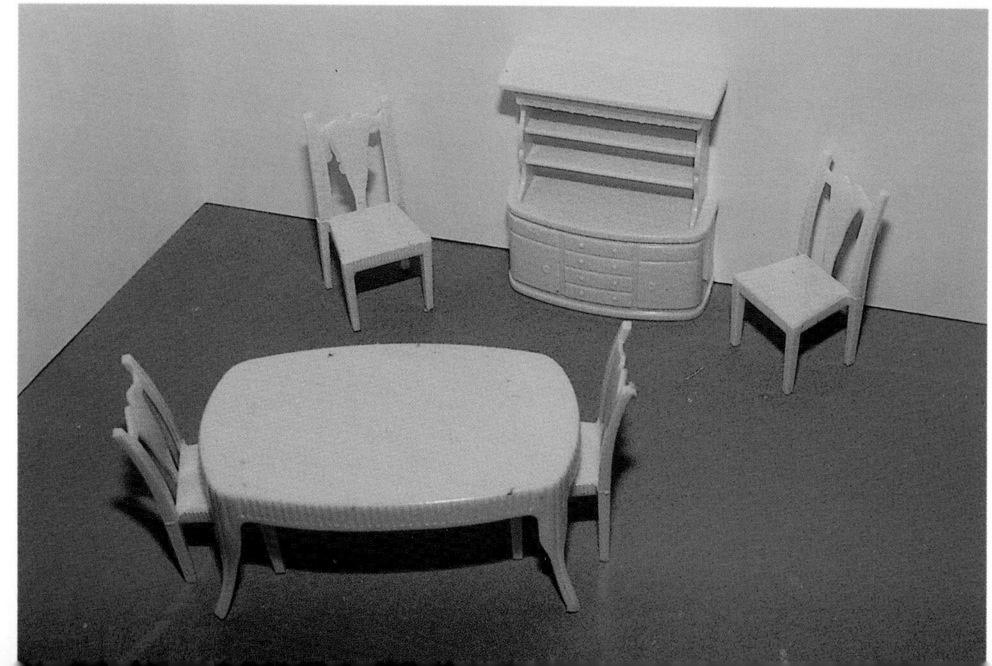

The Superior 3/4" scale dining room furniture came in several different colors, including pink. Photograph and furniture from the collection of Roy Specht.

The Superior kitchen was also produced in white as well as pink. Included was a unique combination sink and stove piece as well as the usual table and chairs. Photograph and furniture from the collection of Roy Specht.

Superior nursery furniture included a crib, chest, and potty chair. Furniture and photograph from the collection of Roy Specht.

The 3/4" Superior line of bathroom furniture was produced for many years and came in white, blue, and yellow. Photograph and furniture from the collection of Roy Specht.

The bedroom included a bed, dresser, stool, and chest. Other furniture produced by Cohn in the 3/4" scale included a washing machine, an umbrella set, and a slide. Most of the furniture came with the Superior mark.

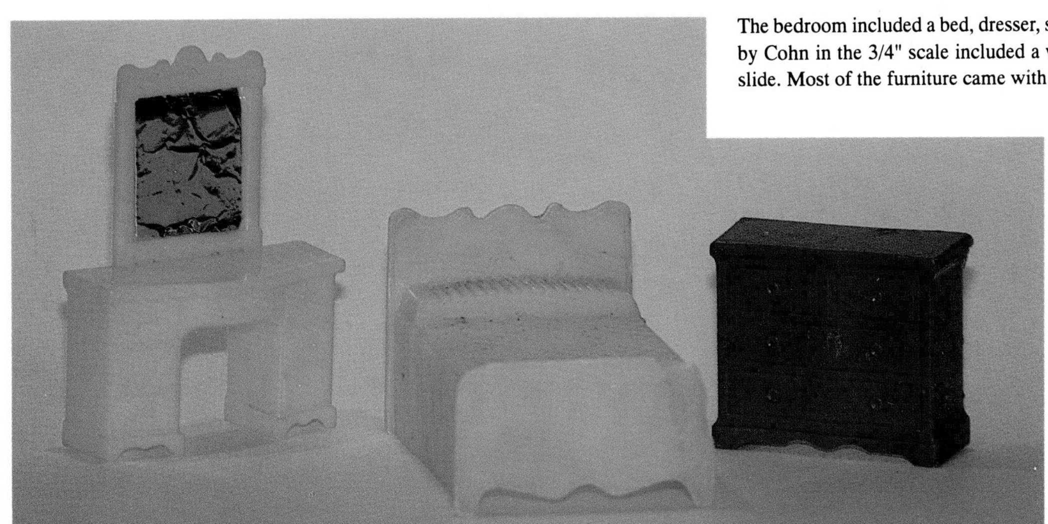

This two-story house was produced by T. Cohn, Inc. in several different colors and models. The early houses included a door that opened as well as one window that could also be opened. The later models featured windows and doors that were only printed on the metal. The houses came in pink, blue, and green.

The inside of the five-room house included a combination living room/dining room. The house measures 23" by 9 1/2" by 13 1/2" tall not including the chimney.

Four room ranch house produced by Cohn in the early 1950s. This house included a large landscaped patio. It is 25" by 16 1/4" by 10 1/2" tall (including patio). The chimney has been replaced. The house came furnished with the 3/4" scale Superior furniture, as well as a swimming pool and slide.

Several years later another ranch house was marketed by the T. Cohn company. It came in either a three- or four-room model. Some of the inside decor was the same as that of the earlier ranch house.

The inside of the three-room ranch house is pictured. The house is missing its chimney. The four-room model of this metal house was still being sold by T. Cohn, Inc. in 1961.

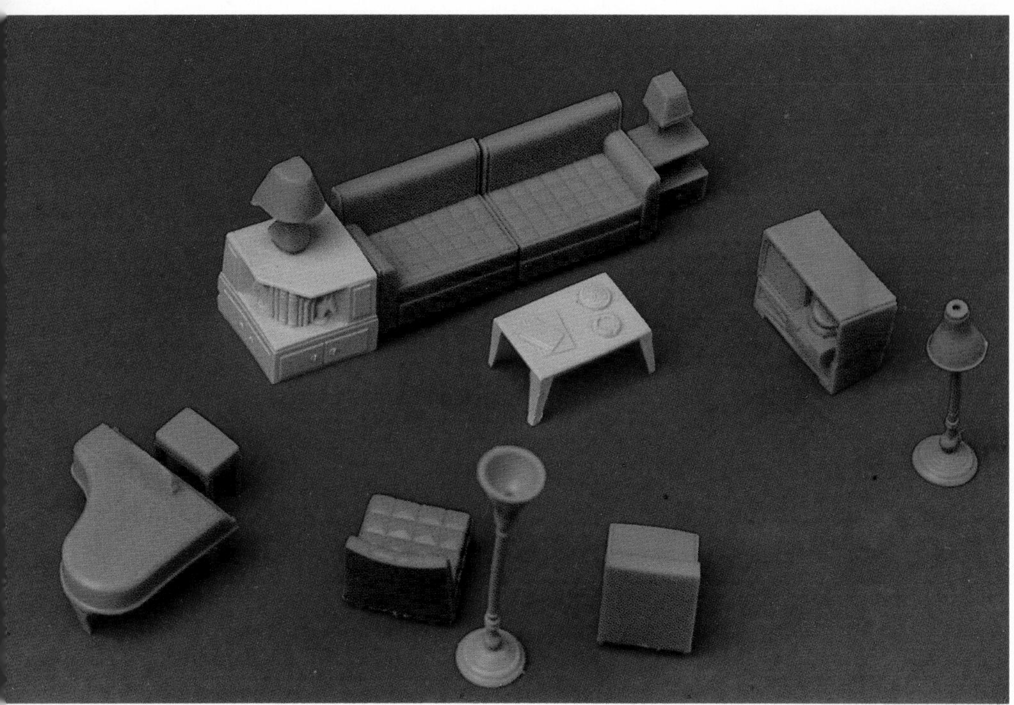

A smaller 1/2" to 1' scale of plastic furniture was also produced by T. Cohn, Inc. These pieces are generally labeled "Superior" inside a pennant. Pictured is the furniture that was made for the living room. Furniture from the collection of Marcie Tubbs. Photograph by Bob Tubbs.

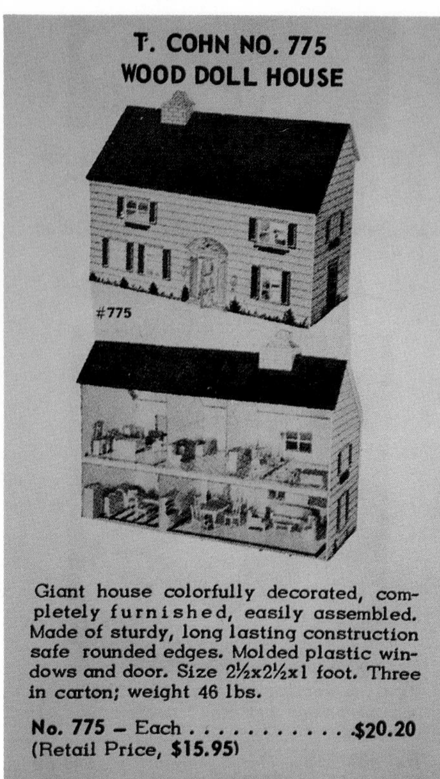

T. COHN NO. 775
WOOD DOLL HOUSE

#775

Giant house colorfully decorated, completely furnished, easily assembled. Made of sturdy, long lasting construction safe rounded edges. Molded plastic windows and door. Size 2½x2½x1 foot. Three in carton; weight 46 lbs.

No. 775 – Each **$20.20**
(Retail Price, **$15.95**)

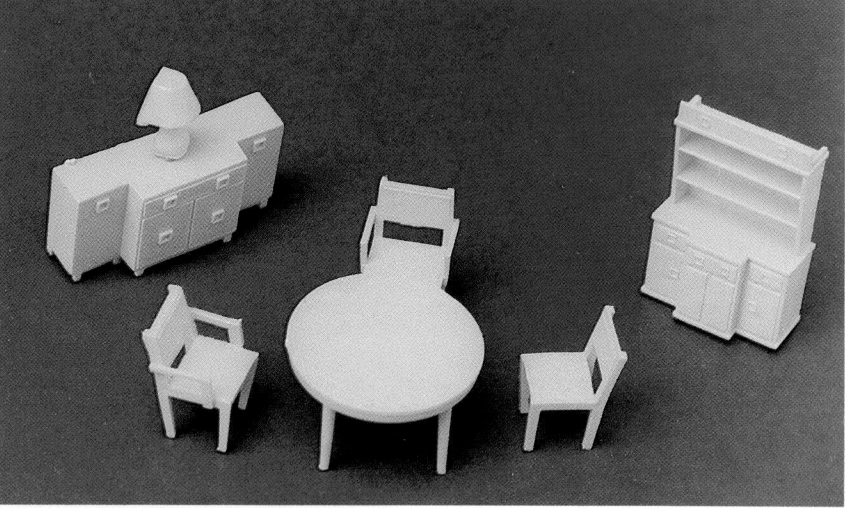

Pictured is the 1/2" scale of Superior dining room furniture. From the collection of Marcie Tubbs. Photograph by Bob Tubbs.

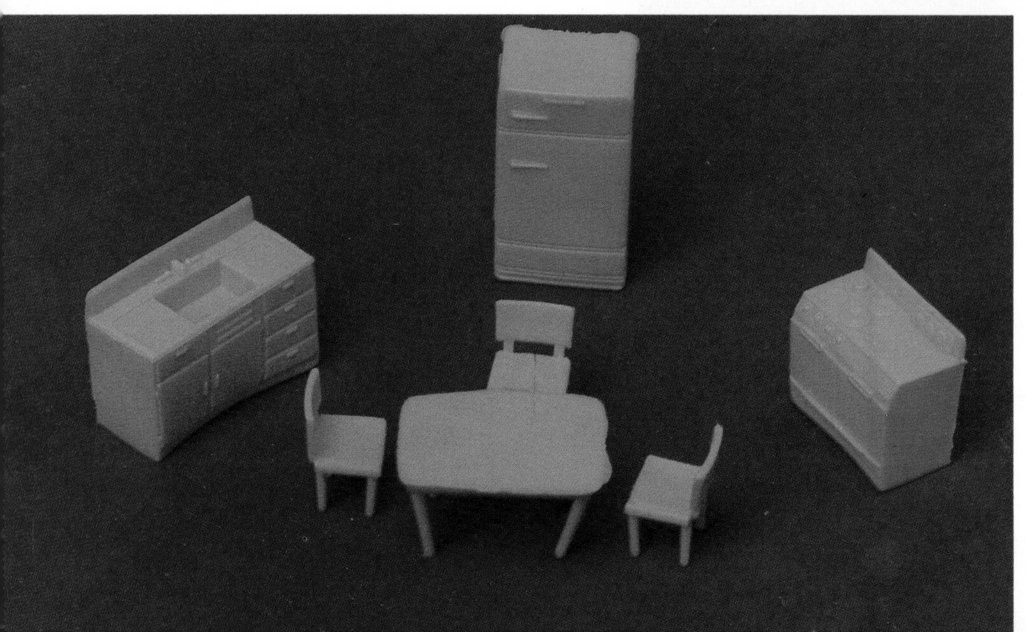

The 1/2" scale of Superior kitchen furniture included a sink, refrigerator, stove, table and chairs. From the collection of Marcie Tubbs. Photograph by Bob Tubbs.

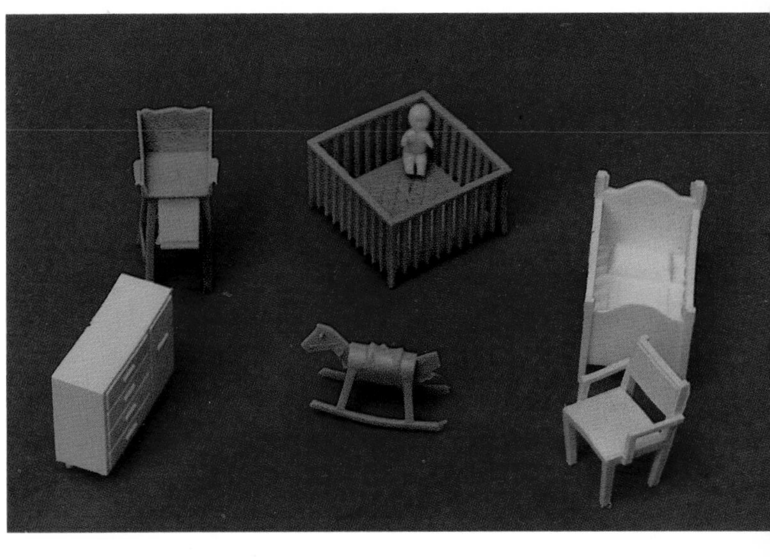

The 1/2" Superior nursery furniture even featured a rocking horse just like the later Marx sets. From the collection of Marcie Tubbs. Photograph by Bob Tubbs.

Opposite page top right:
A Cohn wood (Masonite) dollhouse in a Colonial design was pictured in the Cullum and Boren catalog for 1961. The firm was located in Dallas, Texas. The two-story house contained five rooms and measured 2 1/2' by 2 1/2' by 1'.

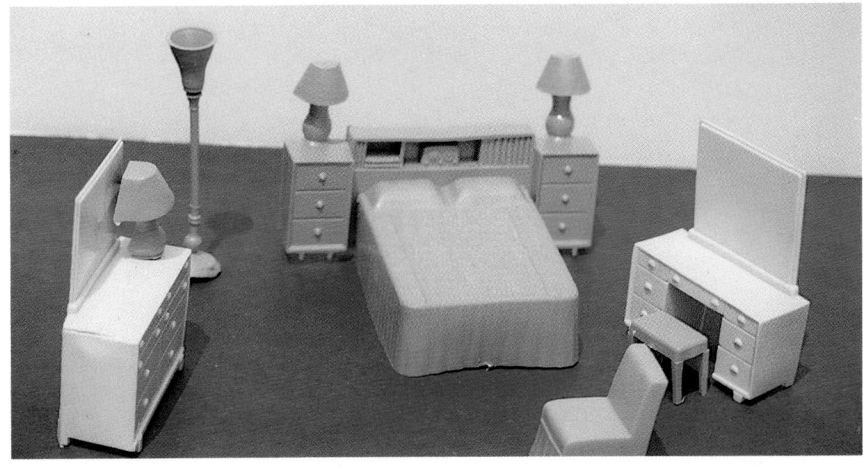

The small Superior bedroom furniture included several "half" lamps along with the other furnishings. From the collection of Marcie Tubbs. Photograph by Bob Tubbs.

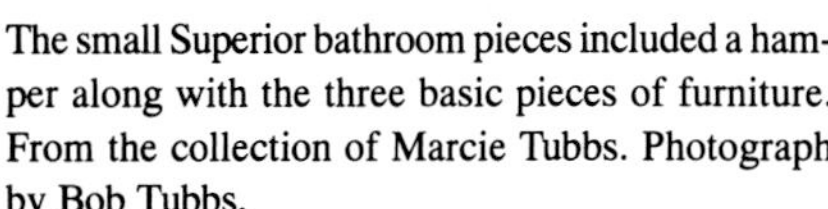

The small Superior bathroom pieces included a hamper along with the three basic pieces of furniture. From the collection of Marcie Tubbs. Photograph by Bob Tubbs.

All of the 1/2" scale of Superior furniture was made of soft plastic. Laundry furniture included a combination washer/dryer, stools, and an ironer. The outdoor furniture included an umbrella table, several lounges, and a small wheeled table. Also pictured are the living room, dining room, bedroom, bathroom, kitchen, and nursery pieces of furniture. Many of these furniture molds are identical to those used by Marx in the later years (see Marx chapter). Furniture from the collection of Marcie Tubbs. Photograph by Bob Tubbs.

Wolverine (Today's Kids)

The Wolverine Supply and Mfg. Co. was founded in 1903 by Benjamin F. Bain. The company was named after the Wolverine mascot of the University of Michigan where Bain had attended school. It began as a tool and die business. One of their early orders was for making the tools to manufacture sand toys. The firm which had placed the order went out of business before they could make use of the tools so Wolverine began the manufacture of the sand toys themselves. By 1913 they were producing gravity-action sand toys, and in the 1920s the company began making housekeeping toys. In 1959 the Rite-Hite toy kitchen appliances were added to their line.

In 1962 the company name was changed to the Wolverine Toy Co. and in 1968 another change was made when the firm was acquired by Spand and Co. The new owners felt that the old factory in Pennsylvania was outdated so new facilities were built in Booneville, Arkansas. The company made the move to Arkansas in late 1970 and early 1971. The company name was changed again in 1986 to Today's Kids and the firm is still producing toys in Arkansas.

The Wolverine Company manufactured dollhouses from 1972 to 1990. The dollhouses were made with two different materials. Beginning in 1972, metal dollhouses similar to the Marx houses were produced. This type of house was marketed by the firm until the end of dollhouse production in 1990. The other type of dollhouse produced by Wolverine was made of masonite, metal, and plastic. Different models of this type of house were produced off and on from 1972 until 1985.

There were seven different models of the Wolverine metal houses. Besides these seven basic models, sometimes different color schemes were used in the houses from year to year. The various designs of the houses are as follows: 1972 Introduction date. No. 810: Two Story Colonial House. The house contained five rooms and a covered patio. The house was made of metal with the patio cover and posts made of plastic. The house measured 23" wide by 16 1/2" high by 9 1/4" deep. The windows and doors on the house were plastic. 1972 Introduction date. No. 800: Ranch House. The one-story three-room metal house also contained a covered patio similar to No. 810. The house measured 23" wide by 10" high by 9 1/4" deep. The door and windows were plastic. 1972 Introduction date. No. 805: Town and Country Doll House. The five-room two-story house was made of metal with a plastic bow window (this was a hallmark on many of the metal houses). The house measured 22 1/4" wide by 17 1/2" tall by 12" deep. The door and windows were plastic. The house was still being sold in the Montgomery Ward Christmas catalog in 1982 at a cost of $19.99 furnished with plastic furniture. 1972 Introduction date. No. 825: Colonial Mansion. The five-room metal house measured 32 1/2" long by 12" deep by 17 1/2" high. It also had a plastic bow window. The door and the windows were plastic. The house also included a two-car attached garage. 1984 Introduction date. No. 808: Augusta Doll House. This two-story metal dollhouse contained four rooms. It measured 17 1/2" tall by 23 1/2" wide by 13 1/2" deep. The difference in this dollhouse and the Town and Country house is that it had an interesting plastic porch across the front. 1986 Introduction date. No. 810: Rosewood Manor. This five-room two-story metal house measured 17 1/2" high by 22 1/2" wide by 12" deep. The outside decoration on this house is much softer in color and gives a pastel look to the decor. This house is marked "Today's Kids." 1986 Introduction date. No. 800: Country Cottage. This is a small metal house made very much like the earlier Ranch House. It is 10" high by 23" wide by 9 1/4" deep. The house contains a combination living room/dining room and a kitchen along with an uncovered patio.

All of these houses were sold complete with plastic furniture. The furniture is not marked as to origin although most of the metal houses are marked. The plastic furniture supplied by Wolverine included the following pieces: Living room: sofa, two arm chairs, large low television, end table, and lamps. Dining room: table, four chairs, and tall china cabinet. Kitchen: round table, two chairs, refrigerator, large one-piece unit that included cabinets, sink, and stove. Bathroom: bathtub, toilet, and vanity sink. Bedrooms: bed with shelf headboard, desk and chair, chest of drawers, lamp, double bed, chest of drawers, and a double dresser with mirror. The furniture was quite cheaply made and doesn't do justice to the very nice metal dollhouses produced by Wolverine. The dolls provided to live in the houses were plastic figures.

The hardboard-masonite dollhouses produced by Wolverine also are very collectible. There were four different designs of these houses. They are as follows: 1972 Introduction date. No. 820: Colonial Doll House. This two-story hardboard house contained six rooms. It measured 24" long, 10 1/2" deep and 19 1/2" tall. It was advertised as being lithographed both inside and out. The house had a steel roof and corner moldings and came complete with furniture. 1980 Introduction date. No. 840: Traditional Doll House. This three-story house was made of masonite wood, plastic, and metal. It contained a living room/dining room and kitchen on the first floor, a bedroom and bathroom on the second floor, and a nursery on the third floor. The house featured a stairway as well as sliding doors on the second floor balcony. The house measured 27 3/4" tall by 23 3/4" wide by 15 1/8" deep. The house was scaled 1" to 1'. 1980 Introduction date. No. 835: Cape Cod. This two-story house was also produced in masonite, plastic, and metal. The house contained two rooms downstairs and two rooms upstairs with a connecting stairway. The house measured 19 1/2" high by 23 3/4" wide by 11 1/8" deep. The house had walls made of masonite, metal flooring on both floors, and the roof, chimney, windows, and door were plastic. The house was scaled 1" to 1'. 1981 Introduction date. No. 850: A-Frame Doll House. The masonite and plastic house had three levels and could be accessed from both sides. The house measured 25" tall by 22 3/4" wide by 24 1/2" deep. The house was scaled 1" to 1'. This house was advertised in the Montgomery Ward Christmas catalog in 1982 at a cost of $54.99. The house came complete with furniture and a family of dolls at this price.

Wolverine provided their own set of dolls plus furniture for their 1" to 1' scale houses. The dolls were No. 845 and were called the Bender family. The family included a father, mother, and baby doll with posable bodies and arms and legs that swiveled.

There were eight different sets of furniture that could be purchased for these larger houses. They included the following pieces: Living room: three-piece sectional sofa with cushions and cocktail table. Dining room: round pedestal table with matching swivel chairs. Kitchen: double basin sink, refrigerator (doors open), and modern range with opening oven door. Patio: adjustible reclining chaise and two chairs. Den: Sofa bed (opens), reclining chair, and console television. Bedroom: bookcase headboard and bed, nightstands, and mirrored dresser (drawers open). Bathroom: corner tub, vanity sink, and toilet. Nursery: crib, playpen, and dresser (drawers open).

Although the metal dollhouses which were produced by Wolverine are still quite plentiful, the larger masonite houses (marketed under the trade name "Rite Scale") are harder to find. One of these houses complete with the company's furniture would, indeed, be a very worthwhile purchase for today's collector.

Two-story Wolverine Colonial house first sold in 1972. The house measures 23" wide by 16 1/2" high by 9 1/4" deep. It is made of metal with a plastic cover and posts on the patio. The windows and doors are plastic.

Inside of the two-story Wolverine Colonial house showing its five rooms.

Wolverine Ranch House. This house was also introduced in 1972. The metal house measures 23" wide by 10" high by 9 1/4" deep. The door and windows were plastic. The patio was covered by a plastic awning.

The inside of the Ranch House showing its three rooms. The partitions are missing.

The Wolverine Town and Country Doll House was a good seller for the company for ten years. It was first introduced in 1972. The plastic bow window was a feature of this house. The metal two-story house measures 22 1/4" wide by 17 1/2" tall by 12" deep. The inside of the Town and Country House featured five rooms, although it is sometimes found without the upstairs partition in a four-room model.

The distinguishing characteristic of the Wolverine two-story Colonial Mansion is its attached two-car garage. This model, too, was first sold in 1972. The inside of the house was the same as the Town and Country model. The house measures 32 1/2" long by 12" deep by 17 1/2" high. Photograph and house from the collection of Rebecca Kepner.

Wolverine Augusta Doll House. This two-story house was made like the Town and Country model with the addition of a plastic front porch. The house measures 17 1/2" tall by 23 1/2" wide by 13 1/2" deep. It was first sold in 1984.

The inside of the Augusta house featured the same decor used on the other two-story metal Wolverine houses.

Rosewood Manor marketed under the new "Today's Kids" name. First sold in 1986. The five-room house measures 17 1/2" high, 22 1/2 " wide, and 12" deep. The practice of making the windows, door, and chimney of plastic continued with this model. Photograph and house from the collection of Marilyn Pittman.

This Colonial dollhouse was first on the market in 1972. It is made of hardboard and was colorfully lithographed both inside and out. The two-story house has six rooms and came completely furnished, as did all the early Wolverine dollhouses. The house measures 24" long by 10 1/2" deep by 19 1/2" high. House from the collection of Susan Jenkins Lacerte. Photograph by Norman R. Lacerte.

Plastic furniture used to furnish the Wolverine metal houses. The furniture is in approximately 3/4" scale but some pieces are a little small for that scale. Pictured is furniture for the living room and dining room. Included is a television that is nearly as large as the sofa.

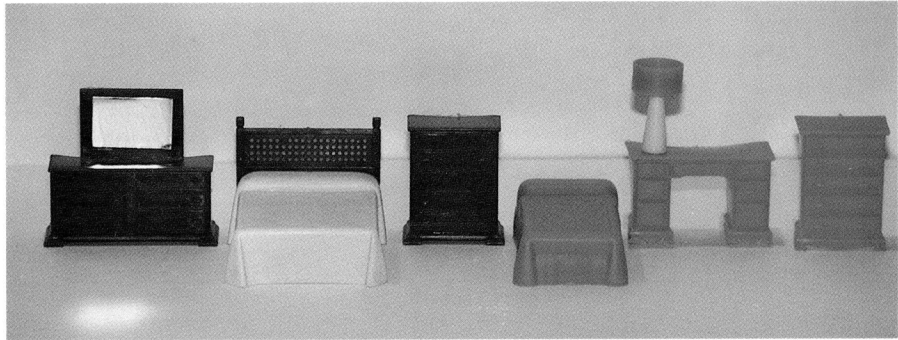

Plastic Wolverine furniture produced for the two bedrooms.

Soft plastic furniture provided by Wolverine for the bathroom and kitchen. The chair could also be used with the bedroom desk.

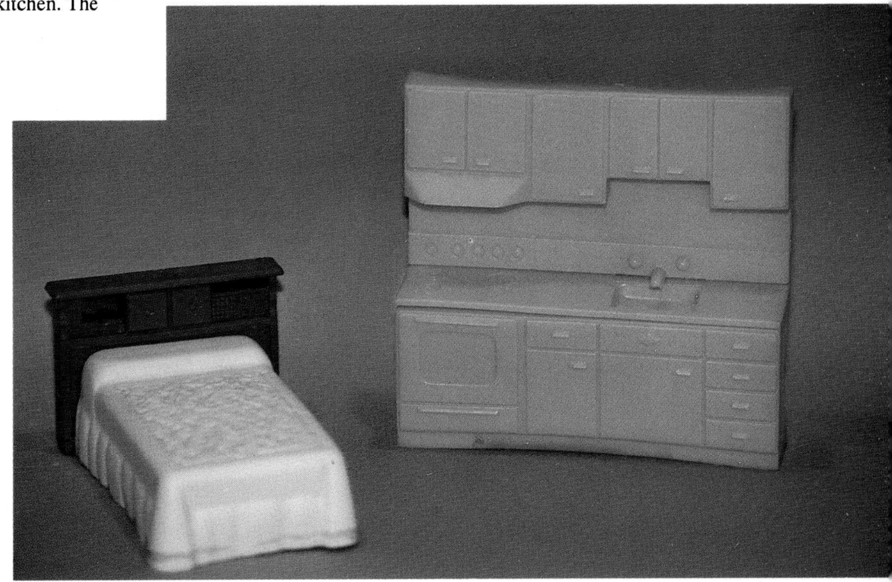

Pictured is the large kitchen unit containing both the sink and the range and a bed featuring a "built-in" headboard. None of the Wolverine furniture was marked as to origin.

Wolverine Cape Cod dollhouse first introduced in 1980. The house has masonite walls, metal flooring and a plastic roof, chimney, windows, and door. The house is scaled 1" to 1' and measures 19 1/2" tall by 23 3/4" wide by 11 1/8" deep. Photograph and house from the collection of Rebecca Kepner.

The inside of the two-story Wolverine Cape Cod house showing its four rooms. A stairway leads to the second floor. Photograph and house from the collection of Rebecca Kepner.

Very special doll houses for very special little girls

A This year the Bender family has super vacation plans that include staying in their very own tri-level vacation home!

THE FAMILY: Father, mother and baby, all poseable.

THE DESIGN: Features a sundeck with planters and shrubbery, tinted "glass" windows and balcony with sliding "glass" doors. Interior is handy to both sides.

CONSTRUCTION: Made of masonite wood/high-impact plastic. Scale: I inch to I foot. 22¾x24½x25 inches high. Easy to assemble.

THE FURNITURE: Includes sets for den, dining room, bedroom, bathroom and kitchen, plus baby crib, 2 patio chairs and chaise.

AGES: 5 years and up.

48 G 62713 M—Ship. wt. 15 lbs. .... set 54.99

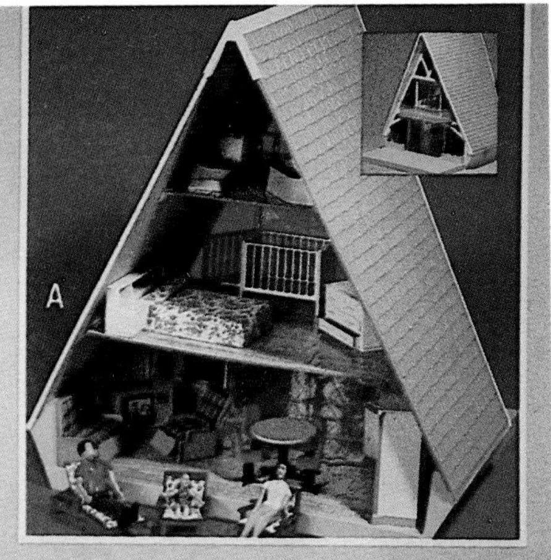

Wolverine A-Frame Doll House as pictured in the Montgomery Ward Christmas catalog for 1982. The house had three levels and could be accessed from both sides. It sold for $54.99 complete with furniture and doll family.

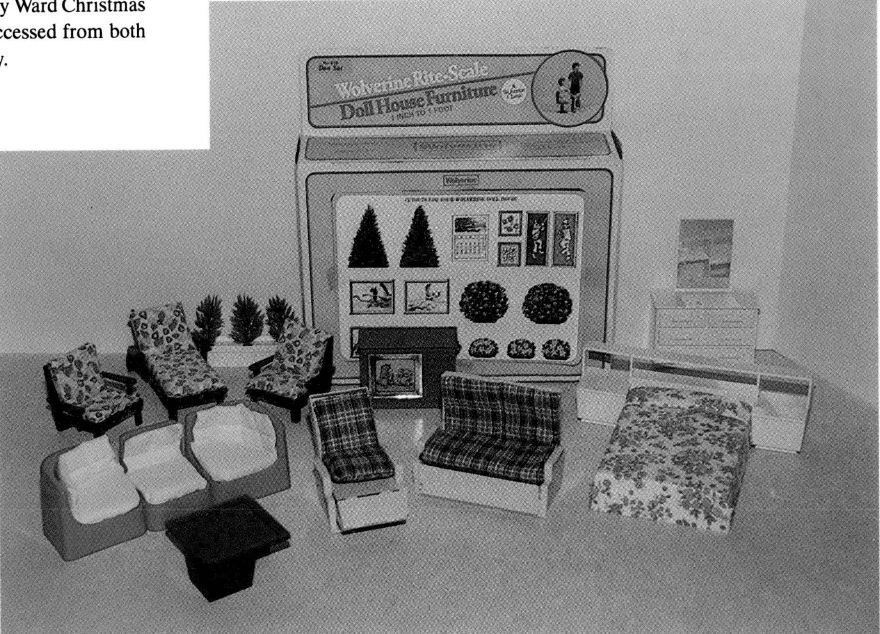

Wolverine marketed eight different sets of 1" to 1' scaled plastic furniture during the 1980s. Pictured are the sets for the bedroom, living room, patio, and den. Photograph and furniture from the collection of Roy Specht.

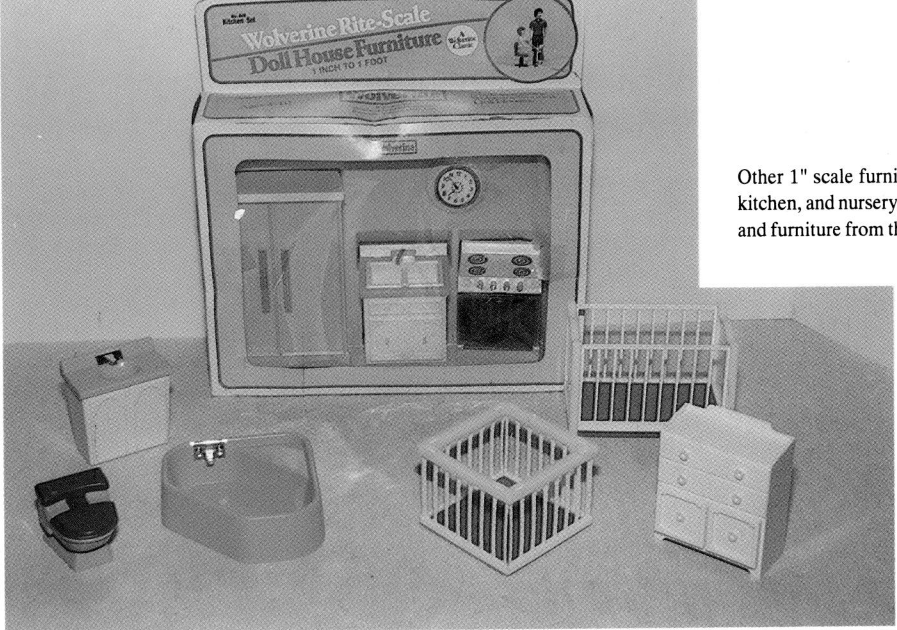

Other 1" scale furniture provided by Wolverine included sets for the bathroom, kitchen, and nursery. Not pictured is the furniture for the dining room. Photograph and furniture from the collection of Roy Specht.

Miscellaneous Plastic (Furniture and Dolls)

Although Renwal, Ideal, Plasco, and Marx produced the most collectible plastic dollhouse furniture, several other companies also manufactured furniture.

Jaydon began production of a full line of 3/4" to 1' scale furniture during World War II. Some of the plastic furniture was also marketed in boxes under the trade name "Best Maid." The dining room set included a table, four chairs, a corner hutch, and a buffet with an opening drawer. The table in this set tends to warp. The living room furniture included a sofa, chairs, coffee table, end table, radio-phonograph, and floor lamp. The kitchen set consisted of a table, four chairs, stove, sink, refrigerator, and cabinet. The bedroom featured an unusual double bed, chest, vanity, and stool. The bathroom pieces included a tub, toilet, sink, hamper, and scale. A piano and bench was also made. Additional pieces like table lamps may have been produced but since the furniture was not marked, it is difficult to identify unless it has been seen in original boxes.

The Multiple Products Corporation from Bronx, New York also produced a line of 3/4" to 1' scale of plastic furniture. This company's product was made of soft plastic during the 1960s. The furniture came boxed and was labeled "Teenettes" Interior Decorators Furniture. The bedroom was advertised as a "Realistic Italian Provincial Bedroom Suite with large double bed and separate headboard." Other sets included an eight-piece Italian Provincial living room suite, a complete Italian Provincial dining room suite, a Country Provincial kitchen and dinnette set and a bathroom with vanity and bathtub. The company also made nursery furniture. Many of these items are marked with the year "1963" as well as the company name. The other furniture is also easy to identify because the pieces are marked with the trademark MPC in a circle.

Irwin Corporation (based in New York) also made a number of small plastic toys that could be used with dollhouses. These included garden and housekeeping items. In addition the company also produced an Interior Decorator Set featured in the Sears Christmas catalog in 1964. The complete set included 236 separate pieces which provided enough furniture for five rooms. These included a dining room, kitchen, bedroom, bathroom, and living room. The furniture came with wall panels and floor coverings so the young decorator could design her own rooms. The furniture was designed to be mixed and matched with chair bottoms and cushions that could be used in several different ways. A doll family was also available to go with the set of furniture. The plastic furniture was in the 3/4" to 1' scale.

The Allied Molding Corporation based in Corona, New York produced several different rooms of furniture using a much smaller scale. The Allied pieces are a little less than 1/2" to 1'. This inexpensive furniture was sold in boxed room sets during the 1950s. It may have been used to furnish some of the more inexpensive dollhouses sold through mail order catalogs. Pictures of furnished metal houses by Jayline look as if they are furnished with Allied furniture. The sets included a kitchen, bedroom, living room, bathroom, dining room, and nursery. The company also produced a small dollhouse to house the furniture. The one-story, four-room house was designed with no roof to make the rooms easily accessible.

Banner Plastics Corp. (New York City) also produced a line of small plastic furniture similar to Allied. The company provided metal rooms to be used with the furniture. There were five rooms which could be fitted together to form a five-room house. Each room was quite small, measuring 7 3/4" by 8 5/8" by 3 3/8". The units were designed to be stacked vertically. This was the same idea used by Marx in the early Newlywed metal rooms. The rooms could be purchased separately or together to make a five-room house. Both the insides and outsides of the rooms were decorated in appropriate designs. The rooms included a bedroom, kitchen, living room, bathroom, and dining room.

The Thomas Manufacturing Corporation in Newark, New Jersey also made interesting small plastic items that can be used to supplement the dollhouse furniture made by other companies. These toys include playground equipment, wagons, strollers, and buggies. The company products are marked "Acme."

A darling set of twin "Bunky Beds" was made by Best Plastics Corporation based in Brooklyn, New York. This set of beds also included twin dolls 3 1/2" tall. The company's business was primarily party favors, toys, and novelties.

Various plastic dollhouse dolls were also produced to accompany the plastic dollhouse furniture. Besides the dolls made by the leading dollhouse furniture manufacturers, the Flagg dolls are probably the most collectible of the dollhouse dolls.

These flexible dolls were made by the Flagg Doll Co. in Jamaica Plain, Massachusetts as early as 1948 and were produced for many years. The dolls have one-piece bodies that can be bent into various positions. The Flagg dolls suitable for dollhouses came in sizes that were appropriate for both the 3/4" to 1' and the 1" to 1' scales of furniture. In the 1" scale, the father doll is 6" tall while the children measure 4 1/2" in height. In the smaller size, the father is 4 1/2" tall while the children measure 3 1/2". The dolls came in various sets and many different models were included in the packages. The basic dollhouse family was made up of parents, boy, girl, and baby dolls. In a more elaborate set, grandparents, nurse, maid, cook, and doctor dolls might be included.

In addition to the dollhouse dolls, other dolls were manufactured to represent other characters such as policemen and postmen. Flagg dolls were produced in various sizes with the largest being 8" tall.

A less well-known set of plastic dolls was made by the EFFanBEE Doll Co. in 1952. These very attractive hard plastic dollhouse dolls have molded hair and painted eyes. The firm produced a father, mother, daughter, and baby doll in the set. The dolls range in size from 2 1/2" to 3 3/4" to 4 3/4". The limbs are joined to the body with string. The baby has curved legs. The dolls are marked on the back "F An B/Made in Canada." Since the dolls are marked, they are easily identified. These dolls are hard to find but they do make very nice additions to a dollhouse collection.

A collector of plastic dollhouse furniture can begin with the major brands of the lines of furniture and then expand to include the more elusive brand names like Jadon, Allied and Banner. This can provide the collector with a hobby that is never ending.

Pictured are double beds and a chest that were part of the Jadon line of furniture. A vanity and stool were also made for the bedroom. Photograph and furniture from the collection of Roy Specht.

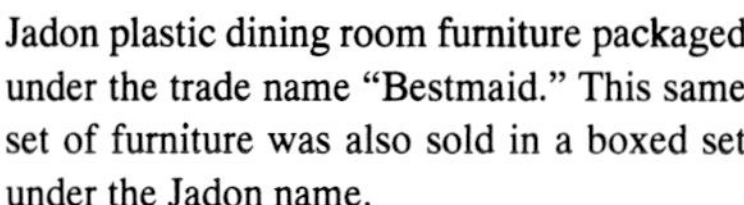

Jadon plastic dining room furniture packaged under the trade name "Bestmaid." This same set of furniture was also sold in a boxed set under the Jadon name.

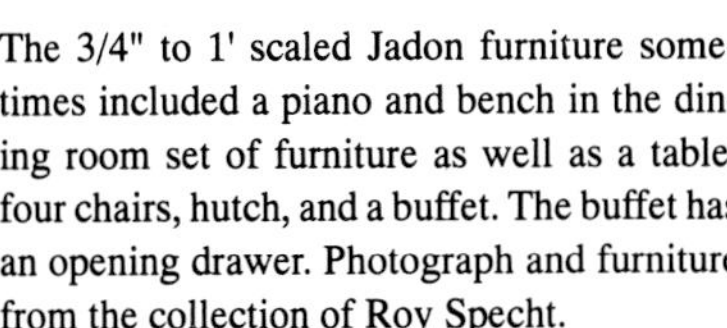

The 3/4" to 1' scaled Jadon furniture sometimes included a piano and bench in the dining room set of furniture as well as a table, four chairs, hutch, and a buffet. The buffet has an opening drawer. Photograph and furniture from the collection of Roy Specht.

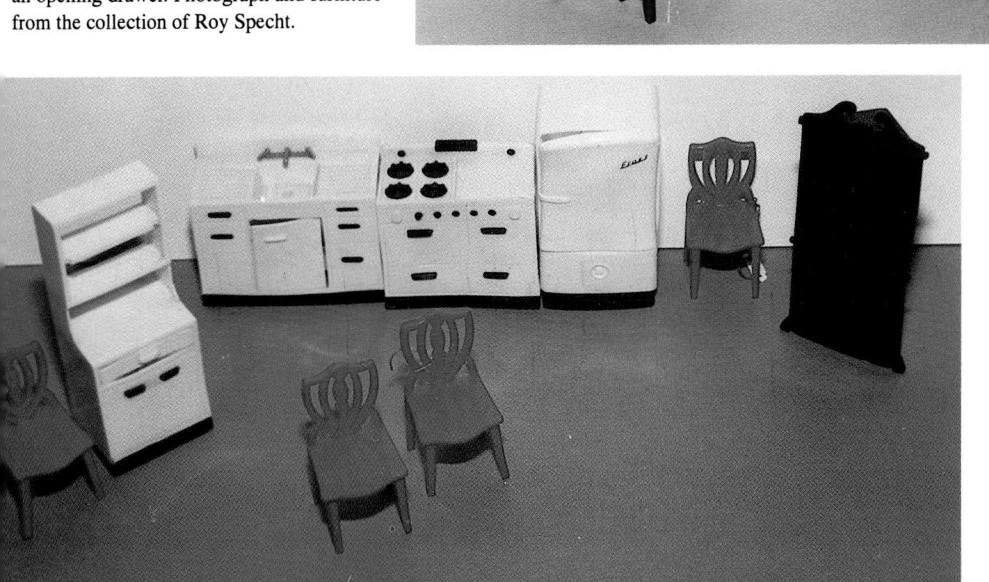

Jadon plastic kitchen furniture which featured several pieces with moving parts. The kitchen also included a table. Photograph and furniture from the collection of Roy Specht.

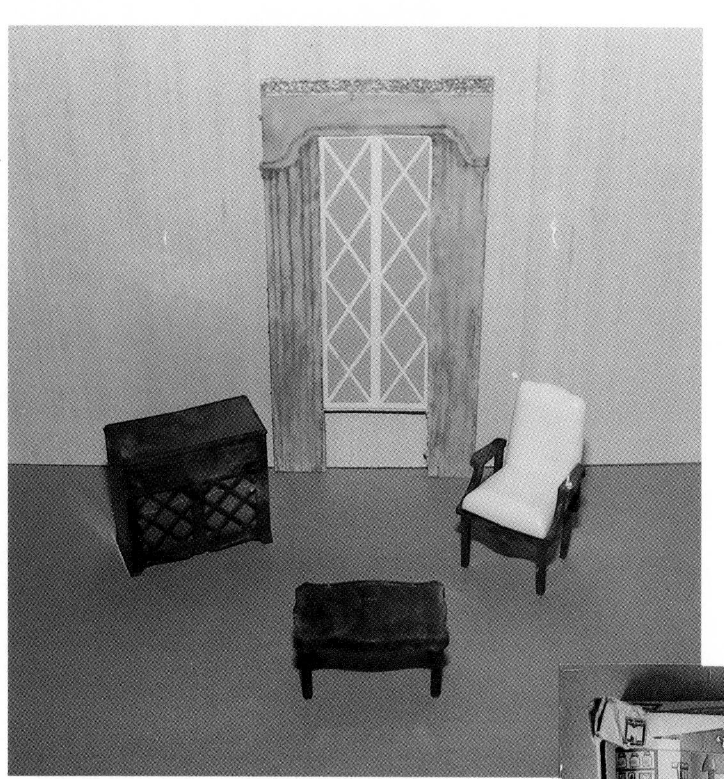

Jadon living room furniture which was first made during World War II. A sofa, lamps, and end table were also produced for this set. None of the Jadon furniture is marked. Photograph and furniture from the collection of Roy Specht.

Boxed Multiple Products Corporation furniture produced in Bronx, New York. This furniture is also in the 3/4" to 1' scale. The furniture is made of soft plastic and dates from the early 1960s. The kitchen pieces include a table, four chairs, sink, stove, refrigerator, and the figure of a woman to "live" in the kitchen.

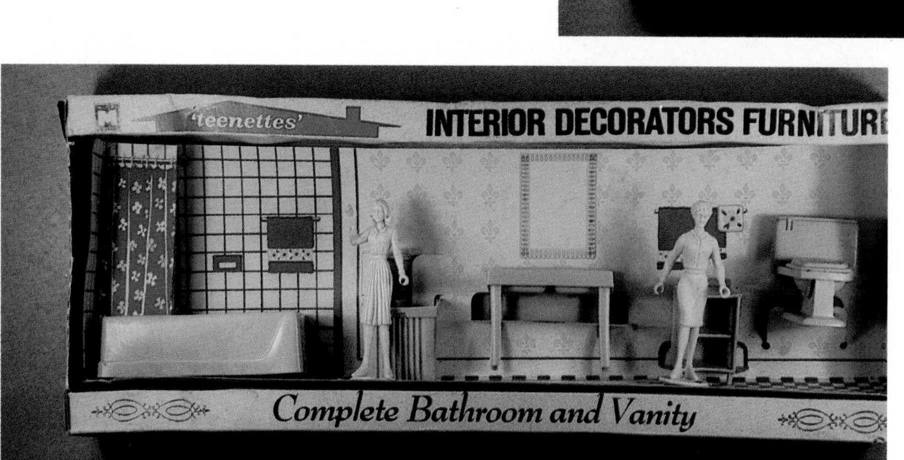

Boxed bathroom from the Multiple Products Corporation. Each piece of furniture is marked "MPC" in a circle.

Living room and nursery pieces from the Multiple Products Corporation. There were originally eight pieces in the living room set of furniture.

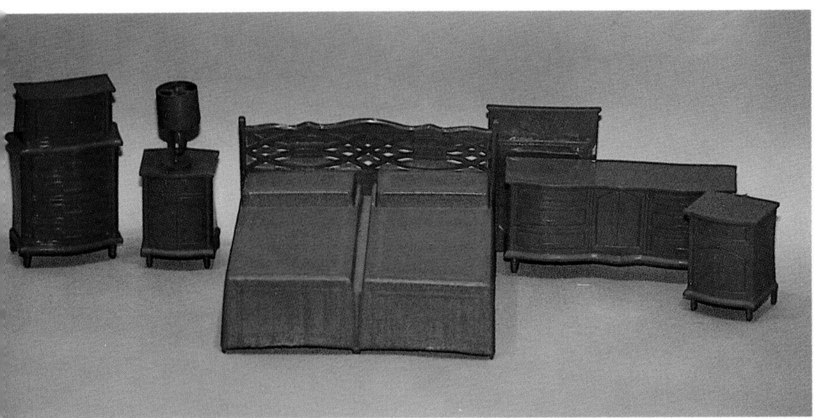

Several of the bedroom pieces of furniture from the Multiple Products Corporation are pictured. There were probably eight pieces of furniture in this set which may have included plastic figures to accompany the furniture.

Pictured is one of the small sets from the Irwin Decorator set from 1964. The dolls could also be purchased to accompany the different rooms.

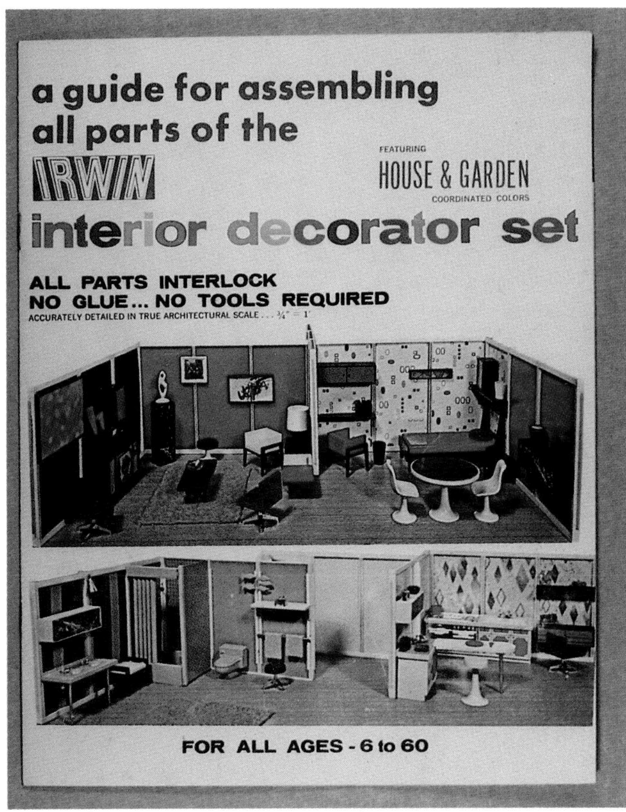

Flyer which accompanied Irwin's Interior Decorator Set from 1964. The set included plastic furniture as well as walls and accessories to make dollhouse rooms.

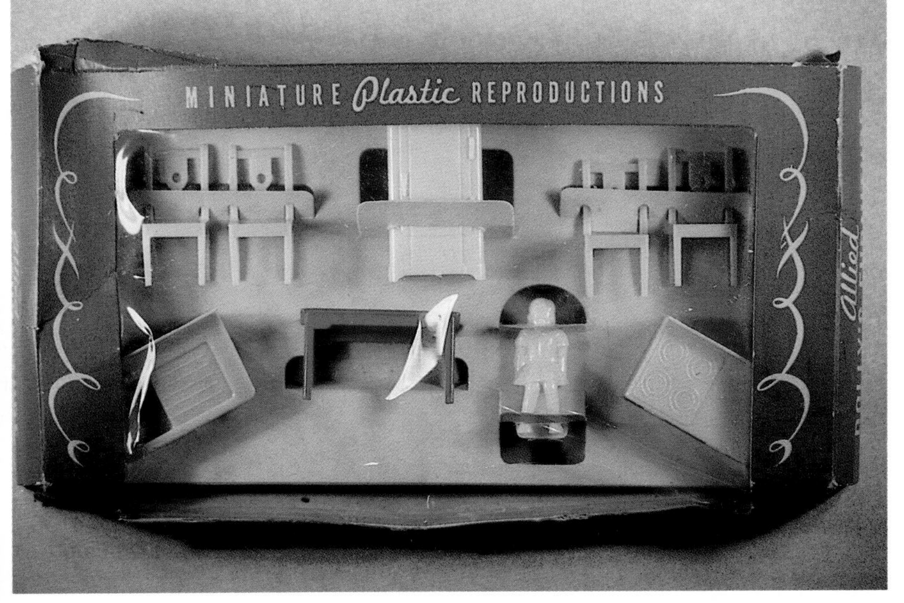

The Allied Molding Corp. also produced very small plastic dollhouse furniture. It was just a little smaller than 1/2" to 1' in scale. Several different designs of boxes were sold with the furniture. The Allied kitchen furniture included a table, four chairs, stove, sink, refrigerator, and small plastic girl.

This early Allied box pictures all the other rooms of furniture, as well as the dollhouse, on its back panal.

Small metal dining room made by Banner Plastics Corp. The room measures 7 3/4" by 8 5/8" by 3 3/8". From the collection of Eleanor O'Neill. Photograph by Betts Darnell.

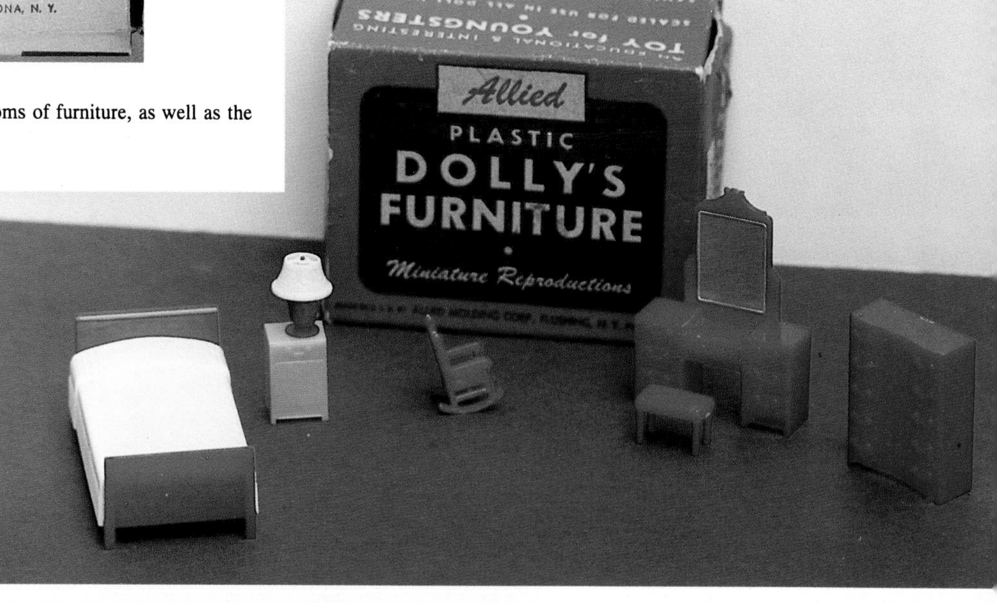

The boxed Allied bedroom furniture included a bed (with removable spread), lamp, nightstand, rocker, vanity, bench, and chest. Furniture from the collection of Marcie Tubbs. Photograph by Bob Tubbs.

Miscellaneous pieces of Allied furniture, including a radio, chair, coffee table, dining room chair, and two dining room cabinets.

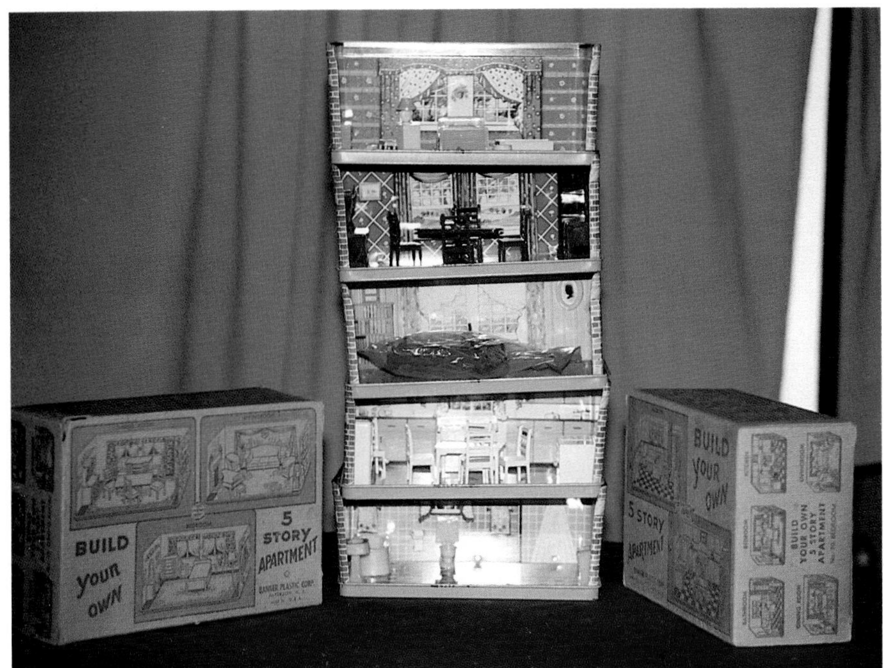

Banner produced five of these small rooms. They included a bedroom, kitchen, living room, bathroom, and dining room. The rooms were furnished with plastic furniture. From the collection of Eleanor O'Neill. Photograph by Betts Darnell.

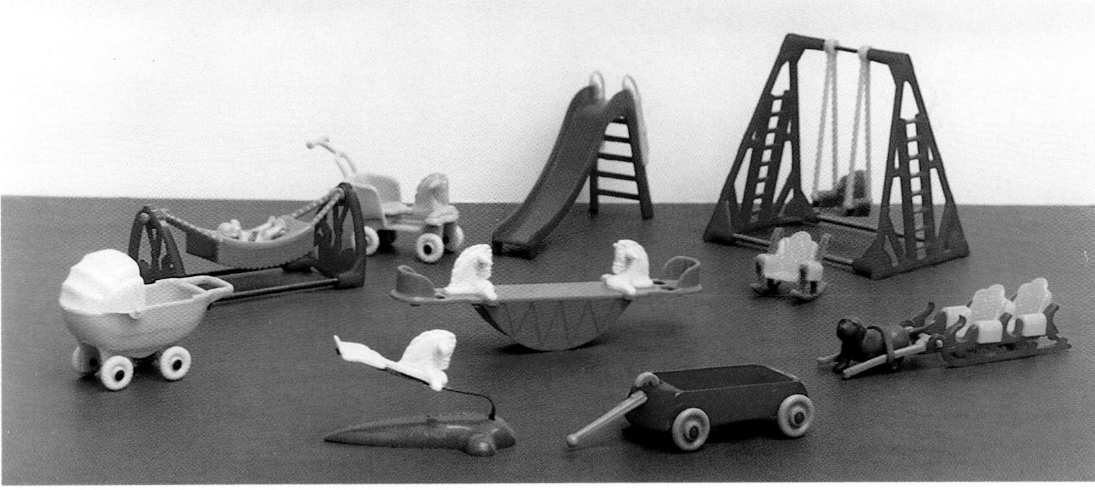

Thomas Manufacturing Corp. produced many different plastic toys that make wonderful additions to the dollhouse furniture made by other companies. The items are marked "Acme." Toys from the collection of Marcie Tubbs. Photograph by Bob Tubbs.

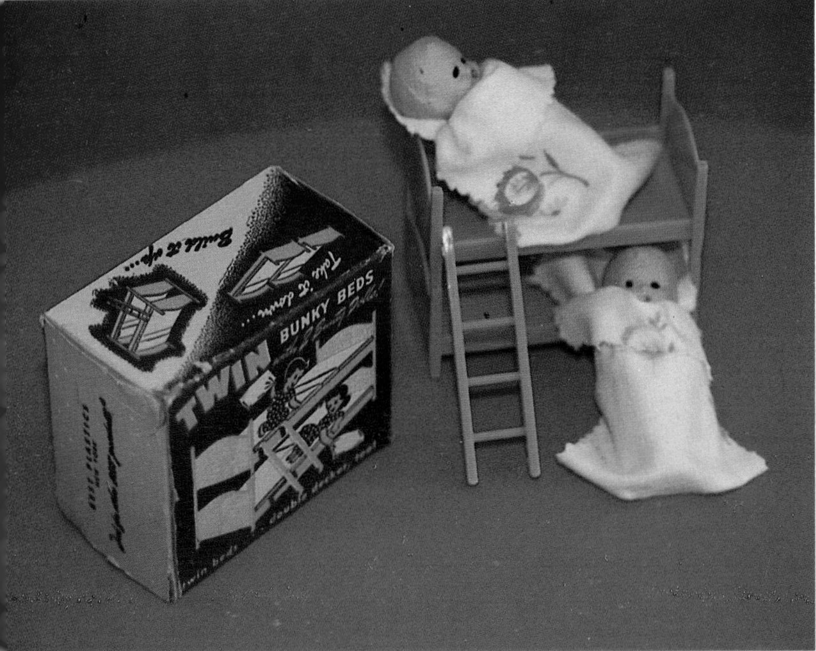

Best Plastics Corporation, best known for the party favors they produced for many years, also was responsible for a set of twin "Bunky Beds" that make nice collectibles for dollhouse collectors. The beds also came with twin babies 3 1/2" tall. Beds from the collection of Marcie Tubbs. Photograph by Bob Tubbs.

These "Bending Doll House Dolls" were advertised in the *House Beautiful* magazine in October, 1948. A five-member family of these dolls sold for $4.25 at that time. The dolls range in size from 2" to 3 1/2" tall and are made of plastic. Probably made by the Flagg Doll Co.

The Flagg Flexible Play Doll was sold from the late 1940s until the 1960s. These dolls are pictured in their original boxes. Although the larger doll is still wearing her original clothing, the smaller dolls have been redressed. Photograph and dolls from the collection of Patty Cooper.

This large assortment of Flagg plastic dolls includes a black doll as well as the more usual models. Circa 1950s. Dolls and photograph from the collection of Patty Cooper.

A more recent set of Flagg family dolls in their original clothing. Photograph and dolls from the collection of Patty Cooper.

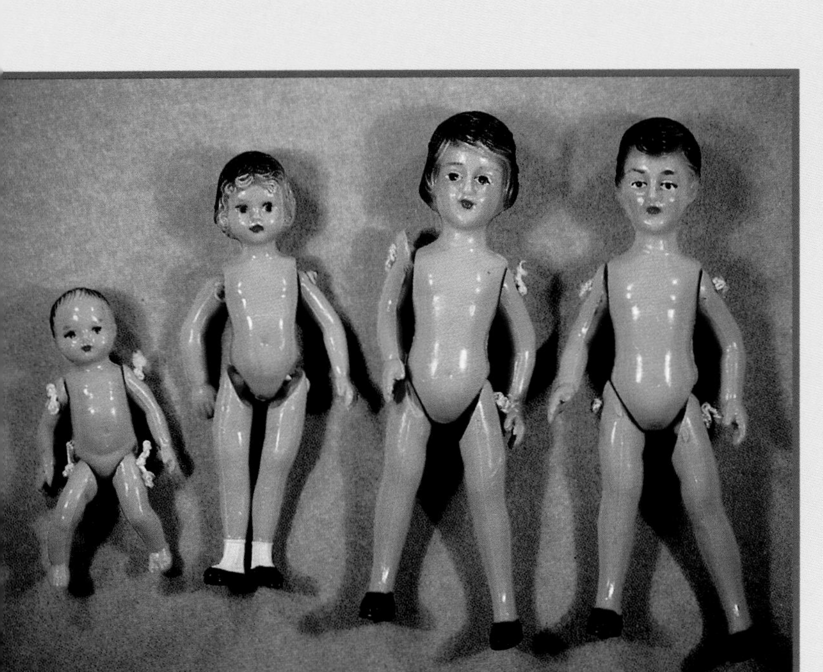

A hard plastic family of small dollhouse dolls was produced by EFFanBEE Doll Co. in 1952. The dolls have molded hair and painted eyes. The limbs are joined to the body with string. They are marked on the back "F AN B/Made in Canada." The dolls measure from 2 1/2" to 4 3/4" tall. From the collection of Marge Meisinger.

Miscellaneous

Miscellaneous Doll Houses

Besides the major companies of Marx, T. Cohn, and Wolverine, many other firms produced metal dollhouses.

One of the earliest of these companies was the Frier Steel Co., located in St. Louis, Missouri. This firm produced at least three different models of dollhouses. The houses are quite heavy but were made so well that they are still found in good condition today. The two-story houses contain from four to six rooms and all include a metal staircase. The front opens for play. The houses apparently date from the 1930s.

At a later date, Jayline Toys, Inc., located in Philadelphia, also produced metal dollhouses. These houses, from the 1950s, were very inexpensive. A furnished five-room house, with garage, was advertised in the Aldens' Christmas catalog in 1950 for $2.69. The house came with forty-one pieces of plastic furniture. The house measured 25" by 7" by 16" tall.

Jayline had earlier produced houses made of Masonite. In a 1945 ad in *Toys and Novelties*, Jayline pictured five different doll houses from their line. Four of the houses were made of masonite. The houses came knocked down and had to be assembled by the purchaser. The houses were priced from $2.00 to $5.00 each. The pictured houses are as follows:

No. 451: Plaza porch front with two trees. Contains two stories with four rooms. Measures 18" by 10" by 16" tall. Has four vertical shuttered windows on front plus a round window or design on the upper story.

No. 453: Colonial house with four columns in front and a triangle decoration above the columns. The house included six rooms in a two-story model. The three upstairs windows and two downstairs windows are vertical in design. The house measures 30" by 12" by 19" high.

No. 450: Tudor style house in a two-story, four-room model. One window is located in the peak and the door is in the middle. The house measurements were 24" by 8" by 16" high.

No. 452: Bungalow design. one-story, five-room house with decorated rooms. The roof lifts off. The house also has a porch. The house is almost square and is two rooms deep. The house measures 24" by 21" by 14" high.

No. 44: This house is made of cardboard and sold for only $1.00. The four-room, two-story house contained two shuttered windows upstairs and one window downstairs. The doorway was covered. The house measured 11 1/4" by 18 1/2" by 16 3/4" high.

One of the most collectible of the Jayline Masonite houses is the Blondie Bumstead Homestead. The house measures 12" by 17" by 22" long. Sold with the house were the cardboard figures of Blondie, Dagwood, Alexander, Cookie, the postman, and the dog Daisy and her pups. Several signs also were included with the house. Over the front bay window was printed "Blondie." At the bottom of the front door was the name "Daisy" (the dog's door). The outside of the house was finished in white siding with red shutters. The roof was shingled in two-tone blue. The inside had four rooms with printed floors. The house was featured unfurnished in the Sears Christmas catalog in 1947 and sold for $3.98.

The National Can Corporation, based in New York, was responsible for other metal houses called the Playsteel Doll Houses. These metal houses were on the market in 1948. The lithographed, two-story houses contained five rooms. Two different models of the houses were made. One had a red roof and the other roof was blue. The inside decoration of both houses was the same. One model contained a door that opened and the other did not. The houses measure 22" by 12" by 19" high.

The Ohio Art Co., located in Bryan, Ohio, is well known for its high quality metal toys which the company manufactured over several decades. The firm, however, produced only one dollhouse. The tiny house was offered in 1949 furnished with twenty-eight pieces of plastic furniture. The house was only 5 1/4" tall by 6" wide by 2" deep. The two-story house contained four rooms and a terrace. Although collectors refer to this house as the Midget Manor, the company's catalog for 1949 does not give the house a name.

During the 1940s, another series of houses appeared to be popular with the public. The Sears Christmas catalogs from both 1945 and 1946 featured these houses made of Tekwood. The unique feature of these dollhouses is the decoration of the inside of the house. The floors and walls are finished with green accents including curtains, pictures, and rugs. The advertising states that the houses are made of Tekwood (wood center with heavy craft paper covering). The decoration of the outside of the houses also includes painted bushes. The casement windows and doors are scored to open and close. The houses contain six rooms and were made in several different designs during these years. They include Colonial, Tudor, and a model with four, two-story columns. Some of the houses were also produced at a cheaper price with no inside decoration (brown natural finish).

Throughout the century, interesting dollhouses have continued to be manufactured. During the middle 1970s, a very nice plastic dollhouse was produced by Tomy. The Tomy Smaller Homes house and furniture were made in the 3/4" to 1'scale. Although these products were manufactured in Japan, the company was headquartered in Carson, California beginning in 1973. The house contains four large rooms. The furniture included kitchen, bedroom, living room, and bathroom pieces. A family was also made to live in the house. The jointed plastic dolls included a father, mother, daughter, and son. The plastic furniture contained working doors and drawers and is very realistic.

The 1980s provided many new and unusual dollhouses for the dollhouse collector. One of the best is the "Sounds Like Home" dollhouse. The house was marketed by Craft Master/ Fundimensions (a Division of General Mills Toy Group) in 1982 through the regular toy outlets.

This wonderful two-story plastic house came with seven sounds, electricity, six rooms of furniture, and a doll. The house also included a cellar door unit which housed an electronic keyboard to activate sound effects. Birds could be heard chirping, a telephone rang, a clock ticked, and a doorbell also rang. The special sound furniture pieces were as follows: a grandfather clock with a random gong sound, an alarm clock with a buzzer alarm, a shower that made the sound of running water, a working music box, a kitchen sink with the sound of running water, a stove that made the sound of eggs frying or a teapot whistling, and a piano that played "There's No Place Like Home." The sounds were powered by either house current or a nine volt battery. The houses could be purchased with seven sounds, electricity, and the Lori doll or in the deluxe model with six complete rooms of furniture.

The furniture came in 3/4" to 1' scale and included the following sets: No. 5-54972: Dining room with electronic grandfather clock, dining table, four chairs, dinnerware, silverware, wine glasses, candles, holders, wallpaper, and carpet. No. 5-54983: Dining room with lighted breakfront, and special pieces to the dinnerware. No. 5-54987: Lori's room with electronic music box, dresser, miniature dollhouse and tennis racket. No. 5-54979: Lori's room with electric lamp, bedclothes, bed, stuffed dog, doll, skateboard, skates, books, wallpaper, and carpet. No. 5-54976: Master bedroom with electronic alarm clock, night table, bed, bedclothes, accessories, wallpaper, and carpet. No. 5-54988: Master bedroom with electric lamp, dresser, mirror, candle, and vase. No. 5-54975: Kitchen with electronic stove, table, four chairs, teapot, fry pan with eggs and bacon, other accessories, wallpaper, and floor covering. No. 5-54986: Kitchen with electronic sink, refrigerator, and accessories. No. 5-54978: Living room with electronic piano, bench, club chair, accessories, wallpaper, and carpet. No. 5-54984: Living room with electric lamp, table, couch, magazine rack, magazines, and candy jar. No. 5-54980: Bathroom with electronic shower, toilet, medicine cabinet, scale, hair dryer, accessories, wallpaper, and floor covering. No. 5-54985: Bathroom with electric light, sink and vanity unit, chair, and accessories. No. 5-54981: Mom and Dad dolls (Dad 4 1/2", Mom 4 3/8"). No. 5-51262: Plants.

The furnishings for the "electronic" dollhouse are very attractive and the pieces were functional with working doors and drawers. The side of the house opens to provide access to four of the rooms. The house measures 16 1/2" deep, 24" wide, and 21 1/2" high (including chimney). The house was apparently only produced for one year. Because of the high price of the completely furnished model, the houses must not have sold well. Because of this scarcity and the high quality of the product, the "Sounds Like Home" house is one of the most sought-after doll houses from the 1980s.

Other interesting houses produced during the 1980s included the Littles House made by Mattel, Inc. in 1981 and 1982. The house was Victorian in style and was 1/2" to 1' in scale. The Mattel Littles dolls are 2 1/2" to 3" tall with large heads and small bodies. The girls have rooted hair while the boys' hair is painted. The dolls are marked "© M.I. 1980." The family included a mother, father, four daughters (Daphne, Flossie, Belinda, and Hedy), a baby, and a son (Kenny). The mother, father, and baby were sold together. The other family members could be purchased accompanied with packages of furniture which contained either one piece or an entire room of furnishings. Many of the furniture items were made of metal. The furniture included the following pieces: bathtub, sink, toilet, bed, dresser, armoire, cradle, sofa, chair, footstool, tilt top table, lamp, plant, drop leaf table, four chairs, stove, and sink/icebox. Accessories included rugs, plants, pictures, lamps, logs, pillows, wagon, rocking horse, and teddy bear. Items apparently added later that are very hard to find include a picnic table and two benches, a porch swing, and a piano and stool.

Fisher Price also produced several interesting dollhouses during the 1980s. A yellow plastic Victorian house was advertised by the company in 1984. The house contained three floors, including the attic. There was also a spiral staircase, a balcony, and lights which worked with batteries. Plastic furniture pieces sold to accompany the house included a toilet, lavatory, shower, refrigerator, stove, sink, cradle, wardrobe, rocking horse, "brass" bed, dresser, table, and chairs. The furniture contained workable drawers and doors. The house measures 15 1/2" by 13 1/2" by 23 1/2" high. Advertising stated the house was 1" to 1' in scale.

Some of the most interesting commercial dollhouses manufactured in the 1990s are the Playmobil 1900 doll houses. The plastic Victorian houses came in several different models. The dolls made for the house are 3" tall. The house that stands 24 1/4" tall contains five rooms plus an attic. It also came with stairs, a balcony, and sun patio. Many different play sets were designed to be sold to furnish the house. Included were the following items: bedroom furniture, a musical piano, patio suite, kitchen suite, child's bedroom, doll assortment, family dolls, and living room. A Model T car and a coach were also available for family transportation. Because the house was quite expensive ($175 unfurnished), it may well be a very desirable collectible for the dollhouse collector of tomorrow. A smaller house measuring 28 7/8" by 8 3/8" by 16 3/8" high was also produced. This three-room house sold for $105. By using an extra component, an even larger Playmobil house could be created which contained seven rooms plus an attic.

Although many other companies continue to produce dollhouses, primarily for the larger fashion dolls, most of the new dollhouse production centers on the kits made for the adult collector. Entire stores are now focused on merchandise for this new breed of collector. The different designs of houses being developed by the various companies, as well as the new miniatures and furniture being sold to furnish these houses, makes this field of collecting a never ending hobby.

Many collectors of the old dollhouses welcome this new phase of dollhouse collecting because it allows them an opportunity to add new accessories to complement the furnishings of an old house.

Perhaps these new collectors will also begin to add the older dollhouses and furniture to their collections. If that happens, the older houses will become even more in demand and the search to find the "perfect house" will become even harder.

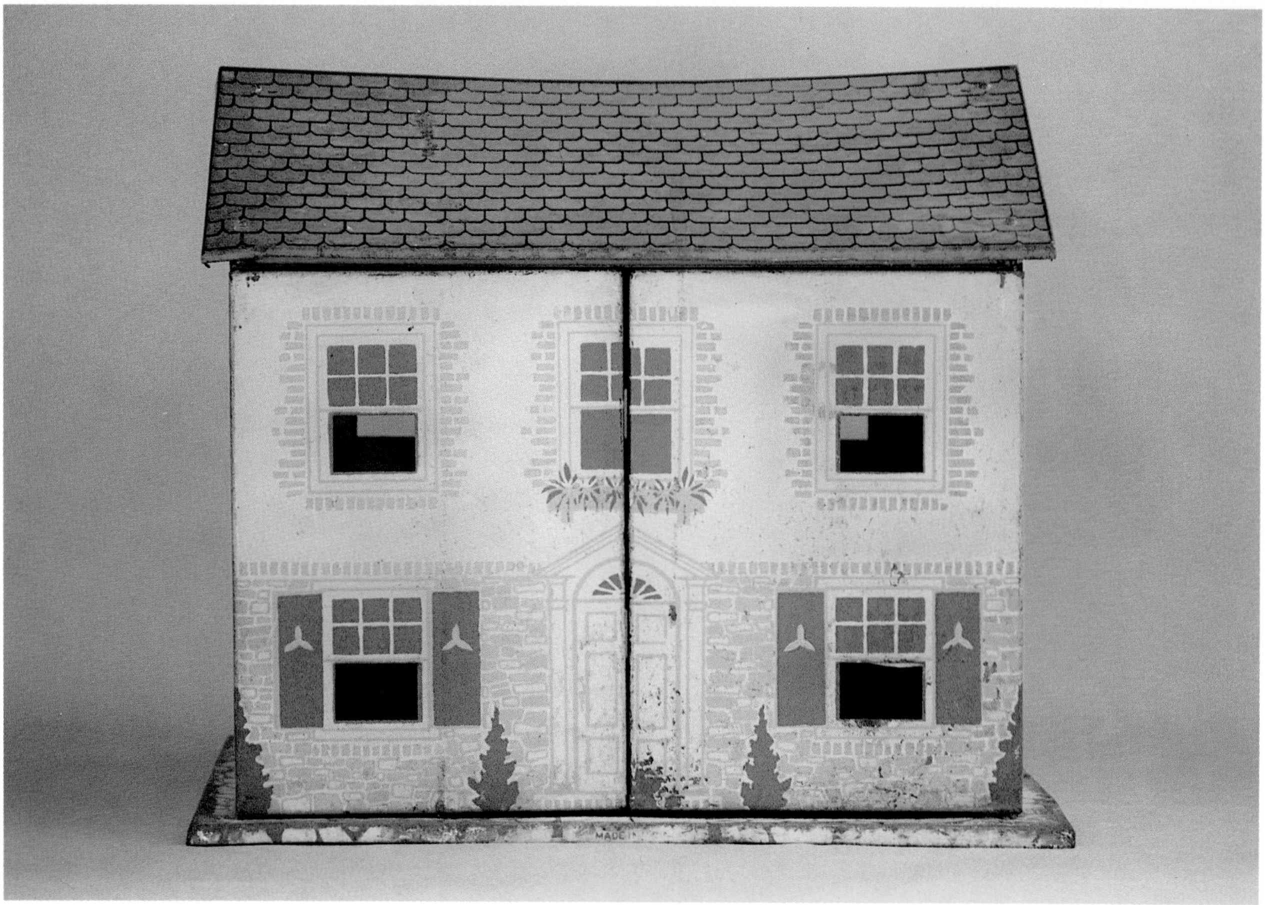

Metal dollhouse made by Frier Steel Co. Marked "Cozy Town Manor/Frier Steel Co./106 Washington Blvd./St. Louis, Mo./Pat. Appl'd For." The house measures 20 1/2" by 16 1/2" by 18" tall.

The Cozy Town Manor contains four large rooms and a metal stairway. The bottom sections of the windows are cut out. The floors are finished with a brown and cream pattern and the walls are also cream. Circa 1930s.

Another model of a Frier Steel house. The front opens to provide access to the rooms inside. This house has additional partitions in order to create a six-room house. Photograph and house from the collection of Patty Cooper.

This Jayline metal house dates from 1950. A similar model was also produced by the company with an added garage. This house measures 18 1/2" by 6 3/4" by 16 1/2" tall. The house is in the 1/2" to 1' scale.

1. COMPLETELY FURNISHED 5 ROOM STEEL DOLL HOUSE WITH PATIO AND GARAGE! 2.69

The small Jayline house contains five rooms. These houses were sold very cheaply through mail order outlets like Aldens. A furnished metal Jayline house could be purchased for $2.69 from Aldens' Christmas catalog in 1950.

New! Blondie Bumpstead Homestead . . . 4 rooms

A doll house just like the Bumpstead home in the comics with Blondie, Dagwood, Elmer, Cookie, postman, Daisy and her pups, and signs! Strong, durable Masonite Presdwood, colorfully decorated. Prefabricated construction—sets up in a jiffy. Side porch, large bay window; other windows of transparent acetate material. Separate door for Daisy and pups. **$3.98**

49 N 2160—About 12x17x22 inches long overall. Shpg. wt. 5 lbs. . . . **$3.98**

The Jayline Blondie Bumstead house was pictured in the Sears, Roebuck and Co. Christmas catalog in 1947. The house was made of masonite and measurered 12" by 17" by 22" long. The house came with figures of the famous comic characters.

This Playsteel metal Colonial dollhouse was marketed in 1948 by the National Can Corporation. The house measures 22" by 12" by 19" high. Photograph and house from the collection of Betty Nichols.

Another model of the Playsteel house featured a blue roof and fieldstone and wood siding on the front. The company advertising referred to this model as the Bucks County House. It sold for $3.98 in 1948. House from the collection of Kathy Garner. Photograph by Bill Garner.

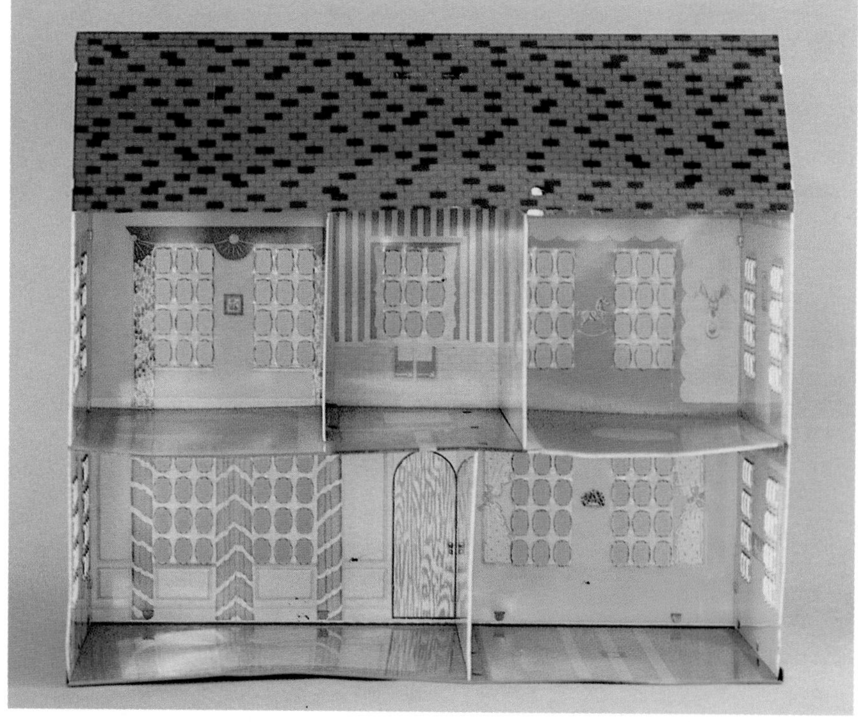

The Playsteel house came in two different models but both houses contained the same decor in their five rooms.

Ohio Art Co. tiny metal house dating from 1949. The house measures 5 1/4" tall by 6" wide by 2" deep. Collectors sometimes refer to this house as the Midget Manor.

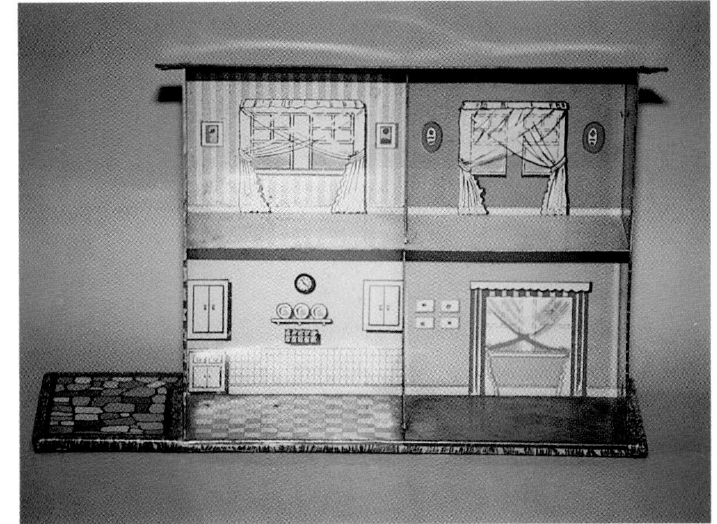

The inside of the Ohio Art Co. house contains four rooms and a terrace.

This hardboard house, made by Brumberger, was featured in the 1975 Sears Christmas catalog. The house was called a Chalet Doll House and measures 20" by 11" by 11" high. The five-room house came furnished with what looks to be the 1/2" scale Superior furniture that dates from the 1950s. Part of the roof lifts up to make access to the inside easier.

This Tudor Tekwood house was featured in the Sears, Roebuck and Co. Christmas catalog in 1945. The house measures 25" by 11 1/2" by 18" tall.

The house contains six rooms decorated with green accents. The casement windows and doors are scored to open and close. The house sold for $4.98 unfurnished. This house was purchased in Clinton, Iowa (home of the Rich Toy Co. during this time period) and the owners thought it was a Rich house. Although the green decoration seems out of character for Rich, one Rich house did feature green floor coverings on the floors (see Rich chapter).

A similar Tekwood house was featured in the Sears Christmas catalog in 1946. This house measures 28" by 12 3/4" by 22" high. The bay windows on this house are similar to the windows on the Jayline Bumstead house. It may be that Jayline was responsible for the production of these Tekwood houses.

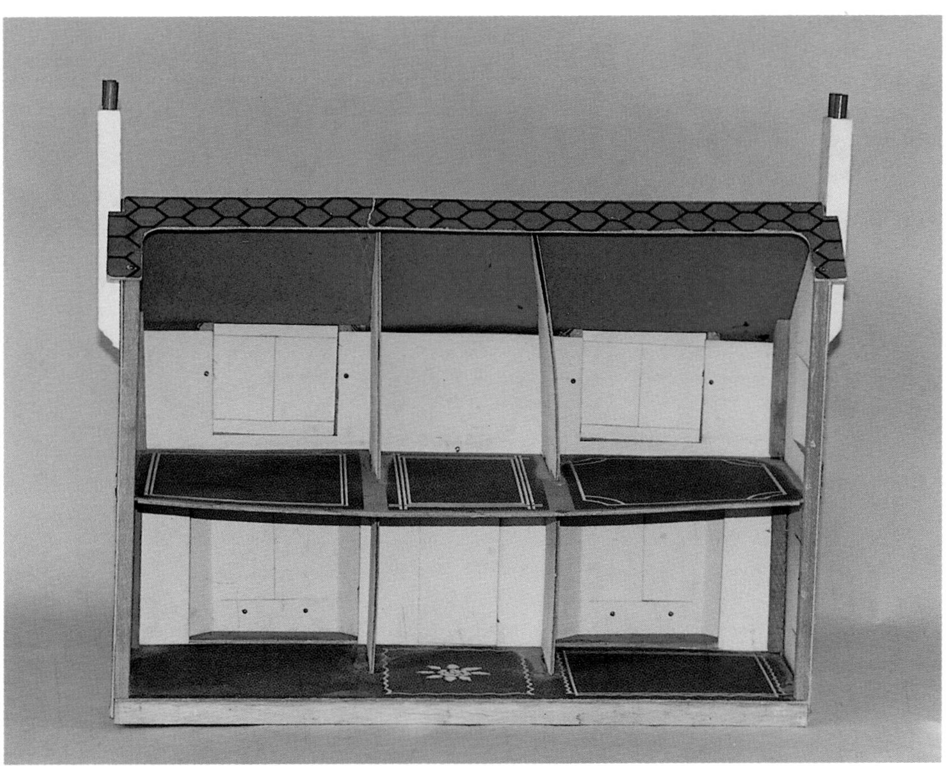

This six-room house sold for $5.98 in 1946. Although the walls were left plain in this house, it did feature the printed green rugs on the floor.

This Tekwood house also appears to have been made by the same company. Jayline advertised a similar house in the mid-1940s but the window arrangement pictured in the ad was slightly different from this house. Several of the scored windows are missing from this house and there may have been a decorative piece on top of the columns. House from the collection of Susan Jenkins Lacerte. Photograph by Norman R. Lacerte.

The inside of the Tekwood house contains six rooms heavily decorated with green. House from the collection of Susan Jenkins Lacerte. Photograph by Norman R. Lacerte.

This unidentified dollhouse dating from the late 1950s or early 1960s is frequently found furnished with Strombecker wood furniture which dates from the last issue of their dollhouse furniture. A television came with the living room pieces. House and photograph from the collection of George Mundorf.

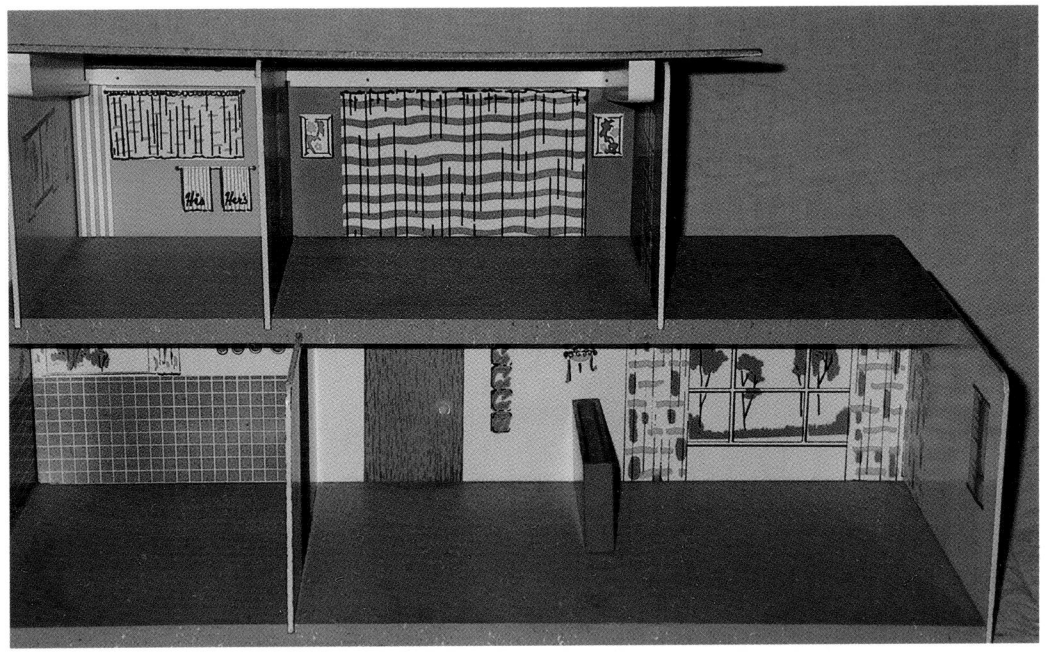

The inside of the house is decorated in keeping with the period. It is possible that Strombecker could have issued this house, furnished with their furniture, as they had in the past with other houses. Photograph and house from the collection of George Mundorf.

This Masonite house features plastic window and door frames. Although the door is missing, it was also plastic. The pink and aqua colors on the house would indicate the house dates from the early 1960s. T. Cohn, Inc. produced a similar house in 1961. This house also has characteristics of the old Keystone houses from the 1950s. A ship and a fireplace are used as decorations on the walls. The chimney has been replaced.

The inside of the house featured five rooms. The rooms are large enough to accomodate the Ideal Young Decorator furniture. The house measures 25" wide, 18 1/2" tall, and 13 1/2" deep.

This Tomy Smaller Homes dollhouse dates from the mid-1970s. The house is made of plastic. Photograph and house from the collection of Roy Specht.

The Smaller Homes house contains four rooms and a stairway. The windows are plastic. Photograph and house from the collection of Roy Specht.

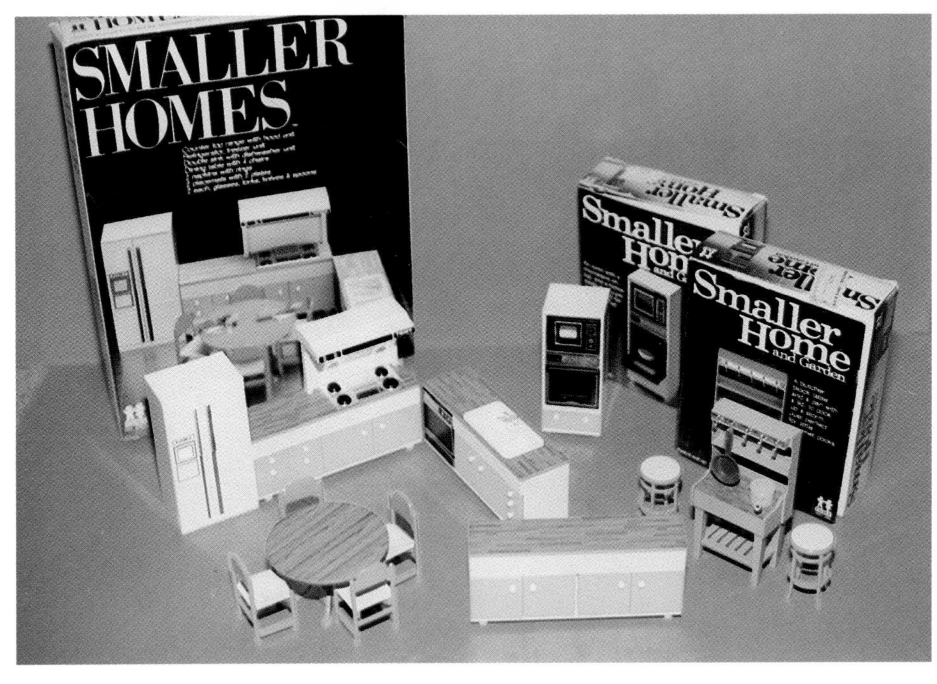

The Tomy kitchen furniture and the original boxes are pictured here. The 3/4" to 1' scale furniture is plastic with working doors and drawers. Photograph and furniture from the collection of Roy Specht.

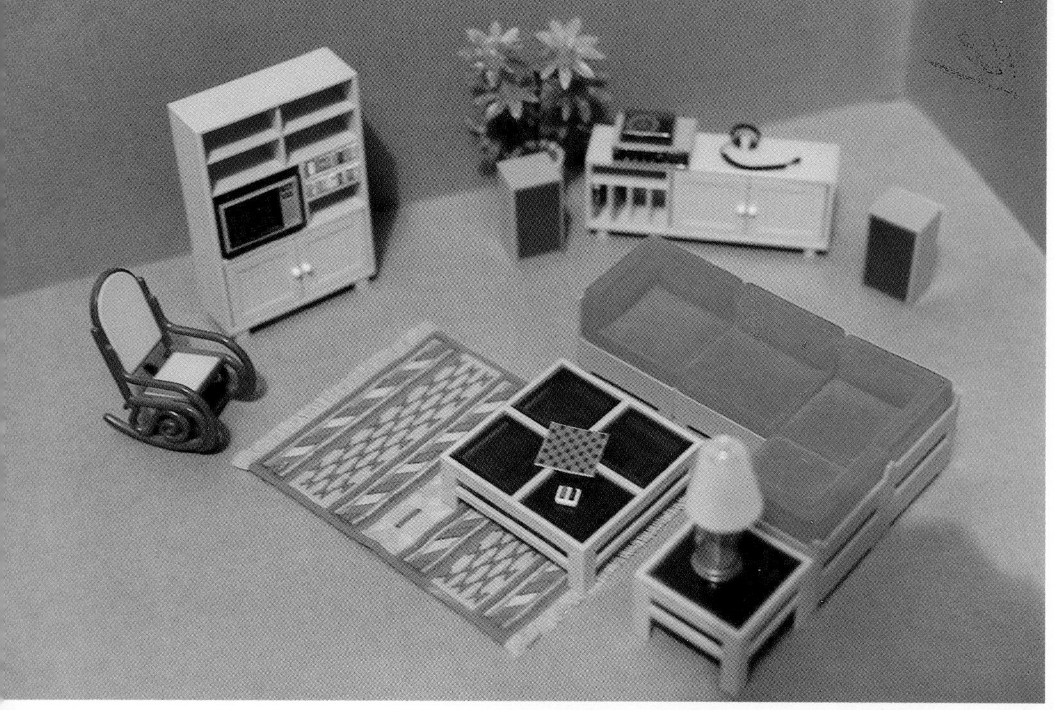

The Smaller Homes living room furniture includes a television plus an entertainment center. Photograph and furniture from the collection of Roy Specht.

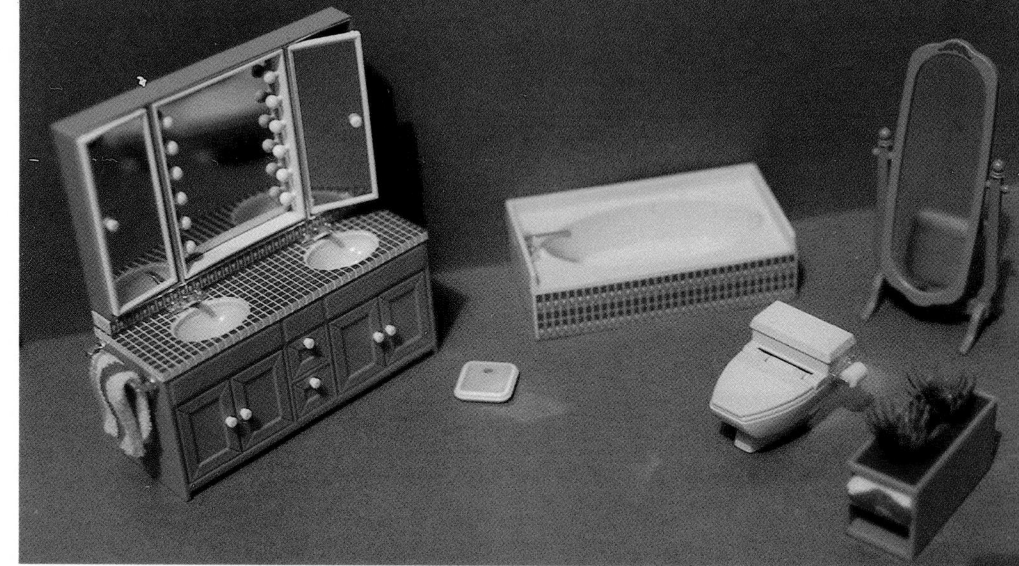

The Smaller Homes bathroom pieces are quite modern and include a large mirrored vanity with two sinks. Photograph and furniture from the collection of Roy Specht.

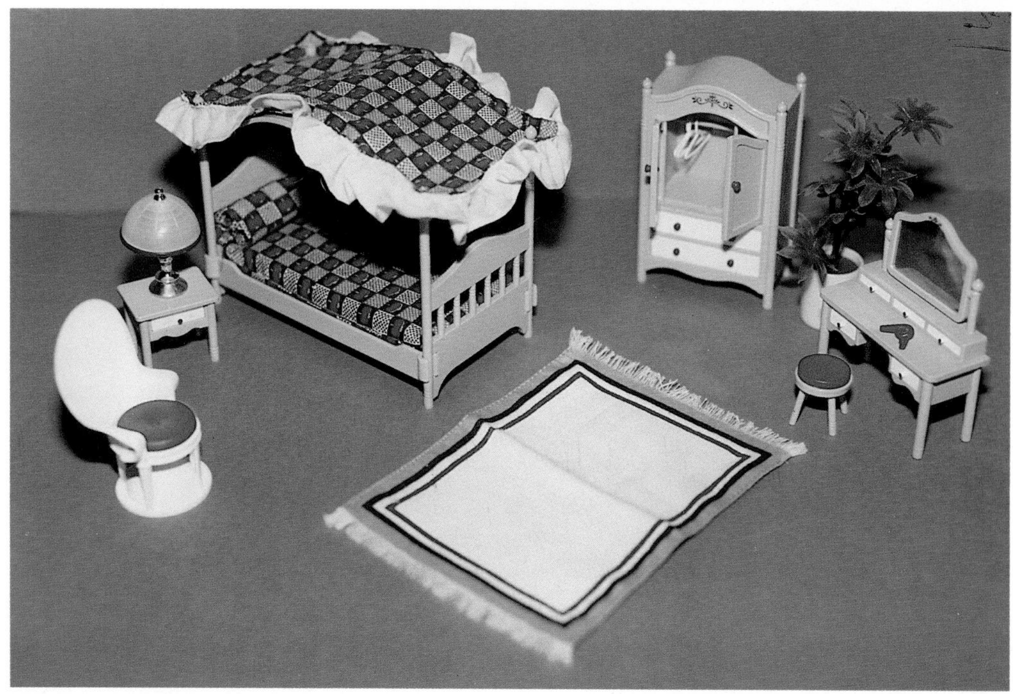

The Smaller Homes bedroom is complete with a canopy bed and hangers to hold the family's clothing. The furniture is marked in a square " © Tomy/Made in Japan." Photograph and furniture from the collection of Roy Specht.

The Smaller Homes plastic jointed family dolls included a father, mother, daughter, and son. Photograph and family from the collection of Roy Specht.

The plastic "Sounds Like Home" dollhouse, along with its electronic accessories, is pictured here. Advertisement from the collection of Kathy Garner.

An advertisement for the "Sounds Like Home" 1982 dollhouse marketed by Craft Master — Fundimensions, a division of General Mills Toy Group. Advertisement from the collection of Kathy Garner.

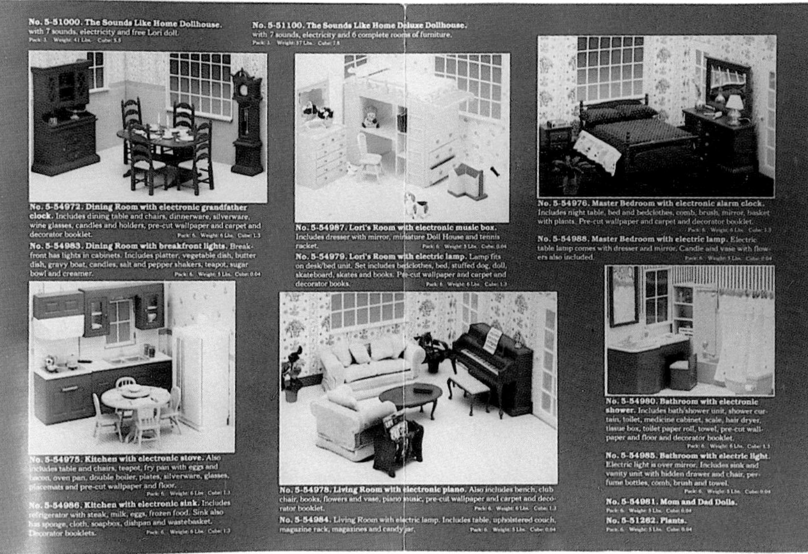

The "Sounds Like Home" dollhouse contained six rooms. The house could be furnished by using fourteen special sets that were produced to accompany the house. Advertisement from the collection of Kathy Garner.

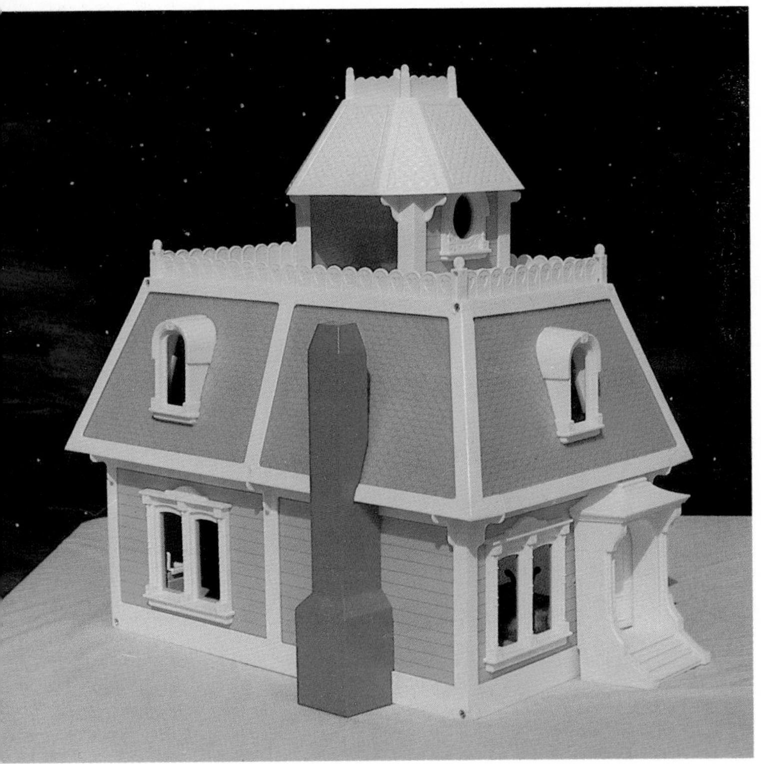

The Littles House was made by Mattel, Inc. in 1981 and 1982. The plastic Victorian house contained four rooms plus a tower room. House from the collection of Kathy Garner. Photograph by Bill Garner.

Flossie and her bed are part of the Littles set produced by Mattel, Inc. in 1981 and 1982. The house and furniture are in the 1/2" to 1' scale. Most of the furnishings were made of metal. The furniture could also be purchased in full room sets.

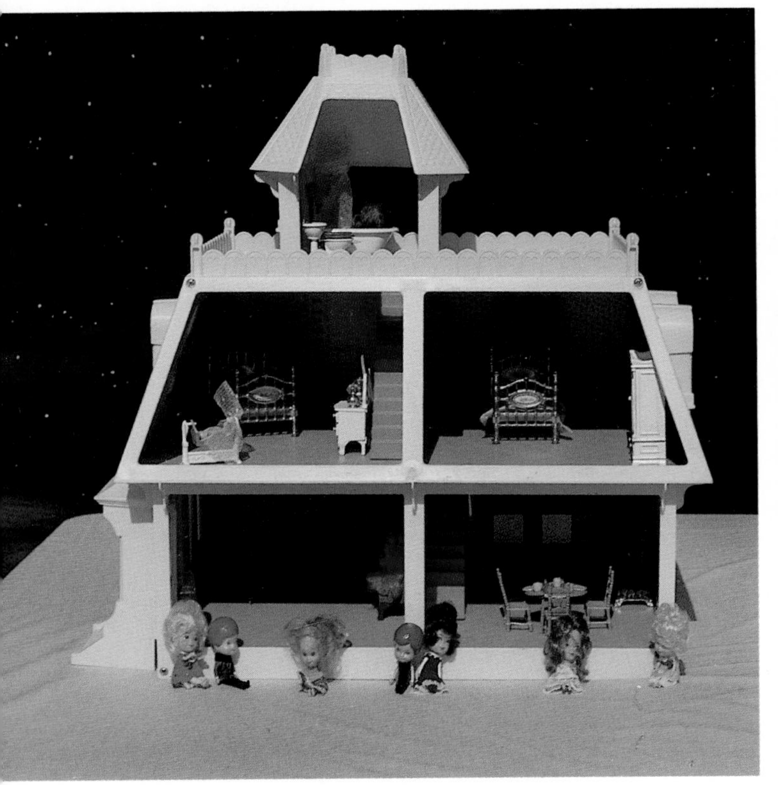

The Mattel, Inc. Littles House was three stories tall. The advertising stated that the house was 1/2" to 1' in scale but the furniture seems to be larger. The dolls made to accompany the house measure from 2 1/2" to 3" tall. House from the collection of Kathy Garner. Photograph by Bill Garner.

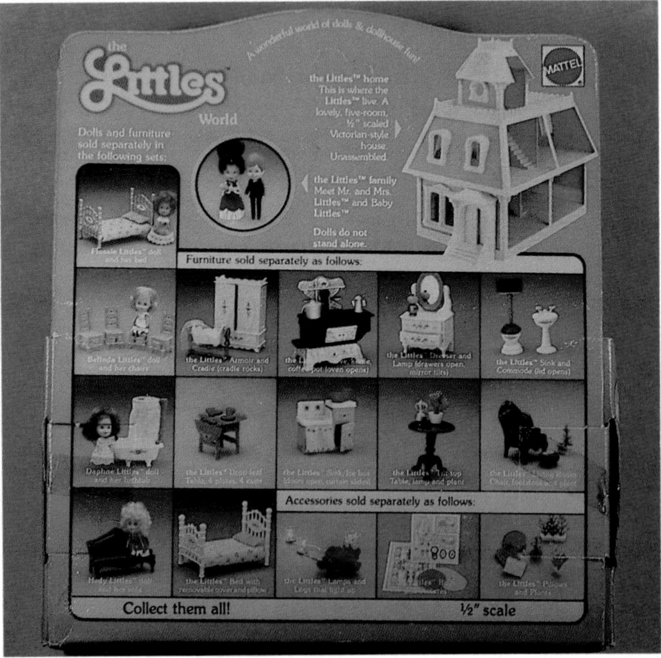

The back of the Mattel furniture box pictures the sets of furniture and accessories that were then available for the Littles House.

Fisher-Price produced this yellow plastic Victorian doll house in 1984. It is shown in the Montgomery Ward Christmas catalog for that year priced for $39.99 unfurnished. The house measures 15 1/2" by 13 1/2" by 23 1/2" high. It was made on the 1" to 1' scale. House from the collection of Kathy Garner. Photograph by Bill Garner.

The Fisher-Price house was three stories tall and could be furnished with a deluxe furniture set. The furniture was priced at $29.99. House from the collection of Kathy Garner. Photograph by Bill Garner.

These Fisher-Price dolls were sold to "live" in the Fisher-Price house. The dolls have moveable heads, arms, and bendable legs. The dolls sold for $7.99.

Playmobil produced a series of large Victorian Houses, along with many accessories, in the early 1990s. Pictured is the house which stands 24 1/2" tall. House from the collection of Kathy Garner. Photograph by Bill Garner.

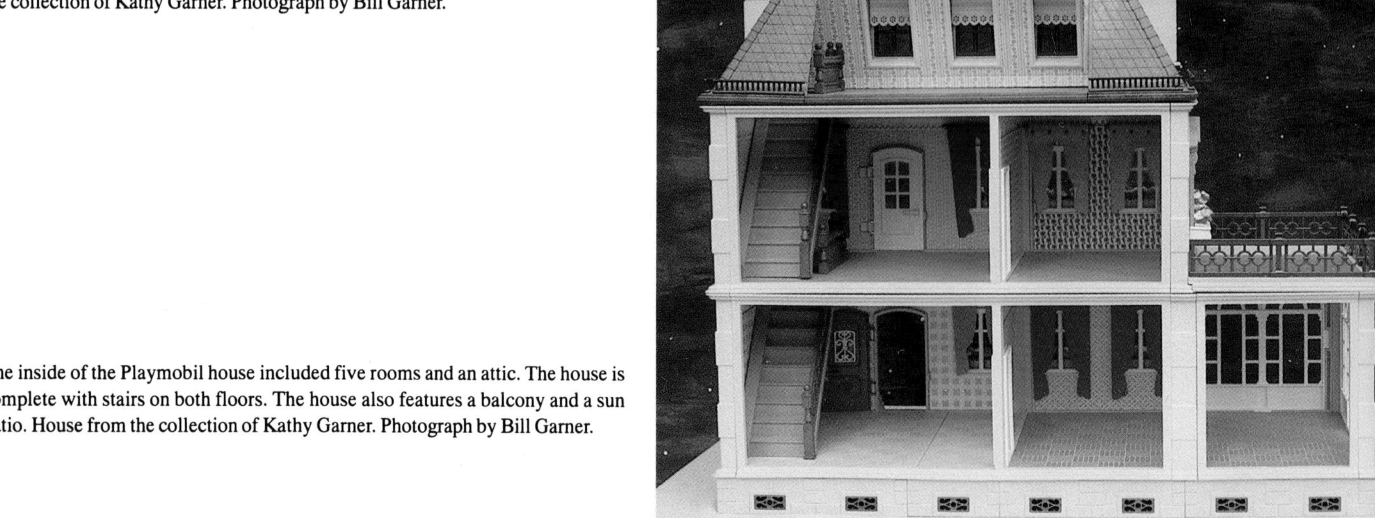

The inside of the Playmobil house included five rooms and an attic. The house is complete with stairs on both floors. The house also features a balcony and a sun patio. House from the collection of Kathy Garner. Photograph by Bill Garner.

Playmobil also made a smaller version of the 1900 house measuring only 16 3/8" tall as well as an even larger version with an extra component added. Besides the houses, Playmobil produced many kits to be used as both furniture and accessories for the houses. All of the pieces could be combined to re-create a 1900 neighborhood. Photograph and collectibles from the collection of Roy Specht.

Heritage (HR-560) Dollhouse kit made by Dura-Craft in 1991. The completed wood house measures 29" wide, 20" deep, and 27" high. The garden was not part of the kit materials. The many dollhouse kits now available may encourage more people to become dollhouse collectors in the future. From the collection of Jeff Zillner and Jim Shivers.

Houses and Furniture For Larger Dolls

Furniture made specifically for dolls has been a staple of the toy manufacturing industry for over a hundred years but furniture designed for a specific doll or set of dolls is a newer innovation.

In the early 1940s when the Nancy Ann Storybook Dolls were meeting with success, the San Francisco based company also began making fabric covered cardboard furniture to be used by their small bisque dolls. Nancy Ann Storybook Dolls, Inc. furniture included the following pieces: wing chair, love seat, settee, arm chair, day bed, bassinet, slipper chair, chaise lounge, dressing table, and bed. This furniture is very hard to locate for today's collector. Many Nancy Ann admirers have to settle for homemade furniture made with the McCalls pattern #811. This old pattern provides instructions for making a bed, sofa, chair, ottoman, and bassinet for dolls 4" to 7" tall.

It was not until the 1950s that new innovations came about in doll production that would change the industry forever. It began with the introduction of the 8" hard plastic Ginny doll in 1948. The small doll was made by Vogue Dolls, Inc. located in Medford, Massachusetts. By the early 1950s, other firms began producing their own versions of Ginny. Nancy Ann Storybook Dolls, Inc. introduced Muffie in 1953 and later Cosmopolitan brought out their Ginger doll. Part of the successful marketing of these small dolls relied on the many different clothing items that were available for each doll. In order to make the dolls appear even more life-like, Vogue soon began selling furniture for Ginny. The wood furniture is quite sturdy and can still be found in good condition today. The furniture was painted pink and white with the Ginny name printed on most pieces. Two different styles of the wood furniture were made. Included in the designs were beds, chest of drawers, rocker, table and chairs, an occasional straight chair, and wardrobe. Vogue also sold the Ginny Gym set which is the most difficult piece to find today.

In 1957 and 1958 Vogue added 8" vinyl baby dolls to their line and these dolls, Ginnette and Jimmy, had their own set of wooden furniture. The pieces included a bed, bathinet, shoofly, and a feeding table.

By the late 1950s, when little girls' doll tastes began to include the 10" high heel Vogue Jill dolls, the company also designed furniture for Jill and her companion dolls, Jeff and Jan. This furniture included a desk and chair, bed, dressing table with bench, and a large wardrobe.

As these 8"-10" dolls became more popular, other companies also designed furniture that could be used for the dolls. The most well known is the furniture made by Strombecker. This firm produced furniture with their own label as well as that with the Betsy McCall logo (see Strombecker chapter). Mattel, Inc. was also active in manufacturing larger doll furniture in 1958. This wood furniture was quite modern and included living room, dining room, and bedroom pieces. The furniture was suitable for the 8" to 10" size dolls.

Richwood Toys, Inc. (Annapolis, Maryland) who produced the 8" hard plastic Sandra Sue doll during the 1950s also provided their dolls with a set of furniture. This lovely mahogany finished wood furniture included a Duncan Phyfe table and matching chairs, a tester bed with canopy, wardrobe, chest on chest, vanity and stool, and a bureau with mirror. Playground equipment was also made for the Sandra Sue dolls. This set of furniture and the Sandra Sue dolls are now in great demand by today's collector.

These new innovations in doll manufacture, (which included the production of lots of different clothing and accessories for the dolls) was carried a step further when Mattel, Inc. designed the Barbie doll in 1958.

Mattel, Inc. carried the concept of making a basic doll with lots of added accessories to new heights and the trend still continues today. Entire books have been written about the Barbie phenomenon but in order to show the influence of Barbie on doll production, this fabulous doll must be included in any discussion of dollhouses and furniture for the larger doll.

When Mattel discovered what a success Barbie was becoming, the company expanded production to include furniture, cars, and houses for its doll. As the Barbie family grew to include other members (Midge, Ken, Allen, Skipper), rooms were also produced for these other Barbie associates. Through the years, Barbie has owned campers, horses, and boats, as well as many different designs of houses, rooms, and apartments.

At first Barbie was only outfitted with several styles of clothing but by 1962 a little girl could own a Barbie car, doll case, and Susy Goose plastic wardrobe and bed. The wardrobe was priced at $3.49 and the canopy bed at $2.94. There was also a portable suitcase house for Barbie in 1962. This corrugated suitcase opened into a one-room apartment and sold for $4.44. By 1964, Barbie had become so popular that ten pages of her merchandise was included in the Sears Christmas catalog for that year. Besides new dolls and clothing, several dollhouse-like units were being sold for Barbie and her friends. These included: Barbie's Campus, Barbie's Little Theatre, Barbie's Fashion Salon, and Barbie's New Dream House. This new chipboard house sold for $4.99 and unfolded to a 44" by 42" size. The house included a living room, bedroom, kitchen, and patio. The furniture was also made of chipboard and came unassembled.

New Barbie furniture pictured in the catalog included a convertible sofa-bed, coffee table, chair, ottoman, end table, lawn swing, chaise lounge, music box piano, dressing table, and a new styled bed and wardrobe. These were all made primarily of plastic.

By 1966, Mattel, Inc. had begun making Barbie houses and rooms using the concept of a vinyl suitcase which opened to provide more space. These vinyl suitcase homes for the Barbie family were made in various models into the early 1970s when more elaborate Barbie multi-storied houses were designed. In 1972 the vinyl suitcase Barbie house measured 12" by 11" by 14" when closed and sold for $8.99

The mid-1960s provided a bonanza for collectors who specialize in the larger dolls and their houses. With the success of Barbie and her accessories, other companies also began making their own 8" to 12" vinyl dolls with extra clothing and accessories. The most popular of these new sets of dolls were the Tammy family dolls made by Ideal in 1963 (see Ideal chapter). Another similar set of dolls was produced by Remco Industries in 1963. This vinyl doll family included Dr. John Littlechap (15" tall), his wife Lisa (15" tall), daughter Judy (13 1/2" tall), and daughter Libby (10 1/2" tall). This family of dolls also was provided with their own living space. Remco produced cardboard houses which included the Master Bedroom Set, Family room Set, and the Dr. Littlechap Office Set. The furniture for all these rooms was made of cardboard. Although these Remco products were not as profitable as those made by Ideal, they are sought after by collectors today.

Tressy, another 11 1/2" vinyl and plastic fashion doll, was also equipped with a place to live. This American Character doll had her own Penthouse Apartment during the mid-1960s. This house was also made of chipboard. The Tressy doll had hair that "grew," which helped her compete against the more popular Mattel Barbie.

By 1966 the collectible vinyl dolls and their accessories had become even smaller. Heidi, the Pocket Book doll made by Remco, was only 5 1/2" tall. This doll had her own house, jeep, and furniture. Her house was made of metal and plastic and came complete with a growing garden. Other plastic furniture that could be purchased for Heidi included a table, chairs, buffet, modern bathroom with corner tub, vanity, bunk beds, dresser, chair, table, and lamp. The living room furniture came with the Heidi house.

The Liddle Kiddles made by Mattel, Inc. also could be purchased with their own furniture and a house. These vinyl dolls were even smaller measuring 2 1/2" to 3 1/2" tall. The Liddle Kiddle Klubhouse was a vinyl suitcase that opened out to provide space for the Liddle Kiddle furniture and toys. In 1967 a Kiddle City was added to the line. It, too, came in a vinyl case. The inside of the case contained small homes, a park, and playground. Other companies made their own tiny vinyl dolls and houses to complete with the most popular Heidi and Liddle Kiddle dolls.

As the 1960s ended, the only small dolls to survive the era were the Mattel, Inc. Barbie family dolls. Tammy, Heidi, Tressy, and all the other new dolls could not compete with the phenomenon that was Barbie.

Other companies did not give up entirely, however. In the early 1970s, another small doll appeared which met with some success. The doll's name was Dawn. She was a 6 1/2" tall vinyl fashion doll made by Topper Corp. There was also a boy doll named Gary in the series and a black doll called Dale. Again, many products were made as accessories for the dolls. These included furniture, a car, and a beauty pageant stage. The furniture was the same design that had been used for the 8" vinyl Penny Brite dolls (made by Deluxe Reading Corp.) in the mid-1960s. The Dawn dolls did not last for very many years but other manufacturers continued to pursue the small doll market.

In the mid-1970s, the Sunshine Family was introduced by Mattel, Inc. This vinyl doll family consisted of a 9 1/2" male doll called Steve, a 9" female doll called Hattie and a 3" baby called Sweets. There was also a black family called Happy Family. These dolls, too, came with various accessories including a house, a surrey cycle, and a Sunshine Van.

Many other small dolls were manufactured during the 1970s to represent famous television personalities. The characters from the Waltons' television program were included in these portrayals. Mego Corp. produced 8" vinyl dolls to represent Grandma and Grandpa Walton, the Walton parents, and children John Boy and Mary Ellen. Two different houses were made in the image of the Walton farm house. The smaller house was made of heavy fiberboard and measured 20" by 13 1/2" by 12". It was to be used with play figures of the family. This house was copyrighted in 1974 by Lorimar Productions, Inc. The other much larger house was made to be used with the dolls produced by Mego and was copyrighted in 1975. This five-room farmhouse measured 24" tall and 35" long. It came complete with several pieces of furniture.

Another character that became popular in the mid-1970s was Holly Hobbie. Several different sizes and models of Holly Hobbie dolls were produced along with many accessories. A Holly Hobbie four-room dollhouse was advertised in the Sears 1975 Christmas catalog. The house was really much like the Sunshine Family house and only provided a back drop for the furniture. There was no roof. The house came with a 5 1/4" vinyl Holly Hobbie doll that could be walked from room to room with the help of a magic wand. The house was made of fiberboard and plastic and measured 18" by 12 1/4" by 8 1/4". It was equipped with plastic furniture.

A much larger and more elaborate Holly Hobbie house was produced during the early 1980s. This house, designed by Judith Massey Cunningham, was quite expensive and was sold in hobby shops. The house came in kit form and had to be assembled. The wood kit was produced by Millie August Miniatures, Inc. and marketed by Plaid Enterprises, Inc. The completed house contained nine large rooms (including the attic). It measured 36" tall, 36 1/4" wide, and 26" deep. The roof could be lifted in back to allow access to the attic rooms. The tower roof could also be removed. The house included a fancy staircase and a simulated stained glass Holly Hobbie window. Four pieces of wood bedroom furniture were also produced to be used in the house. These items included a bed, nightstand, mirrored dresser, and wardrobe dresser. The furniture was made of honey-colored pine. A rug kit and a Holly Hobbie picture were other accessories that could be purchased to accompany the house. This house is currently in demand by both Holly Hobbie and dollhouse collectors so the value of the house continues to rise.

The 1980s brought another unusual collectible when Strawberry Shortcake came on the scene. The American Greeting Corp. had copyrighted the character and Kenner Products produced the dolls and accessories. The dolls with the Strawberry Shortcake name came in all sizes but the most popular ones were the small 5 1/2" and under models. The vinyl dolls could choose from several models of houses but the most elaborate was the house which sold for $150 in 1984. The house was made of molded plastic and measured 27" by 20" by 27" high. The house came furnished with a bedroom, bathroom, kitchen, combination living room, dining room, and an attic. The house was designed for the 5 1/2" and smaller dolls. Since it was so expensive, it is doubtful that the house was a best seller. Houses that were much smaller (10" by 6" by 11") could be purchased for as little as $10.

The current market in furniture and houses for larger dolls revolves mostly around those pieces designed for Barbie and her friends, produced by Mattel, Inc. Since a whole collecting field has grown up around these products, both new and old, the Barbie collectibles continue to dominate the field. The furniture for the Vogue Ginny family dolls continues to be popular with collectors and the new interest being shown in the Richwood Sandra Sue items will also affect the marketplace in the future.

Furniture Series
Wing Chair
1001

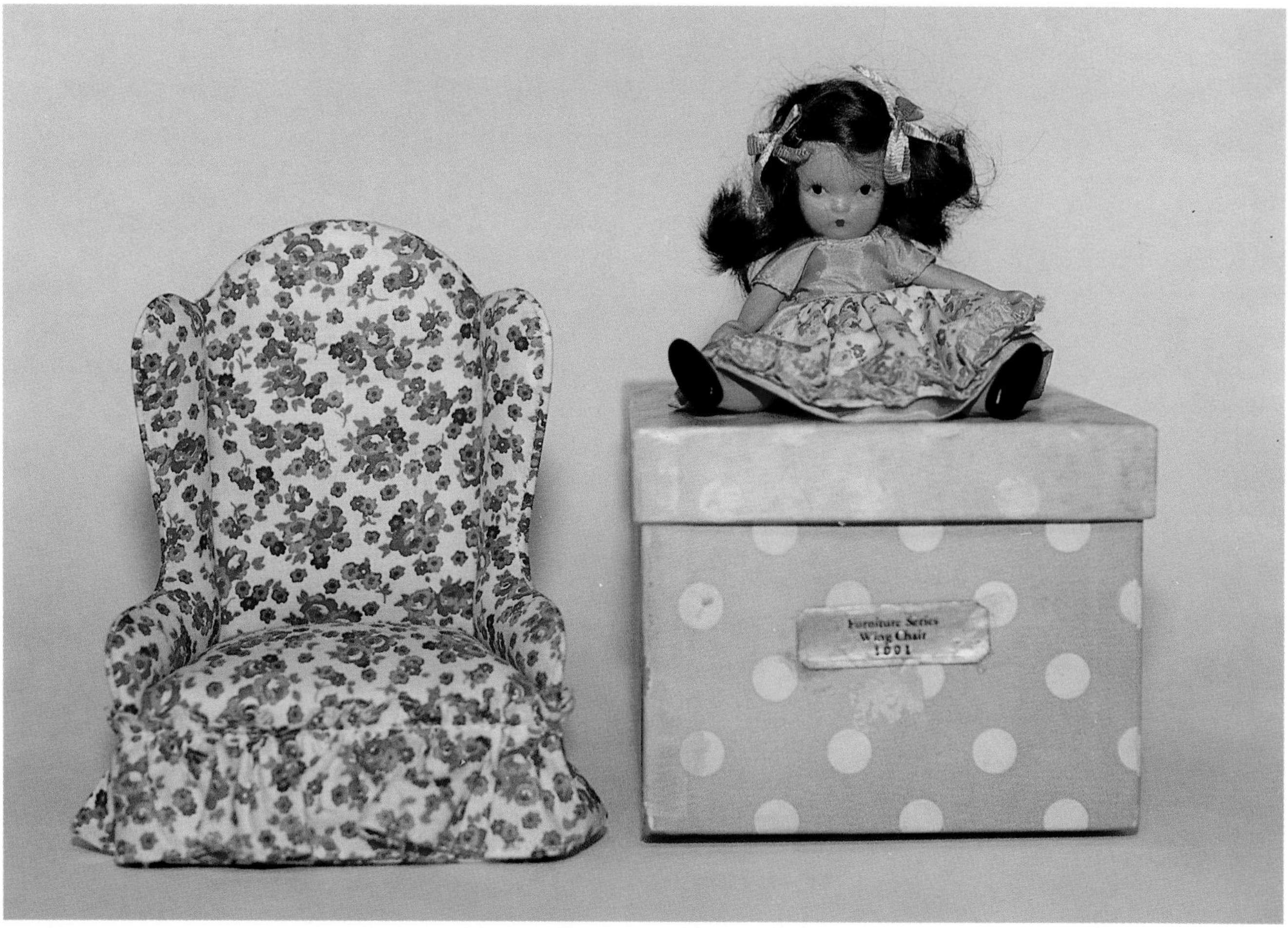

Nancy Ann Storybook Dolls, Inc. fabric-covered cardboard wing chair. The furniture was produced in conjunction with the Nancy Ann Storybook Dolls during the early 1940s. The bisque doll is also a Nancy Ann product from the same period. Photograph, doll, and chair from the collection of Jackie Robertson.

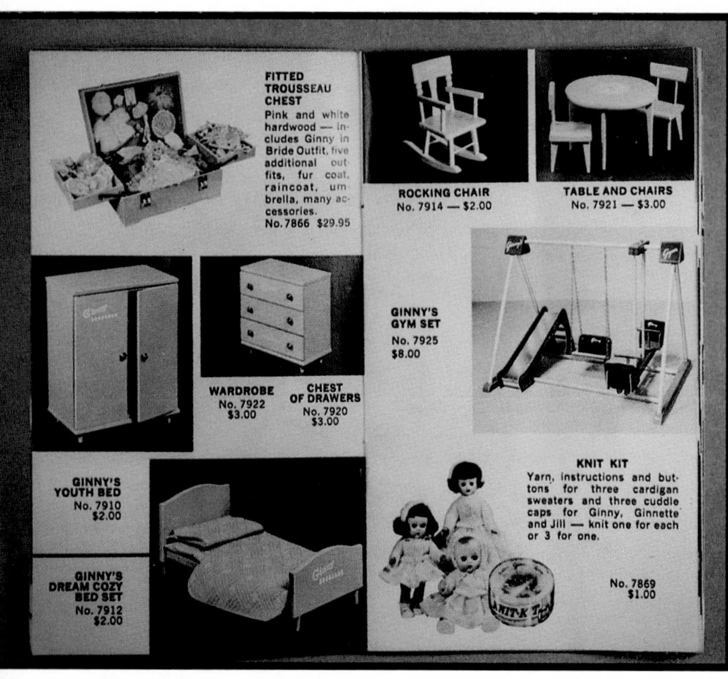

FITTED TROUSSEAU CHEST
Pink and white hardwood — includes Ginny in Bride Outfit, five additional outfits, fur coat, raincoat, umbrella, many accessories.
No. 7866 $29.95
ROCKING CHAIR
No. 7914 — $2.00
TABLE AND CHAIRS
No. 7921 — $3.00
WARDROBE
No. 7922
$3.00
CHEST OF DRAWERS
No. 7920
$3.00
GINNY'S GYM SET
No. 7925
$8.00
GINNY'S YOUTH BED
No. 7910
$2.00
GINNY'S DREAM COZY BED SET
No. 7912
$2.00
KNIT KIT
Yarn, instructions and buttons for three cardigan sweaters and three cuddle caps for Ginny, Ginnette and Jill — knit one for each or 3 for one.
No. 7869
$1.00

Pamphlet copyrighted by Vogue Dolls, Inc. in 1957. It pictures the wood furniture produced for the 8" tall hard plastic Ginny doll. The furniture included a wardrobe, chest of drawers, bed, rocker, table and chairs, and gym set.

Ginny

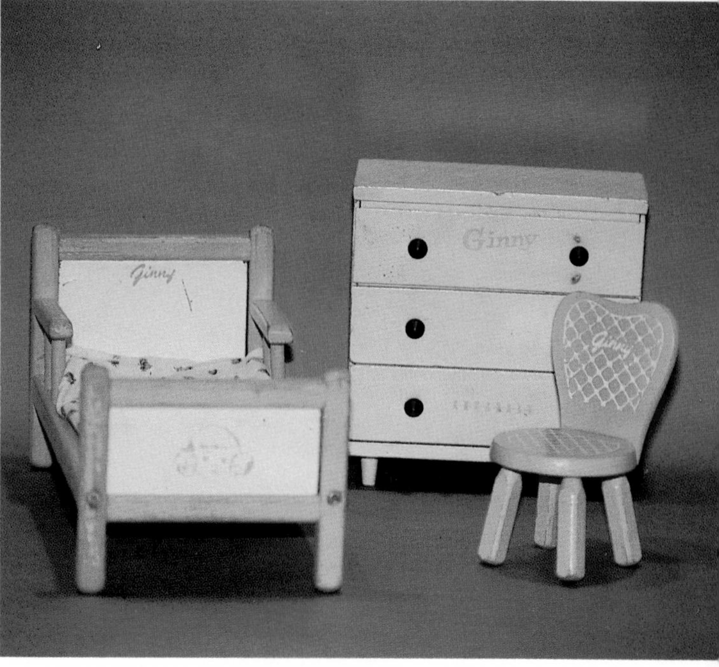

An earlier style Ginny bed is pictured along with the chest of drawers and an occasional chair. All are labeled with the Ginny name. Marketed through Vogue Dolls, Inc.

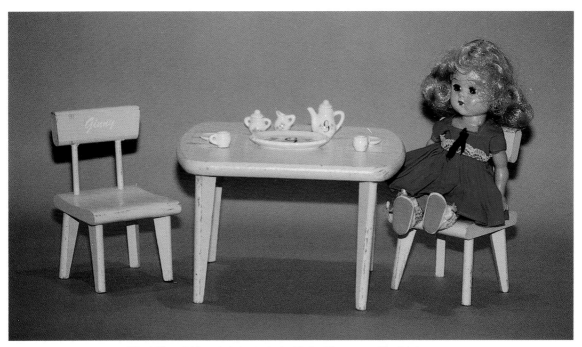

Wood Ginny table and chairs marked with the Ginny name. The furniture was produced to accompany the 8" Vogue hard plastic Vogue Ginny doll pictured here. The tea set is a later Ginny product.

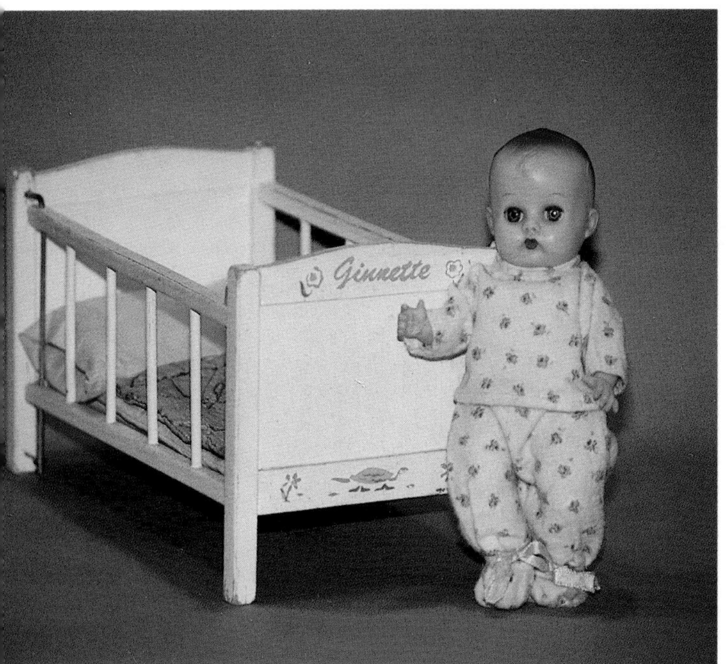

The 8" tall vinyl Ginnette baby doll was also provided with her own furniture by Vogue Dolls, Inc. Pictured is Ginnette with her wooden crib circa 1957.

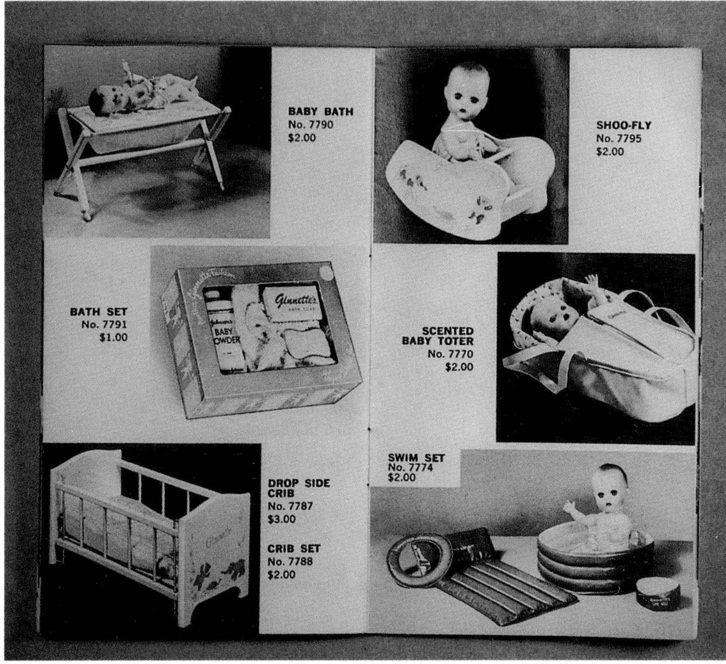

The Vogue catalog from 1957 also pictures a Ginnette Shoo-fly, swim set, and baby toter.

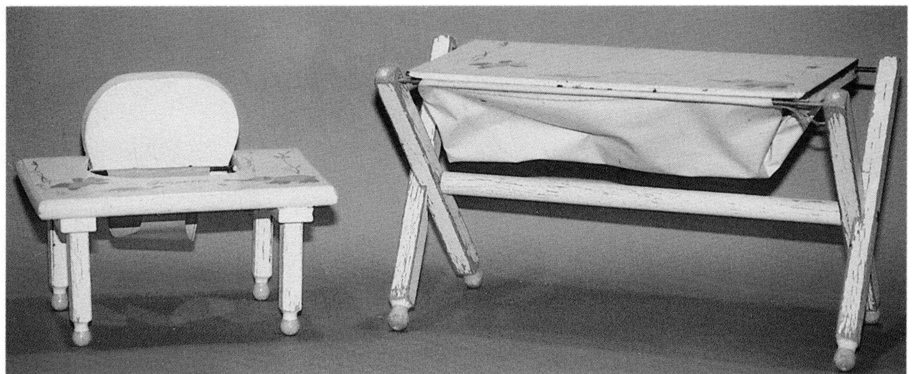

Vogue Ginnette furniture also included a baby bath and a baby tender.

Furniture was also produced specifically for the 10" Jill and Jan dolls. Pictured is a wood desk and chair marketed by Vogue for these dolls.

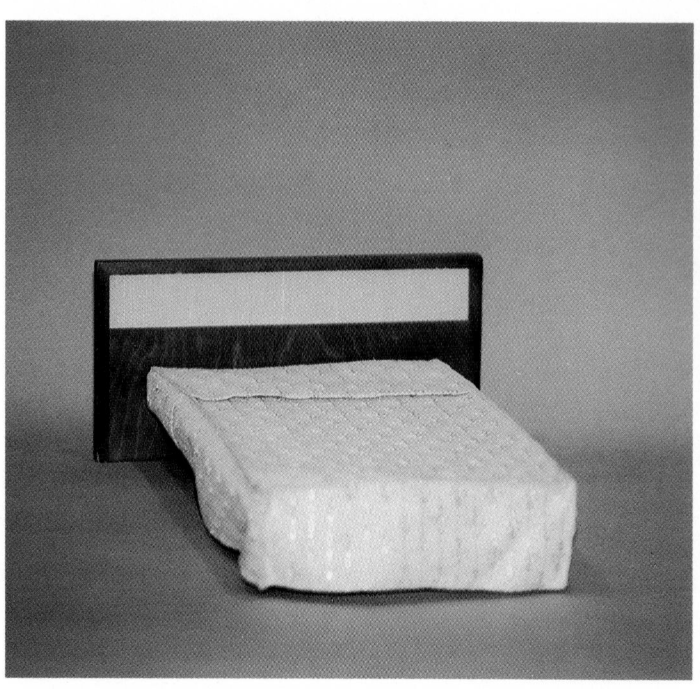

The modern Mattel, Inc. bed, complete with bed spread, as produced circa 1958.

Modern wood dining room furniture produced by Mattel, Inc. circa 1958. This furniture was made to be used with the 8"-10" dolls then popular.

The Mattel, Inc. modern sofa and end table is pictured. These products were produced before Mattel began making the Barbie dolls.

Richwood Toys, Inc. produced mahogany furniture to be used by their 8" tall hard plastic Sandra Sue dolls. This tester bed with canopy is one of their products. Photograph and bed from the collection of Marian Schmuhl.

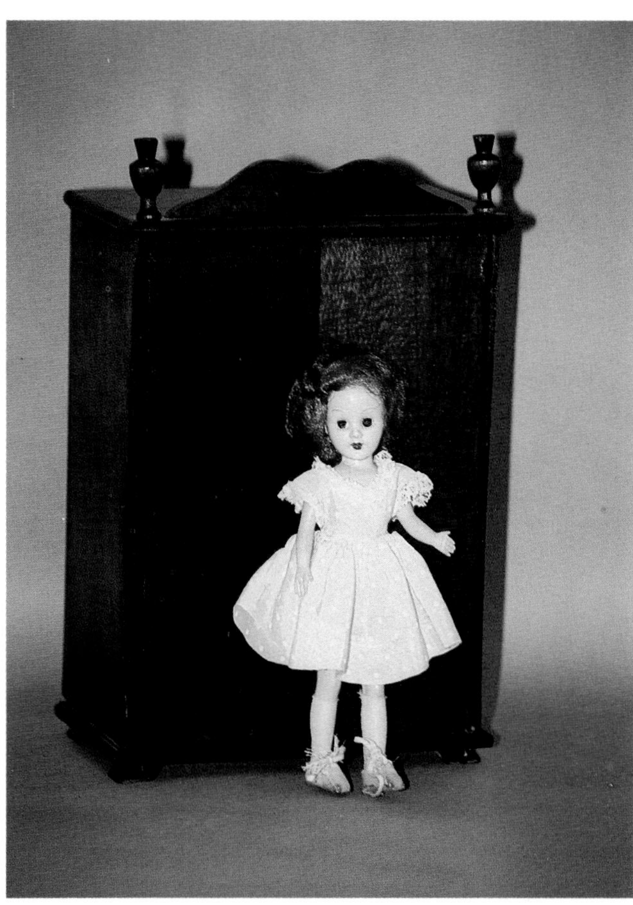

Richwood Toys also marketed this wardrobe as part of its Sandra Sue furniture line. The doll pictured is the Sandra Sue doll with flat feet instead of the high heel model.

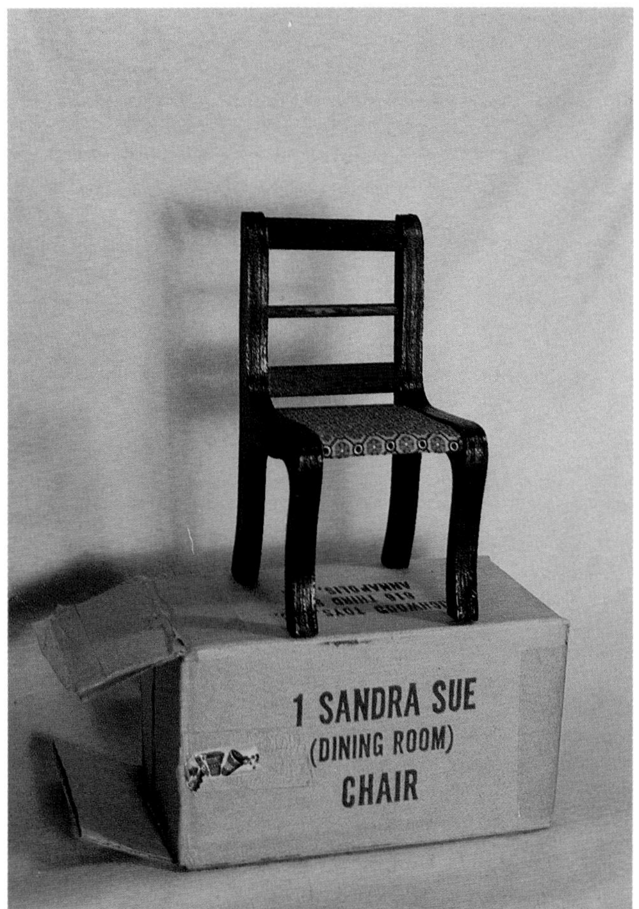

Sandra Sue dining room chair was made by Richwood Toys to accompany the Duncan Phyfe table. Photograph and chair from the collection of Marian Schmuhl.

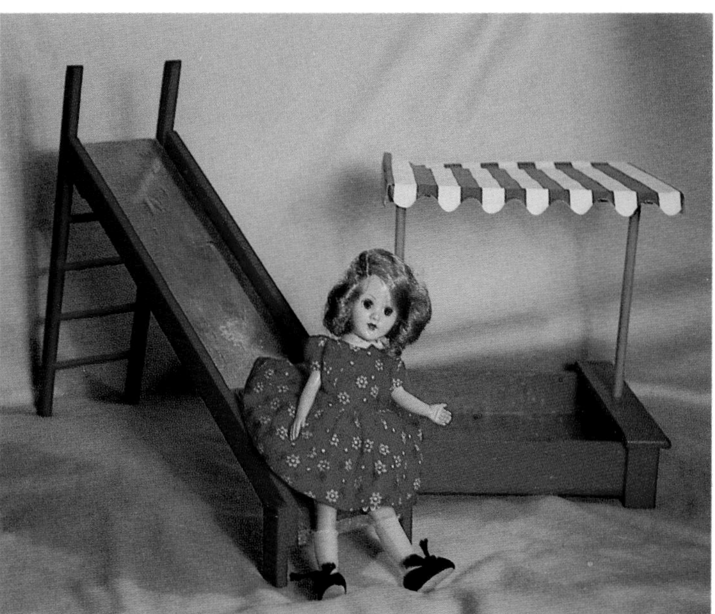

This unusual slide and sand box combination was also produced in the 1950s by Richwood Toys to accompany their line of dolls. Pictured with the slide is a hard plastic Sandra Sue doll. Photograph and collectibles from the collection of Marian Schmuhl.

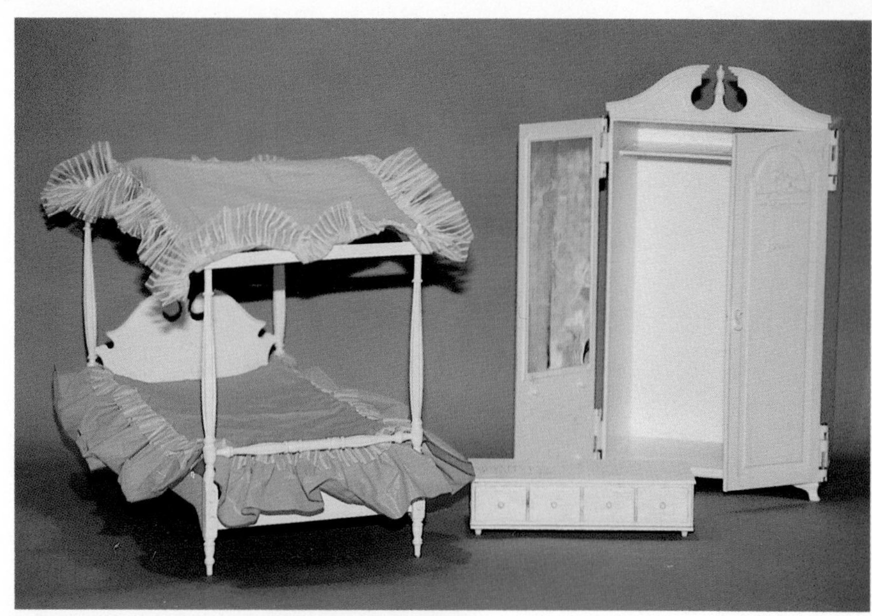

This Susy Goose plastic bed, wardrobe, and storage drawer were marketed in 1962 as accessories for the Barbie doll made by Mattel, Inc.

Barbie's New Dream House (Mattel, Inc.) was advertised in the Sears Christmas catalog in 1964 at a price of $4.99. The chipboard suitcase house unfolded to a size of 44" by 42".

The New Dream House made by Mattel, Inc. included a living room, bedroom, kitchen area, and patio. The furniture was also made of chipboard and came unassembled. Pictured is a Mattel Allen doll (circa 1965) seated in the living room.

Vinyl Barbie suitcase houses were produced by Mattel, Inc. for several years during the late 1960s and early 1970s. This house measures 34" by 11" by 14" high when opened. The house also came with several pieces of plastic furniture including a sofa, table, chairs, and bed.

The Mattel "American Girl" Barbie (1965) stands in Mattel's New Dream House bedroom.

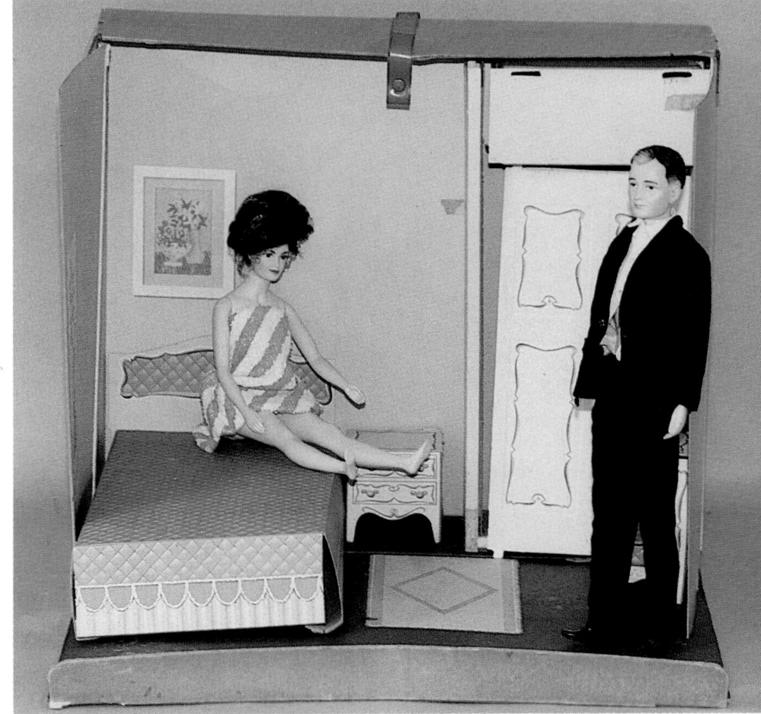

Remco Master Bedroom made for the company's Littlechap dolls in 1963. The chipboard foldaway doll house opens to reveal a room furnished with chipboard furniture which included a bed, storage closet, night table and lamp, chest of drawers, dressing stool, and lounge chair. Pictured in the room are the Mr. and Mrs. Littlechap vinyl dolls.

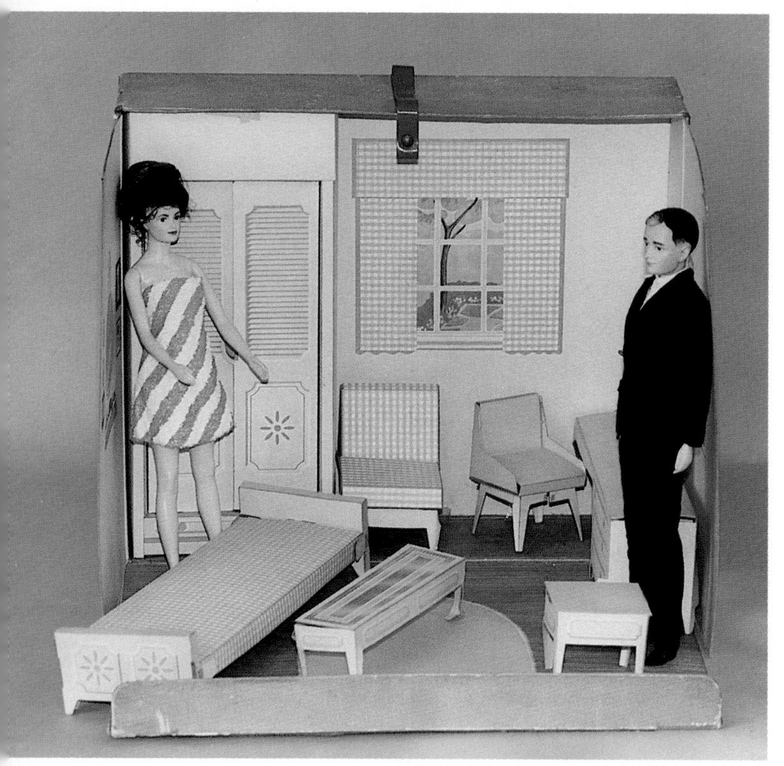

Remco also produced a Family room and an Office for the Littlechap family. Pictured is the foldaway Family room. It is also made of chipboard in a suitcase design. When the suitcase is opened, it contains the Family room complete with furniture. Items included in the original set consist of a convertible sofa bed, storage closet, coffee table, desk, chest of drawers, phonograph, side table, lamp, desk chair, lounge chair, and television. The Dr. Littlechap and Lisa dolls are also shown.

The accessories for the 5 1/2" Remco Heidi doll included a very nice metal and plastic dollhouse. The house was featured in the Montgomery Ward Christmas catalog for 1966. The house contains one large room. The roof of the house is plastic while the walls and floor are metal. A garden also was part of the house setting and it included seeds, a pond, and tools.

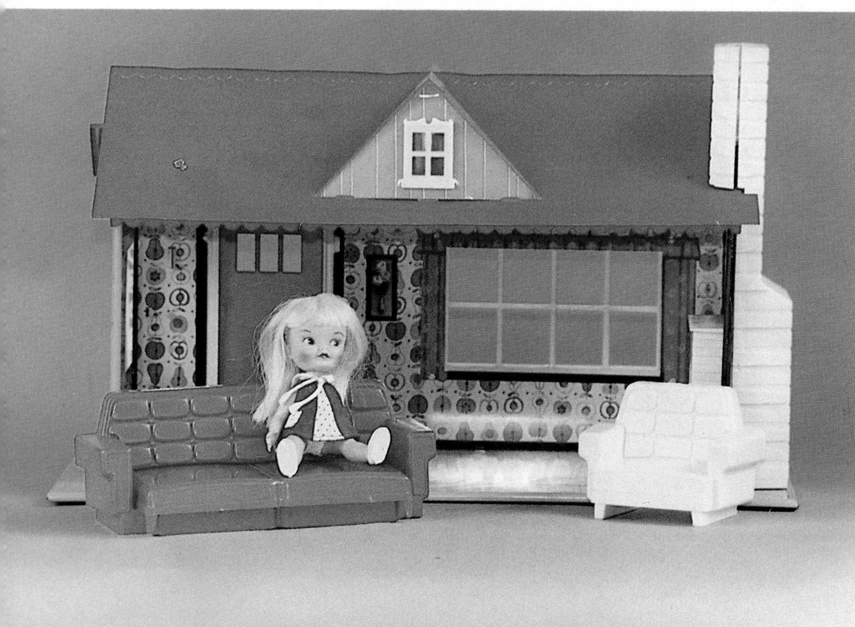

The Remco Heidi house was furnished with a plastic sofa and chair. The house and garden measure 38" long, 10 1/2" high, and 13" deep. The Remco Heidi doll is pictured with the house.

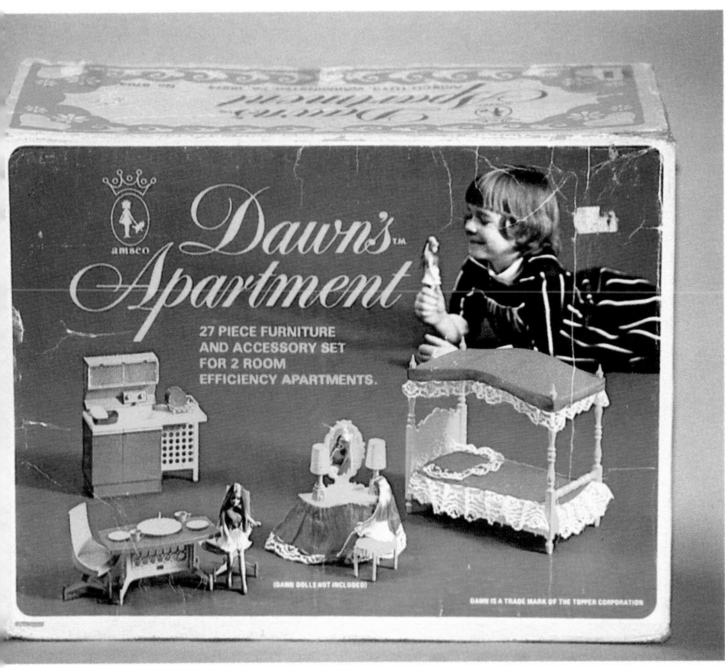

Dawn's Apartment, a twenty-seven piece furniture set for a two-room apartment, was made by Amsco Toys in the early 1970s.

The Plastic Amsco furniture was to be used with the 6 1/2" vinyl Dawn doll made by Topper Corp. Pictured are the kitchen furnishings and accessories.

The Dawn plastic bedroom furniture included a canopy bed, vanity, and bench. The shades are missing from the lamps. All of the furniture in this set is the same design as that used in the 1960s to accompany the 8" vinyl Penny Brite doll, made by Deluxe Reading Corp.

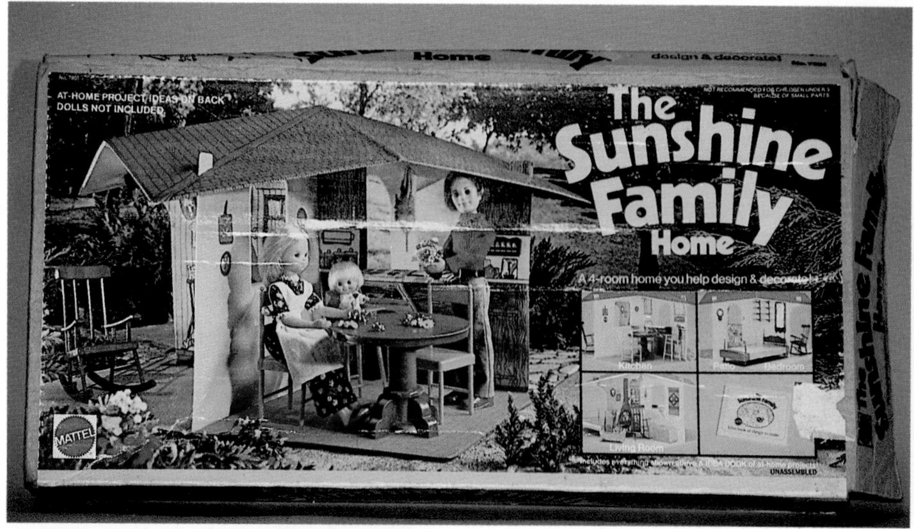

Box for the Sunshine Family House produced by Mattel, Inc. in 1973.

When assembled, the vinyl house containes four open rooms to be used with the Mattel Sunshine Family vinyl dolls. The adult dolls were approximately 9" tall.

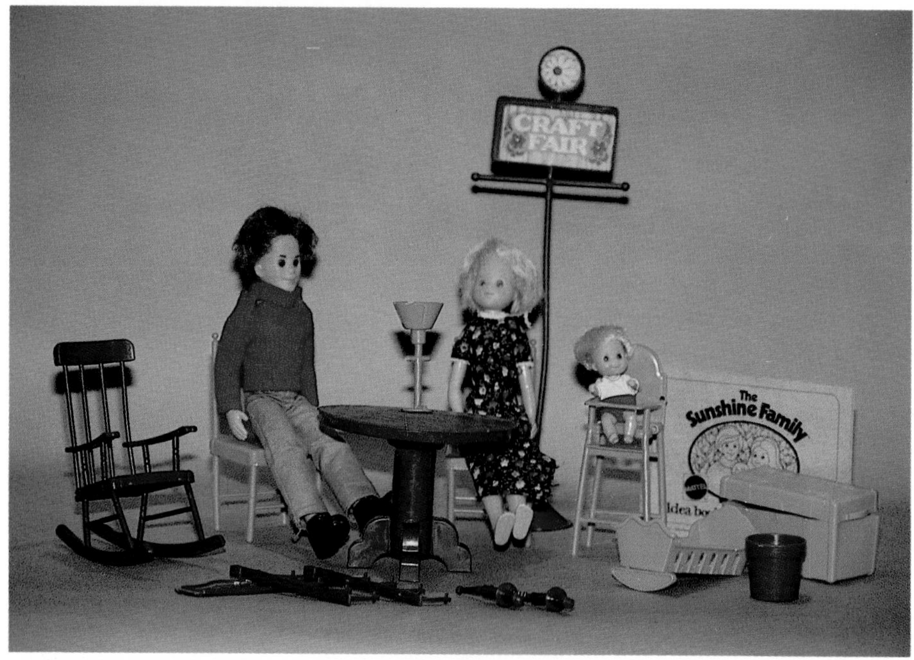

The plastic furnishings, which came with the house, were supposed to be supplemented with items made by the consumer following suggestions included with the house.

This large Walton house was produced by Mego in 1975 to be used with its 8" vinyl Walton dolls. The dolls and house were based on the Walton television program (Lorimar Prod., Inc.).

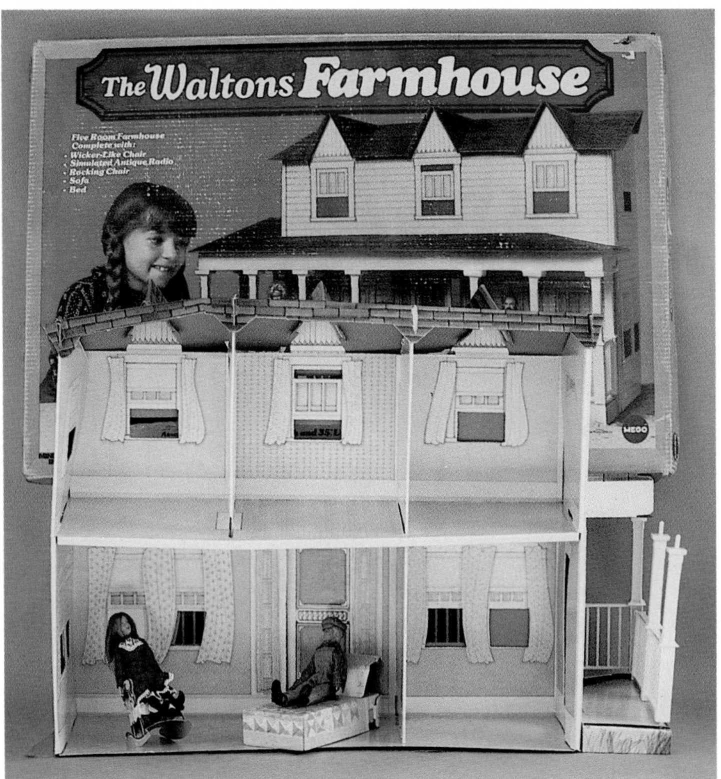

The five-room farm house is made of chipboard and measures 24" tall and 35" long. Several pieces of cardboard furniture also came with the house along with a plastic radio. Pictured with the house are the Mego vinyl Mary Ellen and John Boy dolls.

This large wood Holly Hobbie house was sold in hobby shops during the early 1980s. The house came in kit form and had to be assembled. The kit was produced by Millie August Miniatures, Inc. and marketed by Plaid Enterprises, Inc. The completed nine-room house measured 36" high by 36 1/4" wide by 26" deep. Photograph by Donna Stultz.

Strawberry Shortcake house marketed in 1984. The house is made of molded plastic and measures 27" by 20" by 27" high. Kenner was responsible for most of the toy Strawberry Shortcake products including several different sizes of plastic houses. House from the collection of "Somewhere in Time." Photograph by Pat Collins.

The large Strawberry Shortcake house sold for $150 and included furnishings for a bedroom, bathroom, kitchen, and combination living room/dining room. The house also contained an attic. House from the collection of "Somewhere in Time." Photograph by Pat Collins.

Sources

Dealers

Cobb's Doll Auctions
1909 Harrison Rd. N.
Johnstown, OH 43031
Four auction catalogs yearly
$22 each or $80 year
Send name and address for auction announcements.

Dolls House Antiques
107 Griffin Dr.
Fayetteville, N. Y. 13066
Mail order and shows.

Marilyn's Miniatures of Marshallville
Marilyn Pittman
P. O. Box 246
Marshallville, OH 44645
Dollhouse and dollhouse furniture mail order.

Moreno Valley Miniatures
P. O. Drawer 977
Angel Fire, New Mexico, 87710
SASE for information.

Judith A. Mosholder
R. D. #2 Box 147
Boswell, PA 15531
Send LSASE for list of plastic dollhouse furniture.

Paige Thornton
P.O. Box 669125
Marietta, GA 30066
Send LSASE for each list of plastic, Tootsietoy, or Strombecker furniture, books, or dollhouses.

Publications

Dollhouse and Miniature Collector's Quarterly
Editor: Sharon Unger
P. O. Box 16
Bellaire, MI 49615
$20 per year, (Canada, $25)
For collectors of commercially manufactured dollhouses and related assessories.

International Dolls' House News
P.O. Box 154
Cobham Surrey KT11 2YE
England
Quarterly publication $25 year
A dollhouse magazine for dollhouse collectors and enthusiasts. Established in 1967.

Miniature Collector
30595 Eight Mile Rd.
Livonia, MI 48152
1 year (6 issues) $17.95
Outside USA $23.95 (U.S. Funds)

Nutshell News
21027 Crossroads Circle
Waukesha, WI 53187
Monthly magazine, yearly subscription $34.95
An established miniature magazine, known as "The Bible" of miniatures."

The Tynietoy Preservation Society
In Care of: Ann Meehan
P.O. Box 6686
Portsmouth, N. H. 03802
Two newsletters yearly for $10
Two meetings each year.
Publications: Tynietoy Catalog $10
"I Remember Tynietoy" by Herbert Hosmer $1.00
"The New Model House" $10

Museums

Angels Attic
A museum of antique dollhouses, miniatures, dolls and toys.
516 Colorado Avenue
Santa Monica, CA 90401
Thursday thru Sunday, 12:30 to 4:30 P.M.
310-394-8331

Toy and Miniature Museum of Kansas City
5235 Oak St.
Kansas City, MO 64112
Wednesday-Saturday 10:00 to 4:00, Sundays 1:00 to 4:00
Closed two weeks following Labor Day in September
816-333-2055

Washington Dolls' House and Toy Museum
5236 44th St. NW
Washington D. C. 20015
Tuesday-Saturday 10:00 to 5:00
Sunday 12:00-5:00
202-555-1212

Bibliography

Adams, Margaret, ed. *Collectible Dolls and Accessories of the Twenties and Thirties from Sears, Roebuck and Co.*. New York: Dover Publications, 1986.

Baker, Linda. *Modern Toys: American Toys 1930-1980.* Paducah, Kentucky: Collector Books, 1985.

Becker, R. D. "A History of Strombeck-Becker Manufacturing Company." Moline, Illinois: Circa 1965.

Brett, Mary. "The Baby Boomers' Dollhouse, The Tin Dollhouse of the 1950s." *Toy Collector and Price Guide,* August, 1994.

Callicott, Catherine Dorris and Lawson Holderness. *In Praise of Dollhouses.* New York: William Morrow and Company, 1978.

Christianson, Barbara. "Schoenhut Built Affordable Housing For Dolls." *Antique Week,* June 6, 1994.

Di Noto, Andrea, ed. *The Encyclopedia of Collectibles, Dogs to Fishing Tackle.* Alexandria, Virginia: Time-Life Books, 1978.

Foulke, Jan. *10th Blue Book Dolls and Values.* Cumberland, Maryland: Hobby House Press, Inc., 1991.

Jackson, Valerie. *A Collector's Guide to Doll's Houses.* Philadelphia: Running Press, 1992.

Jacobs, Flora Gill. *A History of Dolls' Houses.* New York: Charles Scribner's Sons, 1965.

Jacobs, Flora Gill. *Dolls' Houses in America.* New York: Charles Scribner's Sons, 1974.

Jenkner, Carol. "Buildings by Bliss". *The Antique Trader Weekly* , May 19, 1993.

Jensen, Julie. "Stuhr Has Handle On Its Industry." *The Dispatch and the Rock Island Argus* , February 6, 1994.

Judd, Polly and Pam. *Hard Plastic Dolls, II.* Cumberland, Maryland: Hobby House Press, Inc., 1989.

The Kilgore-Mfg. Co. *Toys That Last.* Westerville, Ohio: 1931.

King, Constance Eileen. *The Collector's History of Dolls' Houses; Doll's House Dolls, and Miniatures.* New York: St. Martins, 1983.

MacLaren, Catherine. *This Side of Yesterday in Miniature.* La Jolla, CA: Nutshell News, 1975.

Manos, Susan. *Schoenhut Dolls and Toys.* Paducah, Kentucky: Collector Books, 1976.

Mitchell, Donald and Helene. *Dollhouses Past and Present.* Paducah, Kentucky: Collector Books, 1980.

Montgomery Ward Catalogs. Various issues from 1934-1982.

O'Brien, Marian Maeve. *The Collector's Guide to Dollhouses and Miniatures.* New York: Hawthorn Books, Inc., 1974.

O'Neill, Eleanor, ed. *Dollhouse and Miniature Collector's Quarterly.* Advertisements from 1990-1994 issues.

Schroeder, Joseph, ed. *The Wonderful world of Toys, Games and dolls 1860-1930.* Northfield, Illinois: Digest Books, Inc., 1971.

Schwartz, Marvin. *F.A.O. Schwarz Toys Through the Years.* Garden City, New York: Doubleday and Co., Inc., 1975.

Sears, Roebuck and Company Catalogs. Various issues from 1931-1992.

Smith, Michelle L. "Memories From the Marx Toy Box." *Toy Trader* , November, 1993.

Smith, Patricia R. *Effanbee Dolls That Touch Your Heart.* Paducah, Kentucky: Collector Books, 1983.

Snyder, Dee. "The Collectables." *Nutshell News* , "The Newlyweds," August, 1980; "Keystone," October, 1981; "Cast Iron Miniature Toys," March, 1986; "Tynietoys — Fond Recollections," October, 1986; "Colleen Moore's Doll Castle — by Rich Toys," February, 1988; "Nancy Forbes," June, 1989; "Grand Rapids Line,"; August, 1989; "Flagg Folks," May, 1991; "Black Renwal," June, 1992; "The Bumstead Homestead," July, 1992.

Spero, James, ed. *Collectible Toys and Games of the Twenties and Thirties from Sears, Roebuck and Co. Catalogs.* New York: Dover Publications, Inc., 1988.

Strombecker Catalogs 1934, 1936, 1938, and 1950. Moline Illinois: Strombeck-Becker Manufacturing Co.

Towner, Margaret. *Dollhouse Furniture.* Philadelphia, PA: Running Press, 1993.

Whitton, Blair, ed. *Bliss Toys and Dollhouses.* New York: Dover Publications, Inc., 1979.

Whitton, Blair, ed. *The Knopf Collectors' Guides to American Antiques: Toys.* New York: Alfred A. Knopf, 1984.

Whitton, Blair. *Paper Toys of the World.* Cumberland, Maryland: Hobby House Press, Inc.

Whitton, Margaret, ed. *Dollhouses and Dollhouse Furniture Manufactured by A. Schoenhut Co.*

Index

Price Guide

The prices in this value guide should only be used as a guide and should not be used to set prices for dollhouses and doll furniture. Prices vary from one section of the country to another and also from dealer to dealer. The prices listed are the best estimates the author can give at the time of publication, but prices in the collectible field can change quickly. Neither the author nor the publisher assumes responsibility for any losses that might be incurred as a result of consulting this guide.

Dollhouses or furniture that are mint-in-box will be priced higher than the same items in excellent condition without the box. Houses that are worn or have missing parts will be priced lower than a perfect house. Houses that have been redecorated or painted will also be valued for less money than the same house in original condition. Prices are also influenced by the scarcity or popularity of a certain item. Currently, collectors' interest in the major brands of plastic dollhouse furniture has caused the prices for this furniture to escalate. The hard plastic furniture is more collectible than the furniture made of the softer plastic. If a dollhouse is pictured with missing windows or doors, the house is priced accordingly. If a piece of furniture is not included in a boxed set, it is priced separately. If an item is marked with an *, it means that this particular item has not been seen on the market often enough to give a fair price range.

Page	Position	Item	Price
7	TR	BLISS Log Cabin	$2,000-2,500
8	TR	Large Bliss	*
	CL	Bliss House	2,000-3,200
	BR	Bliss House	1,500-1,888
9	TL	Bliss House	500-700
	CL	Garden House	1,500-1,800
	BR	Bliss House	1,500-1,800
10	TL	Bliss House	1,500-1,800
	BR	Bliss House	1,200-1,500
11	TL	Bliss House	2,000-2,500
	BL	Bliss Chair	75-85
		Bliss Sofa	75-100
12	TL	Bliss Piano	75-100
	TR	Chair or table	40-45
		Sofa	75-85
	CR	Chair	35-40
		Stool	20-25
		Sofa	65-75
	BL	Set	600-700
		Bed	125-150
		Washstand	65-85
		Cradle	75-100
		Table	65-75
		Dresser	75-100
		Rocking chair	75-100
13	BL	CONVERSE or CASS	300-400
14	TL	House	350-400
	TR	House	250-450
	BL	Garage	150-200
	BR	House	100-175
15	TL	House	350-450
	BL	House	450-500
16	TL	Kitchen set	85-100
		Chair	10-12
		Other items, each	15-20
	BL	Each	15-20
	CR	Each	15-20
	BR	Chair	10-12
		Other items, each	15-20
19	TL	SCHOENHUT House	575-700
20	TL	House	700-800
	TR	Large house	1,800-2,000
22	TL	Small house	350-450
	TR	House (restored)	400-500
	BL	Tiny house	100-125
23	TL	Large house	1,600-2,000
	BL	House	500-600
24	TL	Tudor house	500-600
	TR	Dutch Colonial	500-600
25	TL	Malibu House	450-500
	BL	Bedroom boxed set	300+
26	TL	Schoenhut D.R. chair	10-12
		Other items, each	20-25
	TR	Shower	30-35
	CL	Bathroom, each	20-25
	BL	Piano/Bench	45-50
		Lamp	18-22
		Clock	15-18
		Fireplace	35-40
27	TL	Sofa	18-25
		Chair	15-20
	CR	Kitchen set	130-140
		Chair	10-15
		Other items, each	20-22
	BL	Boxed incomplete D.R.	150+
		Chair	10-15
		Other items, each	20-25
28	TL	Bedroom chair	10-15
		Other items, each	18-22
	TR	1" kitchen chair	10-15
		Other items, each	20-25
	CR	Tub, toilet, sink each	15-20
		Other items, each	10-15
	BR	Boxed bathroom set	300+
29	TR	Living room each	20-25
	BL	Each item	15-20
30	TL	Piano and bench	25-30
		Other items, each	15-20
	TR	Chair, damaged stove	10-15
		Other items, each	15-20
	CL	Kitchen chair	10-15
		Other items, each	15-20
	BR	Bed, dresser, each	15-20
		Table	10-15
	BL	Each kitchen piece	15-20
31	TL	Chair, stand, each	10-15
		Dresser, bed, each	15-20
	CR	Boxed bathroom set	150+
		Each piece	15-20
	BL	D.R. chair	10-15
		Table, buffet, each	15-20
35	BL	STROMBECKER 1" bed	15-18
		Vanity/bench	15-20
		Chair, nightstand	8-10
36	TL	Bathroom bench	8-10
		Other items, each	15-20
	CR	Shower	25-35
		Other items, each	10-15
	BL	L.R. table, footstool	7-10
		Other items, each	15-20
	BR	Breakfast nook	30-35
37	TL	Piano/bench	30-35
		Other items, each	10-15
	TR	D.R. chair	8-10
	BR	Kitchen chair	9-12
		Cabinet	25-30
		Table, icebox, each	15-20
38	CL	Bedroom chair, lamp, table	5-7
		Bed, dresser, each	10-15
	CR	Stove, icebox, each	15-18
	BR	Bathroom vanity	10-15
		Other items, each	5-10
39	TL	Boxed bathroom set	100-125
	CR	Kitchen chair	4-6
		Other items, each	10-15

40	TR	Piano/bench	15-18
		Other items, each	5-10
	CL	Boxed bathroom set	50-75
	CR	D.R. chair	5-8
		Other items, each	10-15
	BR	Icebox, each	15-20
		Others, stove, as is	10-15
41	TR	Refrigerator, sink	15-20
	CR	Radio, lamp table	18-20
		Mag. tab.,chair & stool	15-20
		Table lamp	8-12
		Floor lamp	10-15
	BL	D.R. chair	8-10
		Other items, each	15-20
42	TR	Bed, chest, each	18-20
		Vanity/bench	18-20
		Stand	10-12
	BR	D.R. chair	3-4
		Other items, each	10-12
43	TL	L.R. boxed set	125-150
		Fireplace	20-22
		Sofa, chair, stool set	25-30
		Floor lamp	15-18
		Other items, each	10-12
	CR	Bedroom chair, stand	4-6
		Other items, each	10-15
	BL	Sofa (wire)	15-20
		Chair (wire)	10-12
		Floor lamp	15-18
		Table (wire)	10-12
		Other items, each	5-10
44	TL	L.R. table	20-25
		Table lamp	8-12
		Floor lamp, cof. table	10-15
		Chair and stool set	15-20
		Sofa	18-20
	TR	D.R. Chair	8-12
		Other items, each	15-20
	BL	Bed, vanity, chest	18-20
		Stand	10-12
	BR	Bed, wardrobe	18-22
		Blanket chest	12-15
		Other items, each	10-12
45	TL	Kitchen items, each	18-22
	TR	Duck rocker	30-35
		Bathroom vanity	18-20
		Lamp	5-10
46	TR	Clock	20-25
		Fireplace	35-40
		Bookcase	35-40
	CR	Radio, piano (musical)	75+
	BR	Radio	18-22
		Tilt top table	30-40
	BL	Secretary	45+
47	TL	Sectional	45-50
	TR	Items, each	5-8
	CL	Set	125-175
		Each item	35-50
	CR	Boxed bedroom	100-125
	BR	Boxed school	150-200
48	TL	Boxed dining room	110-135
	TR	Boxed bathroom	100-110
	CL	Table and chairs	20-25
	CR	Boxed kitchen	100-125
	BL	Boxed living room	100-125
49	CR	D.R. chair	10-12
		Other items, each	20-25
	BL	Bath hamper	10-12
		Other bath items	15-20
50	CL	Boxed bathroom	100-110
	CR	Boxed L.R. (incomplete)	70-85
		Television	12-15
		Lamps	4-6
		Tables	4-8
		Sofa and chair set	15-20
	BL	Sofa	20-25
		Chair and stool	15-20
		Lamp table	20-25
		Table lamp	8-12
		Floor lamp, coffee tab.	10-15
51	TR	Kitchen items, each	15-20
	BL	Strom. unfurnished	125-150
52	TL	Strom. furnished	350-400
	BL	Piano/bench	15-20
		Sofa(sec.), fl. lamp	15-20
		Radio, chair	10-15
		Other items, each	5-8
53	TL	Kitchen chair	3-5
		Other items, each	10-15
	BL	Bathroom items, each	8-10
	BR	Boxed Furn. House	150-175
54	TL	House unfurnished	50-75
55	TL	Boxed chair	18-25
	TR	Wardrobe, bed, feeder	30-35
	CL	Sofa	20-30
		Chair	10-12
		Table	8-10
	CR	Betsy McCall boxed set	200+
	BR	Table and chairs set	25-35
56	TL	Wardrobe	100-115
		Chest	100+
	TR	Canopy bed	150+
	CL	Table and chairs	75-100
	CR	Rocker	65-75
		Single bed	100+
	BL	Outdoor set	45-55
	BR	Kitchen set	65-75
57	TL	Appliances	12-15
		Bowl	3-5
		Table	15-20
		Chair	8-10
	CL	Dining room set	65-75
	CR	Living room set	65-75
	BL	Bedroom set	50-65
	BR	Bathroom set	50-65
58	BR	TYNIETOY unfurn.	*
59	TR	Sideboard	60-70
		Table	45-50
		Chair	40-50
	CR	Bed	85-100
		Dresser	120-125
		Chair	65-70
	BL	Sofa	135+
		Clock	40-50
		Bureau	50-55
60	TL	Crib	60-65
		Rocker	25-30
		Chest	40-45
		Cradle	40-45
	TR	High chair	35-40
		Breakfast nook	190-200
	CL	Chair	40-50
		Table	45-55
	CR	Rocker	65-70
		Settee	80-85
	BL	Chair	60-65
		Mirror	75-85
	BR	Victorian chair	75-85
		Clock	120-130
		Wing chair	125-150
62	TR	RICH cottage	100+
	CL	Cottage	50-75
	BR	Cottage	100+
63	BL	House (no windows)	85-100
64	TL	House (wear)	100-110
65	TL	Tudor house	140-150
66	TL	House (replace wind.)	125-135
67	TL	House (replace wind.)	65-75
68	TL	Rich labeled house	150+
	BL	House	135-150
69	TL	Pink house trim	140-150
70	TL	Williamsburg house	150-200
	TR	Boxed doll	350+
	BL	Booklet Colleen Moore	10-15
71	TR	Castle house	300+
73	TL	KEYSTONE house	140-160
74	TL	Tudor house	200+
75	TL	Keystone House	175-200
76	TR	House (missing parts)	125-135
	CL	Put-A-Way house	200+
77	TL	House (redecorated)	150+

Page	Position	Item	Price
79	TR	Box only NANCY FORBES	10-12
	BR	Each item	5-8
80	TL	Each item	5-8
	TR	Each item	5-8
	CL	Boxed bedroom	30-40
	CR	Boxed bathroom	30-40
	BL	Boxed kitchen	30-40
81	TR	Boxed living room	30-40
	CL	Boxed dining room	30-40
	BR	Larger boxed set	85+
		Each item	3-5
82	TR	Each item	3-5
	CL	DONNA LEE boxed bedroom	25-30
	CR	Boxed bathroom	25-30
	BR	Boxed kitchen	25-30
83	TR	Each item	6-10
	CL	Each item	6-10
	CR	Each item	6-10
	BL	GRAND RAPIDS, each	5-8
84	TR	Bed, dresser	15-20
		Other items, each	5-8
	CL	Rocker, chest, each	12-15
		Magazine rack	5-8
	CR	Hutch	15-20
		Chair	5-8
		Table	12-15
	BL	Each appliance	12-15
86	BL	TOOTSIETOY box only	18-25
	BR	Boxed bedroom	100+
87	TL	Each item	18-22
	TR	Kitchen chair	8-10
		Other items, each	18-22
	CL	Dining room chair	8-10
		Other items, each	18-20
	CR	Boxed Daisy set	100+
	BL	Box only	18-25
		Living room items	18-22
88	TR	Boxed My Dolly not comp.	75+
	BL	Boxed kitchen set	225+
89	TL	Boxed bathroom	225+
	BL	Boxed living room	225+
	BR	Boxed bedroom	225+
90	TL	Boxed music room	250+
	CL	Boxed dining room	225+
	BL	Bench	8-10
		Other items, each	18-22
91	TL	Each	20-25
	TR	Boxed Midget bedroom	65-75
	CL	Phone	20-25
		Sweeper	25-30
	CR	Boxed Midget L.R.	65-75
92	TL	WAYNE house	200-300
	BL	TOOTSIETOY house	450-500
	BR	Daisy house	350-450 93
	TR	Spanish Mansion	650-750
95	TL	ARCADE Kitchen	1600-1800
	BL	Arcade stove	250+
	BL	Arcade cabinet	250+
		Arcade chair	50+
96	TL	Arcade sink	175+
	TR	Breakfast nook	*
	CL	Arcade table	65-75
	BL	HUBLEY furniture, each	150+
97	TL	KILGORE chair	35-45
		Other items, each	50+
	CL	Arcade, each	125-175
	CR	Kilgore, each	50+
	BL	Ladder	45+
		Sink, washer, each	65+
	BR	Bathroom pieces, each	50+
98	TL	Dining room chair	40-45
		Other items, each	50+
	CL	Dining room chair	35-40
		Other items, each	50+
	BL	Bedroom chair, bench	40-45
		Other items, each	60+
99	TL	Highboy	60+
	CL	Each baby item	40-50
	BL	Each item	40-55
	BR	Chair	45+
100	TL	Each baby item	40-50
	CL	Each item	40-50
	BL	Set	*
102	TL	Boxed Warren house	35-45
	BL	BUILT-RITE boxed house	30-40
103	TL	Boxed Country Estate	50-75
104	TR	Art Deco House	100-125
	BL	Dining Room furn., each	3-5
105	TL	Living Room furn., each	3-5
	CL	Bedroom, furn., each	3-5
	CR	Kitchen, furn., each	3-5
	BL	Bathroom, furn., each	3-5
107	TL	Dolly's Playhouse	700-800
	TR	MCLOUGHLIN House	500-600
	BL	House and furniture	250+
108	TR	Boxed L.R. and furn.	65-85
	BL	Boxed D.R. and furn.	50-75
109	TL	Boxed Bedroom and furn.	50-75
	CL	Each boxed room	20-25
110	TL	Mother Goose House	65-85
	CL	Newberg House	35-45
	BR	Sparkle Plenty House	35-45
111	TL	Trixy House	65-75
	CR	Wayne House	100-125
112	TR	Concord House	100-125
	BL	Nels House	85-100
113	TL	House	35-45
	TR	Boxed House	75-100
	CL	Happitime House	75-100
114	TR	Folding house	25-35
	BR	Bakers' Coconut	30-40
115	TL	Little Debbie House	30-40
118	TL	IDEAL boxed D.R.	125-150
119	TL	Chair	8-10
		Arm chair	8-10
		Table	10-12
		Buffet	12-15
		Hutch	8-10
	TR	Boxed bathroom	75-95
	BR	Hamper	4-6
		Medicine cabinet	15-20
		Other items, each	6-8
120	TL	Boxed bathroom	50-75
	TR	Boxed nursery	100-125
	CL	Playpen	25-30
		Buggy, crib, cradle	15-20
		Baby, potty, lamp, each	10-12
	BL	High chair	20-25
		Crib	15-20
		Nightstand	5-6
		Lamp	6-8
		Highboy	10-15
		Folding high chair	15-18
		Bathinet	22-25
		Stroller	40-45
121	TL	Boxed living room	150-200
	CR	Bed, vanity/bench	15-20
		Chair	10-15
		Nightstand	5-6
		Lamp	6-8
		Stools	5-8
		Radiator	15-18
	BL	Fireplace	20-25
		Sofa	12-15
		Chair, lamp, radio, each	10-12
		Tilt top table	15-18
		Coffee table, stool, each	5-8
122	TL	Boxed kitchen	75-100
	CR	Appliances	10-12
		Table	5-10
		Chair	3-6
	BL	Picnic table	18-22
		Lounge	15-20
		Round tab. with umbre.	25-30
		Lawn mower	20-25
		Bench	10-12
		Bird bath	5-10
		Dog-Doghouse	15-18
		Chair	10-15

Page	Position	Item	Price
	BR	Sofa bed	30-40
		Piano/bench	15-18
		Secretary	25-30
		Sewing machine	15-20
		Television	40-55
123	TL	Sofa bed	30-40
		Card table and chairs	125+
	CR	Hamper	4-6
		Tub	15-20
		Toilet	10-15
		Sink	8-9
	BR	Appliances	25-30
		Table	10-15
		Ironer	15-18
124	TL	Washer	20-25
	CL	Bathroom items, each	10-12
	BL	Boxed YOUNG DEC. D.R.	125+
	BR	Chair	8-12
		Table	10-12
		Buffet	15-18
		China cabinet	25+
125	TL	Vanity/bench	25+
		Nightstand	8-10
		Bed	25+
		Armoire	25+
	CR	Sectional	35-40
		Lamp	12-15
		Television	40+
		Coffee table	10-12
	BR	Chair	8-10
		Hamper	10-15
		Toilet, tub, each	15-20
		Sink unit	20-25
126	TL	Appliances	40+
		Chair	8-10
		Table	12-15
	CL	Sink (missing part)	10-15
		Sweeper	18-20
	CR	PETITE PRINCESS TV,std.	200+
		(P.P. furn. prices include access.)	
		Fireplace	20-25
		Clock, screen, each	10-12
		Planter	10-15
		Tables (with access.)	15-20
	BL	Playpen, crib	30-35
		Tricycle	40-50
		High chair	20-25
127	TL	Sofa	15-20
		Tables (with access.)	15-20
		Chairs (large), each	15-18
		Chairs (small)	10-15
		Candelabra	10-15
	TR	Family	35-50
	CR	Piano/bench	25-35
		Chair	10-15
		Candelabra	10-15
	BR	Cabinet, buffet	18-20
		Table	10-15
		Serving cart	15-20
		Chair	8-10
128	TL	Vanity	25-30
		Bed	20-25
		Chest	10-15
		Chaise lounge	15-20
		Tables (with access.)	15-20
	TR	Tub, vanity, each	85-100
		Toilet, linen closet	80-85
		Waste basket	25-30
		Hamper	40-45
		Stool	10-15
	BR	Sink, stove, each	150+
		Refrigerator	150-200
		Table and chairs	150-175
		Hutch	75-100
129	TR	Boxed room	35-45
	BR	Display (missing tops)	400+
130	TL	Vinyl house	40-50
	BR	House (unfurn.)	300-400
131	TR	Furniture, each	5-10
	CR	Tammy House	100+
135	TL	Newlywed furn. house	400+
	TR	Boxed dining room	100-125
	CR	Boxed living room	100-125
	CL	Boxed kitchen	100-125
	BR	Boxed bedroom	100-125
136	TL	MARX house	25-40
137	TL	1/2" furn. (H.P.best) L.R.	3-5
	CL	Dining room chair	1-2
		Other items, each	3-5
	CR	Kitchen chair	1-2
		Other items, each	3-5
	BL	Nursery each	3-5
138	TL	Bedroom each	3-5
	CR	Bathroom each	3-5
	CL	Laundry each	3-5
	BR	House	35-55
139	CR	Disney House	50-75
140	TL	House	75+
	BL	Accessory set	100+
141	TL	3/4" small items each	3-5
		L.R. larger items each	5-8
	CR	D. R. chair	2-3
		Other items each	5-8
	BL	Kitchen appliances each	5-8
		Sweeper, broom	8-10
		Tall cabinet	10-15
		Other items	3-5
142	TL	Nursery chest, potty	3-5
		Other items	5-8
	TR	Umbrella table, chairs	20-25
		Ladder, chaise, each	10-12
		Other items	4-8
	CL	Larger bedroom items	5-8
		Smaller items	3-5
	BR	Larger items	5-8
		Smaller items	3-5
143	TR	Each item	2-3
	CL	Marx house	75-100
144	TL	Babyland	200-225
	TR	School furn. (small)	2-3
		School furn. (large)	5-8
	BL	Ranch unfurnished	75-85
		Ranch furnished	115+
146	TL	Boxed house (incomplete)	45-65
	BL	House with stairs	50-75
147	BL	Marxie Mansion	125+
148	TL	Each piece	5-8
	CR	Ranch house	35-40
	BR	House (missing chimney)	35-45
149	TR	Family	25-35
	CL	Chair	1-2
		Other items	3-6
	BR	Boxed dolls	35+
150	TR	LITTLE HOSTESS sofa	15-18
		Fireplace, screen each	15-18
		Clock, chair	10-15
		Other items	8-10
	CL	Piano	22-25
		Secretary	15-20
		Other items, each	10-12
	BR	Buffet	10-12
		Table	8-10
		Chair	5-7
151	TR	Bed	25-30
		Vanity/bench	15-20
		Highboy, chests each	10-15
		Chaise lounge	10-12
		Nightstand	5-8
		Chair	8-10
	CL	Shower, sink	15-20
		Medicine cabinet	12-15
		Other items	8-12
	BR	Appliances	15-20
		Table	12-15
		China cabinet	20-25
152	TL	Boxes only	2-4
	CR	Package	50-75
		Television	20+
	BL	Boxed furnished house	100+
153	TR	Boxed furnished house	75-100
	CL	Furniture, each piece	1-3
	BR	Furniture, each piece	1-3

154	TL	Unfurnished house	25-35
155	TL	Cardboard house	35-50
	BL	SINDY Scene Setter	40-50
		Entertainment pieces	15-20
		Sofa and chair, each	8-10
156	TL	Appliances, each	12-15
	CL	Chairs	3-5
		Table	10-12
		Other items, each	12-15
	BR	Furniture, each	8-10
157	BR	PLASCO boxed bathroom	75+
		Toilet	7-9
		Other items, each	5-8
158	TL	Boxed dining room	90+
		China cabinet	10-12
		Tables	8-10
		Buffet	5-8
		Chairs	3-5
	CR	Boxed set fabric finish	125+
		Television	15-20
		Fireplace	12-15
		Clock	10-12
		Sofa (fabric)	15-20
		Chair (fabric)	10-15
		Desk and chair	5-8
		Coffee table	3-5
	BL	Boxed set living room	100+
		Sofa	8-12
		Chairs	6-8
159	TL	Boxed set bedroom	100+
		Beds	8-10
		Highboy	10-12
		Mirrored dresser, vanity	8-10
		Nightstand, stool	3-5
	BR	Boxed kitchen	75+
160	TL	Boxed garden	75-100
		Chaise lounge	15-20
		Umbrella table/chairs	20-25
		Table	5-7
		Birdbath, fountain	5-10
		Chairs	5-8
	CR	Appliances, cabinet	5-8
		Chairs	2-4
		Table	4-6
		Cabinet with shelves	8-12
	BL	Crib	20-25
		Bathinet	25+
		Vanity/stool	12-15
		Dresser with mirror	15-20
		Highboy	15-18
		Baby	15-20
161	TL	Boxed set with record	125-150
	CR	Unfurnished house	200+
162	TR	Bubble pack dining room	35-50
	BL	Bubble pack bedroom	35-50
	BR	Another style dining room	35-50
164	BR	RENWAL boxed bedroom	125+
165	TL	Bed	8-12
		Vanity/bench	15-18
		Highboy	10-12
		Dresser with mirror	15-18
		Table lamp, nightstand	5-8
	CL	Rocker	8-12
		Baby, potty	9-12
	CR	Boxed nursery (room)	175+
		Playpen, high chair	18-20
		Cradle, bathinet	18-20
		Highboy	10-12
		Buggy	20-25
		Nightstand, lamp	5-8
	BR	Boxed dining room	125+
166	TR	Buffet, hutch	8-12
		Server	5-7
		Table	10-12
		Chair	4-6
	CL	Boxed kitchen	125+
	CR	Appliances (open doors)	15-18
		Table	8-12
		Chair	3-6
	BL	Boxed bathroom	85-100
		Hamper	3-5
		Bathroom pieces	6-9
167	TR	Boxed living room	150+
	TL	Washer	25-30
		Iron and ironing board	20-25
	CR	Smaller boxed L.R.	85-100
	BL	Sofa	10-15
		Chair	8-10
		Floor lamp	10-12
		Radio-Phonograph	14-18
		Tables, table lamp	5-8
168	CR	Slide, teeter-totter	15-20
		Swing	25-30
		Tricycle	20-25
		Kiddie car	30-35
	BL	Boxed accessories	100+
		Sweeper	20-25
		Garbage can	12-15
		Scale, stool	6-8
		Phone, smoking stand	8-10
		Dust pan	3-5
		Alarm clock	5-8
		Mantle clock	8-10
169	TL	Broom	50+
		Table radio	8-10
		Sewing machine	25-30
		Radio-phonograph	14-18
	TR	Mop	25-35
		Stroller	25-30
	CL	Piano/bench	20-25
		Card table/chairs	50-75
	BR	Cook 'n Serve Set	*
170	TL	Busy Little Mother Set	*
	CR	Boxed School	200+
171	TL	Boxed small nursery	175+
	CR	Boxed large nursery	200+
		Cribs	6-7
		Babies	3-4
		Scale	8-10
	BL	Black finish chair	5-8
		Other items	10-15
172	TR	Family dolls, each (box)	35-50
	CL	Doll displays	150+
	BR	Furniture display	500+
174	BL	T. COHN house	65+
175	BL	House missing window	40-50
176	CR	Superior 3/4" furn.	3-5
	BL	Superior chair	1-2
177	TL	Kitchen furniture	3-5
		Chair	1-2
	TR	Nursery	3-5
	CR	Bathroom	3-5
	BL	Bedroom	3-5
178	TL	Green house	45-55
179	TL	Ranch house	30-40
	CR	Small house, no chim.	30-35
180	TL	1/2" Superior furn.	1-3
	CR	Superior dining room	1-3
	BL	Superior kitchen	1-3
181	TR	Nursery	1-3
	CL	Bedroom	1-3
	CR	Bathroom	1-3
	BL	Furniture	1-3
183	TL	WOLVERINE house	35-50
184	TR	Small ranch	20-30
	BR	Town and Country	25-35
185	TR	Attached garage	35-50
	CL	Augusta	30-40
186	TL	Rosewood	30-40
	BL	Colonial	40-50
187	TL	Furniture, each	1-2
	TR	Furniture, each	1-2
	CL	Furniture, each	1-2
	BR	Furniture, each	1-2
188	TL	Cape Cod	45-65
189	CR	Boxed sets	25-35
	BL	Unboxed sets	20-22
		Each piece	8-10
191	TL	Boxed Bestmaid	35-45
	TR	Bed JAYDON	15-20
		Other	5-10
	CR	Piano/bench	8-12

Page	Position	Item	Price
		Table, hutch, buffet	5-8
		Chair	3-4
	BL	Appliances	6-10
		Chair	3-5
192	TL	Radio	10-12
		Chair	5-8
		Coffee table	3-5
	CR	Boxed MULTIPLE Prod.	20-22
	CL	Boxed bathroom	20-22
	BR	Furniture each	1-2
193	TL	Bedroom each	1-2
	CL	IRWIN set	15-20
	BL	ALLIED boxed kitchen	20-25
194	TR	BANNER boxed room	35+
	CR	Allied boxed bedroom	20-25
	BL	Furniture each	1-3
195	TL	Banner boxed room	35+
	CR	ACME items	6+
	BL	Bunk beds (BEST)	20+
196	TL	Bending dolls each	12-15
	BL	FLAGG dolls boxed orig.	20-30
197	TL	Flagg dolls each	12-15
	CR	Flagg dolls each	12-15
	BL	EFFanBEE Dolls	*
200	TL	FRIER Cozy Town House	125+
201	TL	Frier house (redec.)	85+
	BL	JAYLINE house	45-65
203	BR	PLAYSTEEL House	75+
204	TL	Playsteel no chimney	50+
205	TL	OHIO ART house	45-65
	BL	Brumberger (no chim.)	20-25
206	TL	Tudor Tekwood	85-125
207	TL	Tekwood 1946 (damage)	75-100
208	TL	Tekwood missing windows	60-75
209	TL	Four-room house	75-100
210	TL	Masonite (missing pieces)	50+
211	TL	TOMY Smaller Homes	65-75
	BL	Boxed refrig.,stove, sink, table, chairs, access.	70-75
		Single items	10-12
		Smaller boxed items	20-25
		Counter/stools	15-18
212	TL	Sofa, hutch, phono., each	15-18
		End table, coffee tab.	3-5
		Rocker	8-10
	CR	Tub, toilet	10-12
		Mirror, planter	8-10
		Vanity	15-20
212	BL	Bed	15-18
		Wardrobe, vanity	10-15
		Chair, lamp, nightstand	8-12
213	TL	Family set	18-22
	BL	SOUNDS LIKE HOME Furn.	1800-2000
		Room boxes	75-200+
214	TL	LITTLES unfurnished	75-100
	TR	Boxed furniture	12-15
215	TR	FISHER PRICE unfurn.	75-100
	BR	Furnished	250+
216	TR	Fisher Price Family	20-25
	CL	PLAYMOBIL House	175+
217	TL	Playmobil layout	1,000-1,200
220	TL	NANCY ANN boxed chair	200+
	BR	GINNY Bed	45-55
		Chest	55+
		Chair	40-50
221	TL	Table/chairs	45-55
	CL	GINNETTE bed	30-40
	BR	Feeder, bathinet	30-40
222	TL	JILL desk, chair	50+
	TR	MATTEL, INC. bed	20-25
	CL	Table/chairs	30-35
	BL	Sofa/table	30-35
223	TL	RICHWOOD bed	100+
	TR	Wardrobe	100+
	BL	Boxed chair	50+
	BR	Slide, sand unit	100+
224	TR	SUSY GOOSE bed, wardrobe	25-35
	CL	New Dream House (used)	45-55
225	TR	Vinyl suitcase (used)	30-35
	BR	LITTLECHAP bedroom (used)	30-35
226	TL	LITTLECHAP fam. rm. (used)	30-35
	CR	HEIDI house	45-55
227	TL	Boxed DAWN'S Apartment	55-75
228	TR	Boxed SUNSHINE house	25-35
229	TL	Boxed WALTON Farmhouse	100+
230	TR	HOLLY HOBBIE House	600+
	BL	STRAWBERRY SHORTCAKE	200+